Vehicular Sensing for Improved Urban Mobility

Vehicular Sensing for Improved Urban Mobility

Guest Editors

Constantin-Florin Caruntu
Ciprian-Romeo Comsa

Basel • Beijing • Wuhan • Barcelona • Belgrade • Novi Sad • Cluj • Manchester

Guest Editors

Constantin-Florin Caruntu
Gheorghe Asachi Technical
University of Iasi (TUIASI)
Iasi
Romania

Ciprian-Romeo Comsa
Gheorghe Asachi Technical
University of Iasi (TUIASI)
Iasi
Romania

Editorial Office

MDPI AG
Grosspeteranlage 5
4052 Basel, Switzerland

This is a reprint of the Special Issue, published open access by the journal *Sensors* (ISSN 1424-8220), freely accessible at: https://www.mdpi.com/journal/sensors/special_issues/5114L2PB0L.

For citation purposes, cite each article independently as indicated on the article page online and as indicated below:

Lastname, A.A.; Lastname, B.B. Article Title. *Journal Name* **Year**, *Volume Number*, Page Range.

ISBN 978-3-7258-2501-1 (Hbk)
ISBN 978-3-7258-2502-8 (PDF)
https://doi.org/10.3390/books978-3-7258-2502-8

Contents

About the Editors

Constantin-Florin Caruntu

Constantin-Florin Caruntu is a professor and researcher at the Gheorghe Asachi Technical University of Iasi (TUIASI), Romania, specializing in automatic control and applied informatics. Since starting his academic career in 2003, his work has spanned smart mobility, networked control systems, and predictive control. Caruntu has held several leadership roles, including Vice Dean with Research and Innovation Activities, Full Professor, and Doctoral Supervisor, and has played a key role in mentoring students.

In 2016, he also transitioned into industry roles at Continental Automotive Romania, where he served as Business Owner, Team Leader, and Technical Project Leader in AI and Intelligent Infrastructure. His leadership in developing AI-driven automotive technologies has been widely recognized, earning him international accolades.

Caruntu's research contributions, particularly in vehicle platooning, cooperative automated vehicles, and control strategies for automated systems, have made a significant impact on intelligent transportation. He has led numerous national and international research projects, working with global experts to advance model predictive and automotive control systems. His active participation in scientific conferences and project leadership further underscores his dedication to advancing smart mobility solutions. He is the author/coauthor of more than 150 publications in journals, books, and conference proceedings.

Ciprian-Romeo Comsa

Ciprian-Romeo Comsa is a professor and researcher at the Gheorghe Asachi Technical University of Iasi (TUIASI), Romania, with over two decades of experience in wireless communications, signal processing, and smart mobility. His academic career is complemented by his industry role at Continental Automotive Romania, where he contributes to innovations in smart mobility.

Comsa holds a Ph.D. in electrical engineering from the New Jersey Institute of Technology, where he focused on wireless communication systems and object detection. His leadership roles have involved overseeing significant projects like the CONSENSUM-AIR 5G Lab, and he has been nominated as Vice-Rector for Innovation at TUIASI, as of 2024.

His research interests span automotive control systems, inter-vehicular communications, and AI applications in mobility, making him an essential figure in advancing both academic and industry initiatives in smart mobility.

Preface

The evolution of urban mobility has seen unprecedented technological advances, particularly in the fields of vehicular sensing and intelligent transportation systems (ITS). This Reprint, "Vehicular Sensing for Improved Urban Mobility," brings together key research contributions that explore the critical role of sensor technologies, connected vehicles, and communication systems in reshaping how we navigate urban environments.

The subject of this Reprint centers around the integration of vehicle-to-everything (V2X) technologies, including vehicle-to-vehicle (V2V) and vehicle-to-infrastructure (V2I) communications, enabled by emerging 5G networks and artificial intelligence (AI). These technological advances are leading to safer, more efficient, and environmentally friendly transportation systems. Our aim is to present the latest developments in vehicular sensing, highlighting how these innovations contribute to urban mobility by addressing challenges such as traffic congestion, road safety, and environmental impact.

The motivation behind this compilation lies in the rapidly growing interest and research in smart cities, autonomous driving, and connected vehicles. As global urban populations expand, the need for improved mobility solutions becomes ever more pressing. This collection showcases cutting-edge studies that propose solutions to these pressing challenges through advancements in vehicular sensing.

The Reprint is addressed to researchers, engineers, policymakers, and professionals working in the fields of transportation engineering, automotive technology, urban planning, and smart infrastructure. It serves as a comprehensive resource for those interested in the intersection of vehicular sensing technologies and urban mobility.

We extend our gratitude to the contributing authors, whose rigorous research and innovative thinking form the backbone of this Reprint. Additionally, we acknowledge the invaluable assistance and support of the reviewers, editors, and collaborators who helped ensure the quality and impact of the work presented here.

We hope this Reprint serves as a valuable resource and inspiration for future advancements in vehicular sensing and urban mobility, contributing to safer, smarter, and more connected cities.

Constantin-Florin Caruntu and Ciprian-Romeo Comsa
Guest Editors

sensors

Editorial

Vehicular Sensing for Improved Urban Mobility

Constantin-Florin Caruntu [1,*] and Ciprian-Romeo Comsa [2]

1 Department of Automatic Control and Applied Informatics, "Gheorghe Asachi" Technical University of Iasi, 700050 Iasi, Romania
2 Department of Telecommunications and Information Technologies, "Gheorghe Asachi" Technical University of Iasi, 700506 Iasi, Romania; ccomsa@etti.tuiasi.ro
* Correspondence: caruntuc@tuiasi.ro

In recent years, advancements in the automotive industry have accelerated the development of connected and autonomous vehicles (CAVs). While fully autonomous vehicles still require significant progress to meet all safety and security standards, there are emerging opportunities to enhance traffic safety using environmental sensor data and connectivity with other vehicles and smart infrastructure, especially in urban settings.

Environmental sensors such as cameras, radars, and lidars provide critical information about objects and other traffic participants around vehicles. The latest CAV developments enable the sharing of this information between vehicles and infrastructure, known as collective perception. Vehicles can now create and improve their environment models (EMs) with data from their own sensors and from vehicle-to-everything (V2X) communication technologies. This allows vehicles to transmit their heading, position, and speed through cooperative awareness messages (CAMs) and enhance precision with collective perception messages (CPMs).

Urban infrastructure, equipped with various sensors, like cameras, radars, and GNSS, plays a crucial role in traffic safety. Smart infrastructure employs computer vision and artificial intelligence (AI) for object detection, classification, pose estimation, tracking, and behavior prediction, thus enhancing traffic participants' awareness. Thus, this Special Issue explores all aspects of vehicular sensing in urban mobility, including architecture, emerging sensors, communication technologies, advanced applications, and deployment issues, such as developing smart infrastructure systems for information sharing and safety.

The rapid expansion of urban environments necessitates innovative solutions for improving mobility and reducing congestion. Vehicular sensing technologies hold significant promise in addressing these challenges by enhancing traffic management, ensuring safety, and optimizing vehicle performance. This Special Issue, titled "Vehicular Sensing for Improved Urban Mobility", showcases ten cutting-edge studies that advance the state of the art in vehicular sensing and its applications in urban settings.

Citation: Caruntu, C.-F.; Comsa, C.-R. Vehicular Sensing for Improved Urban Mobility. *Sensors* **2024**, *24*, 5134. https://doi.org/10.3390/s24165134

Received: 2 August 2024
Accepted: 6 August 2024
Published: 8 August 2024

1. Overview of the Contributions

In their study, Fernández et al. [1] develop a dual-slope path loss model tailored for vehicular sensing applications in diverse urban and suburban environments. This model aims to improve the reliability and accuracy of vehicular communications, which are crucial for the deployment of intelligent transportation systems (ITSs) and vehicular ad hoc networks (VANETs).

Herbers, Doerzaph, and Stowe [2] explore how line-of-sight (LOS) sensors and connected vehicle technology (CVT) can mitigate and prevent crash and near-crash scenarios. Their research demonstrates the potential of these technologies to enhance vehicle safety through advanced driving assistance systems (ADASs).

Heyer-Wollenberg et al. [3] propose a cooperative method to improve the accuracy of Turn Movement Count (TMC) by incorporating contextual observations from surrounding areas. This method significantly enhances the identification of vehicle movements under challenging conditions, contributing to more accurate traffic analysis and management.

In their paper, Silva et al. [4] present a model-based approach to quantify the dependability of VANETs, particularly in urban advanced mobility (UAM) contexts. By leveraging virtual machine migration, they aim to enhance the reliability and availability of VANETs which are essential for integrating UAM into urban infrastructures.

Lazar et al. [5] discuss a comprehensive control architecture for connected vehicle platoons, utilizing vehicle-to-everything (V2X) communication. This architecture improves road safety, traffic flow, and fuel efficiency, offering a promising solution to contemporary traffic problems.

Shopovska et al. [6] address the challenge of detecting vulnerable road users (VRUs under varying lighting conditions. They introduce a high-dynamic-range tone mapping technique for intelligent automotive systems, enhancing the performance of imaging sensors in extreme lighting scenarios.

Achirei et al. [7] present a model predictive control framework for omnidirectional mobile robots, emphasizing the use of convolutional neural networks (CNNs) for object detection. Their approach significantly improves the navigation and operational efficiency of mobile robots in urban logistic environments.

Frej et al. [8] conducted an experimental study on the longitudinal acceleration of urban buses and coaches, analyzing vehicle motion dynamics and driver behavior. Their findings provide valuable insights for enhancing passenger comfort and safety in urban public transportation.

Park et al. [9] propose a secure mutual authentication and key agreement scheme using physically unclonable functions (PUFs) for Internet of Drones (IoD) applications This scheme ensures robust security in UAV operations, which is crucial for urban traffic surveillance and environmental monitoring.

Zhao and Zhao [10] developed an algorithm for online stochastic error modeling of inertial sensors used in urban navigation systems. Their approach enhances the accuracy of inertial navigation during GNSS outages, which is vital for reliable vehicle positioning in dense urban areas.

2. Conclusions

This Editorial provides an overview of each paper's contributions to the field of vehicular sensing and emphasizes their impact on improving urban mobility. The papers in this Special Issue highlight the diverse applications and significant advancements in vehicular sensing technologies. By addressing key challenges in urban mobility, these studies contribute to safer, more efficient, and sustainable transportation systems.

We, the Editorial Team, appreciate all the innovative research endeavors presented in this Special Issue. We extend our thanks to the authors for their diligent incorporation of feedback, critical assessment of their work, and adherence to timelines, which have enabled the successful publication of this Special Issue. The Guest Editors are pleased with the conclusive outcomes of the published papers and anticipate their utility for researchers, engineers, designers, and other professionals engaged in various aspects of advanced analytical and numerical simulation approaches, as well as experimental studies applied to vehicular sensing and urban mobility. We also express our gratitude to the reviewers for their crucial contributions and the dissemination of scientific findings. Lastly, we thank the Editorial Board of *Sensors* for their patience, support, and exceptional contributions. We hope the readers feel inspired by and can learn from the research articles in this Special Issue.

Funding: This research received no external funding.

Conflicts of Interest: The authors declare no conflicts of interest.

References

1. Fernández, H.; Rubio, L.; Peñarrocha, V.M.R.; Reig, J. Dual-Slope Path Loss Model for Integrating Vehicular Sensing Applications in Urban and Suburban Environments. *Sensors* **2024**, *24*, 4334. [CrossRef]
2. Herbers, E.; Doerzaph, Z.; Stowe, L. The Impact of Line-of-Sight and Connected Vehicle Technology on Mitigating and Preventing Crash and Near-Crash Events. *Sensors* **2024**, *24*, 484. [CrossRef] [PubMed]
3. Heyer-Wollenberg, P.; Lyu, C.; Jovanov, L.; Goossens, B.; Philips, W. Improving Turn Movement Count Using Cooperative Feedback. *Sensors* **2023**, *23*, 9772. [CrossRef] [PubMed]
4. Silva, L.G.; Cardoso, I.; Brito, C.; Barbosa, V.; Nogueira, B.; Choi, E.; Nguyen, T.A.; Min, D.; Lee, J.W.; Silva, F.A. Urban Advanced Mobility Dependability: A Model-Based Quantification on Vehicular Ad Hoc Networks with Virtual Machine Migration. *Sensors* **2023**, *23*, 9485. [CrossRef] [PubMed]
5. Lazar, R.-G.; Pauca, O.; Maxim, A.; Caruntu, C.-F. Control Architecture for Connected Vehicle Platoons: From Sensor Data to Controller Design Using Vehicle-to-Everything Communication. *Sensors* **2023**, *23*, 7576. [CrossRef] [PubMed]
6. Shopovska, I.; Stojkovic, A.; Aelterman, J.; Van Hamme, D.; Philips, W. High-Dynamic-Range Tone Mapping in Intelligent Automotive Systems. *Sensors* **2023**, *23*, 5767. [CrossRef] [PubMed]
7. Achirei, S.-D.; Mocanu, R.; Popovici, A.-T.; Dosoftei, C.-C. Model-Predictive Control for Omnidirectional Mobile Robots in Logistic Environments Based on Object Detection Using CNNs. *Sensors* **2023**, *23*, 4992. [CrossRef]
8. Frej, D.; Grabski, P.; Jurecki, R.S.; Szumska, E.M. Experimental Study on Longitudinal Acceleration of Urban Buses and Coaches in Different Road Maneuvers. *Sensors* **2023**, *23*, 3125. [CrossRef] [PubMed]
9. Park, Y.; Ryu, D.; Kwon, D.; Park, Y. Provably Secure Mutual Authentication and Key Agreement Scheme Using PUF in Internet of Drones Deployments. *Sensors* **2023**, *23*, 2034. [CrossRef] [PubMed]
10. Zhao, L.; Zhao, L. An Algorithm for Online Stochastic Error Modeling of Inertial Sensors in Urban Cities. *Sensors* **2023**, *23*, 1257. [CrossRef] [PubMed]

sensors

MDPI

Article

An Algorithm for Online Stochastic Error Modeling of Inertial Sensors in Urban Cities

Luodi Zhao [1,2,3] and Long Zhao [1,2,3,*]

1 School of Automation Science and Electrical Engineering, Beihang University, Beijing 100191, China
2 Digital Navigation Center, Beihang University, Beijing 100191, China
3 Science and Technology on Aircraft Control Laboratory, Beihang University, Beijing 100191, China
* Correspondence: buaa_dnc@buaa.edu.cn

Abstract: Regardless of whether the global navigation satellite system (GNSS)/inertial navigation system (INS) is integrated or the INS operates independently during GNSS outages, the stochastic error of the inertial sensor has an important impact on the navigation performance. The structure of stochastic error in low-cost inertial sensors is quite complex; therefore, it is difficult to identify and separate errors in the spectral domain using classical stochastic error methods such as the Allan variance (AV) method and power spectral density (PSD) analysis. However, a recently proposed estimation, based on generalized wavelet moment estimation (GMWM), is applied to the stochastic error modeling of inertial sensors, giving significant advantages. Focusing on the online implementation of GMWM and its integration within a general navigation filter, this paper proposes an algorithm for online stochastic error calibration of inertial sensors in urban cities. We further develop the autonomous stochastic error model by constructing a complete stochastic error model and determining model ranking criterion. Then, a detecting module is designed to work together with the autonomous stochastic error model as feedback for the INS/GNSS integration. Finally, two experiments are conducted to compare the positioning performance of this algorithm with other classical methods. The results validate the capability of this algorithm to improve navigation accuracy and achieve the online realization of complex stochastic models.

Keywords: GMWM; stochastic error; inertial sensor; sensor calibration; error model; Allan variance

Citation: Zhao, L.; Zhao, L. An Algorithm for Online Stochastic Error Modeling of Inertial Sensors in Urban Cities. *Sensors* **2023**, *23*, 1257. https://doi.org/10.3390/s23031257

Academic Editors: Constantin Caruntu and Ciprian Romeo Comşa

Received: 23 November 2022
Revised: 9 January 2023
Accepted: 17 January 2023
Published: 21 January 2023

1. Introduction

Modeling and estimation of inertial sensor errors are generally challenging tasks, especially for low-cost inertial micro-electromechanical system (MEMS) sensors, since the error model has complex spectral structures. For a global navigation satellite system (GNSS) and inertial navigation system (INS) integrated system, it is usually performed through a general Kalman filter, e.g., an extended Kalman filter (EKF), which is closely related to the inertial sensor modeling. When the GNSS signals are partially or completely unavailable, the INS operates in coasting mode, i.e., the navigation parameters can be estimated completely independently of the GNSS. Consequently, the overall navigation performance depends greatly on the accuracy of the inertial signal, or more precisely, on the errors of the inertial signal. These errors are integrated into the INS, and their impact increases dramatically over time. In conclusion, accurate modeling and estimation of the error of inertial signals are crucial for improving the quality of navigation performance.

The errors of inertial sensors can generally be divided into deterministic errors and stochastic errors. Most deterministic errors can be compensated for by physical models and have been widely studied [1–3], while stochastic errors are difficult to model. This is because there are many influence factors and normally the model is too complex to estimate correctly. Traditional estimation methods, such as Allan variance (AV) and power spectral density (PSD) analysis methods, have obvious disadvantages when the stochastic error structure is complex [4]. AV is currently the most widely used method in engineering to identify and calibrate inertial sensors [5–9]. Although this method was originally intended to study the stability of oscillators, it has been successfully applied to problems in a large number of different types of sensors, among which is the modeling of inertial sensor errors. However, AV is only suitable for stochastic processes that can be clearly identified and separated in the spectral domain. More strictly, stochastic processes are thought not to be affected by spectral ambiguity, while spectral ambiguity is common for low-cost MEMS inertial measurement units (IMU) [10,11]. Since human judgment is required in the identification of noise when performing Allan variance, it will absolutely cause deviation in estimation parameters in most cases. For PSD analysis, the periodogram is an inconsistent estimator of the power spectral density function and can be badly biased even for large sample sizes (because of frequency leakage effects). Moreover, when the PSD has large variability over a very narrow frequency band, it will make the least squares optimization problem based on the difference between the empirical PSD and the model-based PSD more difficult to solve [12]. However, reference [13] associated the wavelet variance (WV) with the PSD, and WV can be calculated from the samples using the wavelet transform estimator. Accordingly, reference [14] proposed a generalized wavelet moment estimation method, which identifies the time series to be estimated as a combination of stochastic processes. In such cases, the GMWM estimator is asymptotically consistent, and the empirical WV of the series corresponds to the WV implied by the assumed model. The generalized least squares method has been adopted to minimize the discrepancy between the two and to estimate the parameters of the latter [15]. This method can effectively avoid the disadvantages of the above traditional methods and has significant practical application value. In addition to the modeling of a given sequence of IMU data, other researchers have studied the modeling of IMU error within filters. Reference [16] introduced the stochastic IMU error models within a Sage Husa adaptive robust Kalman filter. Reference [17] developed an adaptive Kalman filter with colored noise for gyroscope random drift. However, these researchers calibrated inertial sensors before or after the experiments, i.e., offline calibration. Moreover, they hardly considered the time-taken for the calibration process. For these reasons, we propose an algorithm for online modeling of inertial sensor errors based on GMWM and mainly focus on the modeling of the stochastic error of the vehicle-mounted inertial sensors in urban areas.

The algorithm in this work is designed under the frame of INS/GNSS integration within KF. The main contributions of this paper are as follows: Firstly, this paper realizes autonomous stochastic error modeling by constructing a complete stochastic error model and a model ranking criterion. Secondly, this paper proposes a static state detecting algorithm with an adaptive threshold. It collects and accumulates static data when the vehicle stops. Finally, the frame of INS/GNSS integration is developed, combining autonomous stochastic error modeling and detecting module as feedback within the EKF. It realizes online modeling of the stochastic errors and provides a more accurate navigation solution as the vehicle runs. Two experiments, the GNSS denial experiment and GNSS available experiment, are designed to validate the feasibility of the proposed algorithm.

The rest of the paper is organized as follows: Section 2 introduces the main contributions of this paper. It introduces the principles of the system frame first. Then, it explains the basic principle of GMWM in Section 2.1 and presents the autonomous modeling in Section 2.2. The static state detection with adaptive threshold is developed in Section 2.3. Sections 2.1–2.3 detail the whole algorithm for online modeling of inertial errors. Section 3 details two conducted experiments to compare the algorithm proposed in this paper with other classical methods, and analyzes the positioning performance. Section 4 reveals the conclusions and further research directions.

2. Online Stochastic Error Modeling of Inertial Sensors

Generally, measurements of inertial sensors have various errors which contaminate the true measurements from the gyroscopes and the accelerators, decreasing the performance of the navigation solutions. For inertial sensors, the angular rate of the gyroscope and the specific force observation equation of the accelerometer are calculated as follows:

$$\begin{aligned} \omega &= \omega_{true} + b_\omega + S_\omega \omega + c_\omega t + \varepsilon_\omega \\ f &= f_{true} + b_f + S_f f + c_f t + \varepsilon_f \end{aligned} \quad (1)$$

where ω is the gyro output measurement, ω_{true} is the true rotation rate, b_ω is the gyroscope bias, S_ω is the gyroscope scale factor, c_ω is the gyroscope temperature coefficient, t is the temperature, f is the accelerometer output measurement, f_{true} is the true specific force, b_f is the accelerometer bias, S_f is the accelerometer scale factor, c_f is the accelerometer temperature coefficient, and ε_ω and ε_ω are the sensor noises. For IMU calibration, the six-position static calibration test is generally adopted to determine both the deterministic bias and scale factor of the gyros and accelerometers. Moreover, the temperature variation effect is neglected here in this paper due to the relatively short duration datasets. Thus, this research mainly explores the modeling and estimation of sensor noise, ε_ω and ε_f.

GMWM is a recently proposed technique to model the sensor noise ε_ω and ε_f. We make further improvements to GMWM to realize online modeling and design a novel navigation solution. It is designed within an INS/GNSS integrated system with feedback to provide information on IMU stochastic errors. The layout of the frame is shown in Figure 1. Here, we first introduce the basic principles of the system, then more details and some supporting conclusions will be further explained in Sections 2.1–2.3 of each module accordingly.

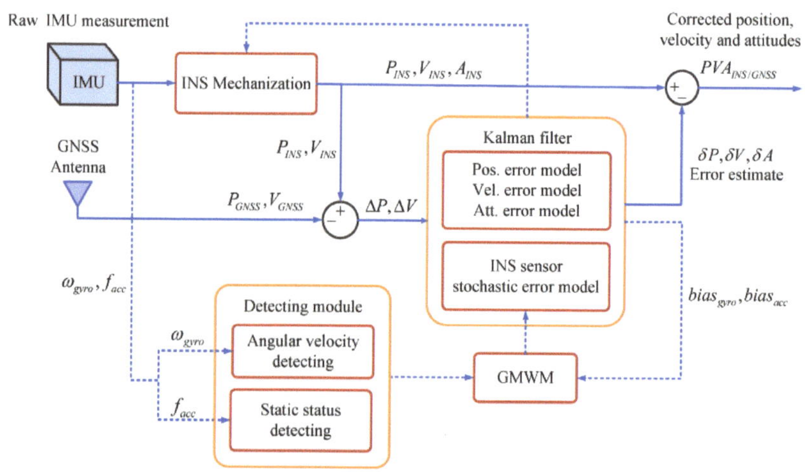

Figure 1. Layout of online stochastic error modeling of inertial sensors in INS/GNSS integration.

- When GNSS signal is received, the INS/GNSS integration within EFK obtains the navigation solution. Differing from traditional INS/GNSS integration, this system has an additional feedback to the EKF and has a stochastic error model as an augmented error vector in the EKF. When GNSS signal is blocked, the INS works in coasting mode;
- The feedback consists of autonomous stochastic error modeling of inertial sensors and the detecting module. The raw observations from the IMU will go to the detecting module. The static state detecting with adaptive threshold judges the motion state of the vehicle. If the vehicle is static, e.g., waiting before traffic lights or temporarily avoiding pedestrians or other vehicles, the IMU data accumulate during this static duration. After bias removal, accumulations will go to autonomous stochastic error modeling where the GMWM will prepare the best model of inertial stochastic error. Meanwhile, the angular rate detector will judge if the angular rate of the vehicle is higher than 30°/s. If not, the best model can work as an augmented vector together

with the navigation error models in a Kalman filter. The Kalman filter outputs the bias of gyros and accelerators fed to autonomous stochastic error modeling in return;
- The IMU data for autonomous stochastic error modeling only accumulate during the static duration; therefore, the amount of data is quite small, and the calculation is generally completed within a few seconds. As the vehicle runs, the IMU data accumulation increases with extension of the static duration; consequently, the accuracy of the stochastic model improves.

2.1. Generalized Method of Wavelet Moments

GMWM is an estimation method based on the idea of generalized method of moments (GMM) estimators and the wavelet variance (WV) [16]. The GMWM makes use of the relationship between the WV and the parameters of a latent process, estimating the latter by minimizing the distance between the empirical WV and model-based WV [14]. The calculation process for the GMWM can be illustrated as follows:

The wavelet coefficients are built using wavelet filters $\left\{\widetilde{h}_{j,l} : j = 1, \cdots J\right\}$, where the j-th level wavelet filter of length is $L_j = (2^j - 1)(L_j - 1) + 1$. In the stationary or non-stationary process, we get the maximum overlap discrete wavelet transform (MODWT) coefficients, $W_{j,k}$

$$\overline{W}_{j,k} = \sum_{l=0}^{L_1-1} \widetilde{h}_{j,l} Y_{k-1}, k \in Z \tag{2}$$

WV is defined as the variance of the wavelet coefficients, $W_{j,k}$, at the dyadic scales $\tau_j = 2^{j-1}$.

$$v^2(\tau_j) = \text{var}\left[\overline{W}_{j,k}\right] \tag{3}$$

For a finite observed process, the MODWT-estimated WV can be calculated as follows:

$$v(\tau_j) = \frac{1}{M_j} \sum_{k=L_j}^{N} \overline{W}_{j,k}^2 \tag{4}$$

where $W_{j,k} = \sum_{l=0}^{L_j-1} \widetilde{h}_{j,l} y_{k-l}, k \in (L_j; N)$ and $M_j = N - L_j + 1$.

The PSD of the wavelet coefficient, $S_{W_j}(f) = \left|\widetilde{H}_j(f)\right|^2 S_{F_\theta}(f)$, supports a direct relationship between WV and PSD, where the variance of the mentioned series of wavelet coefficients are the direct integral of its PSD as follows:

$$v(\tau_j) = \int_{-1/2}^{1/2} S_{W_j}(f)df = \int_{-1/2}^{1/2} \left|\widetilde{H}_j(f)\right|^2 S_{F_\theta}(f)df \tag{5}$$

where $H_j(f)$ is the transfer function of the filter $h_{j,l}$, F_θ is the model built using one or more stochastic processes that describes the dynamics of the observed sensor error sequence, and S_{F_θ} is the PSD implied by the model F_θ. Therefore, there is an implicit connection between the WV and the parameters of the data generating model, F_θ. We exploit this connection by defining an estimator for θ, namely by matching a sample estimate of the WV together with the model-based expression of the WV. The GMWM estimator is used to minimize the distance between the empirical and estimated WV in order to estimate the parameters of the latent composite processes as follows:

$$\hat{\theta} = \underset{\theta \in \Theta}{\text{argmin}} (\hat{v} - v(\theta))^T \Omega (\hat{v} - v(\theta)) \tag{6}$$

where θ represents the time series model parameter that we intend to estimate belonging to the compact set Θ, and Ω is a symmetric positive definite weighting matrix chosen in a suitable manner to make the GMWM estimator is as efficient as possible. It is also important to mention that this method could also be based on the AV since the aforementioned Haar WV is simply twice the AV with additional benefits. A detailed mathematical background on the GMWM can be found in [12,14].

The next step is the parameter estimation of the possible model, F_θ, and ranking these models using specific criterion to determine the best one.

2.2. Autonomous Stochastic Error Modeling

2.2.1. Complete Stochastic Error Model

Within our research interests, the problem of modeling and estimation focuses on the stochastic error components affecting gyroscopes and accelerators. Hence, we restricted the possible models, F_θ, to a complex model which is defined as a combination of independent basic stochastic processes. These basic stochastic processes are widely used within the design of navigation filters and can precisely describe the behaviors of inertial sensors: gaussian white noise (WN), random walk (RW), drift (DR), quantization noise (QN) and finite auto regressive model (AR) [18–23].

In order to cover as many basic stochastic processes as possible, this paper defines the complete model of IMU stochastic error as

$$error = 4 \times AR + DR + WN + QN + RW \tag{7}$$

The complete model consisting of these basic stochastic processes is universally suitable to a variety of inertial sensors. When identifying the structure of the stochastic error, all the combinations of these basic stochastic processes within the complete model are regarded as the candidate models. Then, parameters of all the candidate models are estimated by GMWM and later, the best or most suitable model is selected by a designed model ranking criterion.

2.2.2. Model Ranking Criterion

After using GMWM to estimate the parameters of all candidate models, it is necessary to establish a model ranking criterion according to the actual requirements of noise modeling and estimation in the practice appliance. According to this criterion, candidate models are evaluated and ranked to select the most suitable one. Ref. [15] gave a ranking criterion called wavelet variance information, which weighs the model fitness and computational complexity to evaluate the trade-off between the model accuracy and the estimation of the time-taken. Ref. [15] statically collected IMU data for several hours, leading to a significant amount of data. It makes the parameter estimation quite time-consuming, taking even up to several hours, especially when the model contains a large number of stochastic processes. However, for the online modeling problem of stochastic errors studied in this paper, a relatively small amount of IMU data are processed. Accordingly, the computational calculation is small, which leads to tiny difference between the time-taken for estimating different candidate models. Consequently, only the model accuracy is considered in this research here.

The objective function given by Equation (6) can be regarded as a mismatch between the WV calculated by the model F_θ and the WV calculated by observed measurements, and its purpose is to minimize this difference and make the model more closely match the observations. Based on this physical explanation, the ranking criterion can be defined as goodness of fitness (GOF) as follows:

$$GOF = (\hat{\nu} - \nu(\hat{\theta}))^T \Omega (\hat{\nu} - \nu(\hat{\theta})) \tag{8}$$

After the parameter estimation of all candidate models is completed by GMWM, all candidate models are evaluated by the GOF criterion, and the model with the smallest GOF value is selected as the optimal model.

A summary of the overall flow of autonomous stochastic error modeling of inertial sensors is shown in Figure 2.

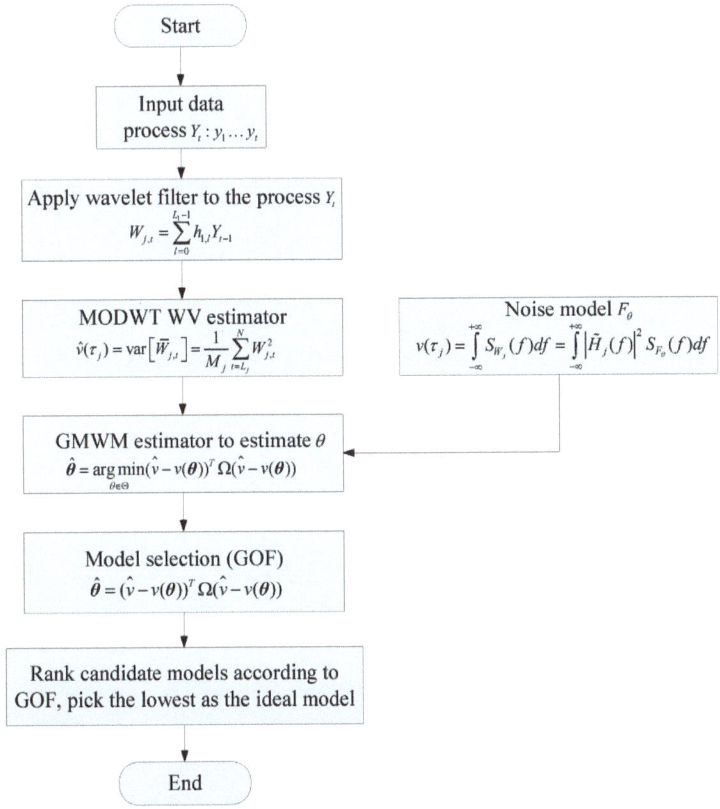

Figure 2. The overall flow of autonomous stochastic error modeling of inertial sensors.

2.3. *Static State Detecting with Adaptive Threshold*

For the vehicle navigation system, the inertial outputs contain some specific constraint information under different motion states. Under the premise of adding no extra cost and devices, the information can provide additional constraints for the navigation system, which is helpful to improve the accuracy and stability of the integrated navigation system. In particular, when the vehicle is in a static state, the inertial sensor is not affected by vehicle maneuvers, so the accelerations kept stable and the velocity remains close to zero. Nevertheless, the stability analysis of the accelerometer output can be performed to detect the static states. A common method is to use the standard deviation of the accelerometer output in a fixed time window as the test statistic [24]:

$$\begin{cases} T_i(\text{ Accel. } X) < \lambda \\ T_i(\text{ Accel. } Y) < \lambda \end{cases}, T_i = \sqrt{\frac{1}{N-1}\sum_{j=i-N+1}^{i}(A_i - U_i)^2} \quad (9)$$

where A_i is the accelerometer output at epoch i, U_i is the mean value of the data in the fixed time window at epoch i, N is the number of data in the fixed time window, and $N = 100$ in this paper. T_i is the standard deviation of the data in the fixed time window at epoch i. Empirically, $\lambda = 0.02$.

Due to the different characteristics of different inertial sensors, the standard deviation of the output in the static state is different as well. Hence, adopting a fixed empirical threshold is prone to misjudgment. Moreover, for the circumstances studied in this paper it is crucial to ensure that the data for GMWM estimation modeling are from a static state False detections have a more significant impact on the accuracy of modeling estimation than missed detections. Aiming at solving this problem, this paper proposes an adaptive method to determine the detecting threshold as follows:

$$\lambda_i = \begin{cases} 2\left(\frac{k-1}{k}\right)\lambda_{i-1} - \frac{|T_i - T_{i-1}|}{k} & , T_{i-1} < \lambda_{i-1} \\ \lambda_0 & , T_{i-1} \geq \lambda_{i-1} \end{cases} \quad (10)$$

where k is the number of the static states that has been detected at epoch i. λ_i is the test statistic at epoch i, λ_0 is the initial value, and $\lambda_0 = 0.02$, empirically.

The first term in Equation (10) physically means that when the standard deviation of the data is smaller than the test threshold, it is more inclined to assume that the vehicle stays in a continuous static state, so the detection threshold increases to make it easier to detect the static state. The second item physically means that if the standard deviation of the data between the current epoch and the previous epoch is quite large, the motion state of the vehicle has changed. It is more inclined to assume that the vehicle is not in the static state, so the detection threshold decreases to make it more difficult to detect the static state. The effect of this static state detection method with adaptive threshold will be experimentally verified in Section 4.

The static state detection with an adaptive threshold works together with the angular rate detection as the detecting module to provide extra guidance for autonomous stochastic error modeling. The detecting module is supported by two important conclusions given in [21]. One is that although stochastic errors do depend on the dynamic characteristics, for one specific IMU, the structure of the stochastic error is not affected by the applied dynamics. Only parameter values differ according to dynamic variations. The other is that for the general range of MEMS-IMU, the largest factor among various dynamic characteristics affecting stochastic errors is the angular rate. Moreover, a relatively low angular rate, normally below $30°/s$, does not cause an evident change in the parameters. Fortunately, this is most commonly the case when turning in urban cities. Therefore, the angular rate detector will judge if the angular rate of the vehicle is higher than $30°/s$. If not, the stochastic error model estimated under the static state can replace the dynamic model.

In practical applications, ranking all candidate models to identify the error structure will take up the most majority of the calculation time. However, identifying the structure will be conducted only once, i.e., after the first static duration, since the structure will not differ with the dynamic variations. Once the error structure is determined, only the parameter estimation of this fixed model structure will be conducted later through the experiment, and it can be processed within several seconds. It means that the autonomous stochastic error modeling lasts only a few seconds after the vehicle starts moving, then the KF will be able to adopt the stochastic error to obtain the navigation solution.

3. Experiments

This section is split into two main parts:

(1) GNSS denial experiments, which are designed to verify the feasibility of the autonomous stochastic error modeling based on GMWM;

(2) GNSS available experiments, which are designed to verify the feasibility of the proposed algorithm for online stochastic error modeling of inertial sensors. We compare the navigation performance of the proposed algorithm with the other traditional methods and further analyze the performance of the proposed algorithm in different trajectory sections as the car runs.

3.1. GNSS Denial Experiment

As previously mentioned, AV is probably the most commonly used method for model identification and sensor calibration. In 1998, the IEEE standard officially put forward this technique as a noise identification method to determine the characteristics of the underlying random processes that perturb data. In general, AV only considers five basic stochastic processes: QN, WN, BI, RW, and DR. These processes correspond to the linear regions in a log–log plot, which will present a typical U/V-shaped curve in ideal circumstances. Therefore, parameters are usually estimated by performing linear regression of (visually) identified linear regions in such log–log plots. Further research on this principle can be found in [8–10,25,26]. A GMWM-based algorithm can be considered as a further extension of AV because they both identify and quantify the different noise terms that exist in inertial sensor data. Hence, in this experiment, we compare the three models with different GOF values generated by the autonomous stochastic error modeling method, AV method, and loosely coupled navigation solution with a reference trajectory.

The trajectory was produced by a car with SPAN NovAtel-CPT driving around an urban area of Beijing on 18 June 2021. The reference is provided by SPAN NovAtel-CPT under the post-processed solution. The raw IMU data were collected statically for 2 h at 125 Hz by SPAN NovAtel-CPT, as shown in Figure 3. After the data were fed to the autonomous stochastic error modeling method, we adopted three models with the minimum GOF values to analyze the performance as an example. Three artificial GNSS outages, each lasting 60 s, were designed to cover two turns and one straight line, as shown in Figure 4. The planimetric navigation drift in the earth centered earth fixed (ECEF) coordinate system at the end of GNSS outages was analyzed to judge the quality of the five different models.

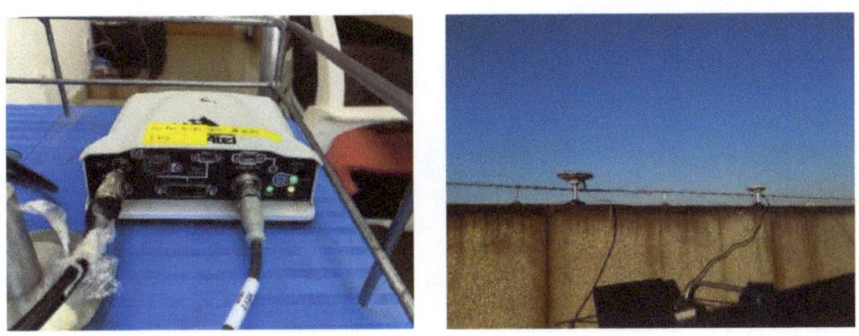

Figure 3. SPAN NovAtel (**left column**) is placed statically indoors and the antenna (**right column**) is set up outdoors to collect GNSS signals.

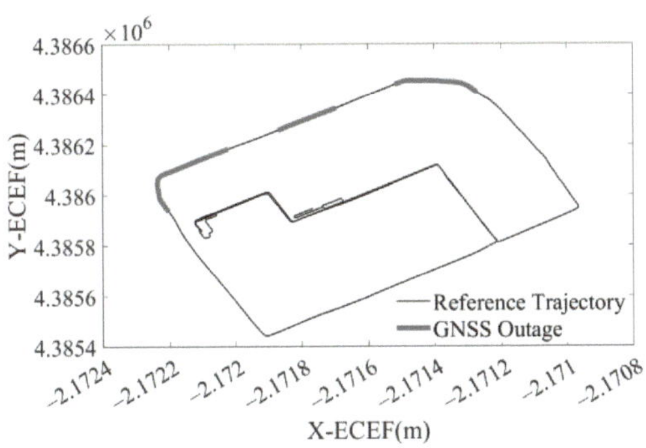

Figure 4. The reference trajectory and three artificial GNSS outages.

3.1.1. Stochastic Error Modeling

Based on the optimal model autonomous selection method proposed in Section 3.2, the 2 h static IMU data after bias removal were processed by the GMWM-based method and the AV-based method. A GMWM log–log plot of the best model with the minimum GOF and an AV log–log plot is shown in Figure 5. It implies that the GMWM-based method proposed in this paper can achieve a good fit to the static data. The curves in the AV log–log plot represent the characteristics of three stochastic processes, i.e., WN, BI, and RW. The WN parameters can be deduced from a slope of $-1/2$ at $\tau = 1$ on the left part. BI due to flicker noise in the measurements can be identified at the lowest point in the curve. The RW parameters can be deduced from a slope of $1/2$ on the right part.

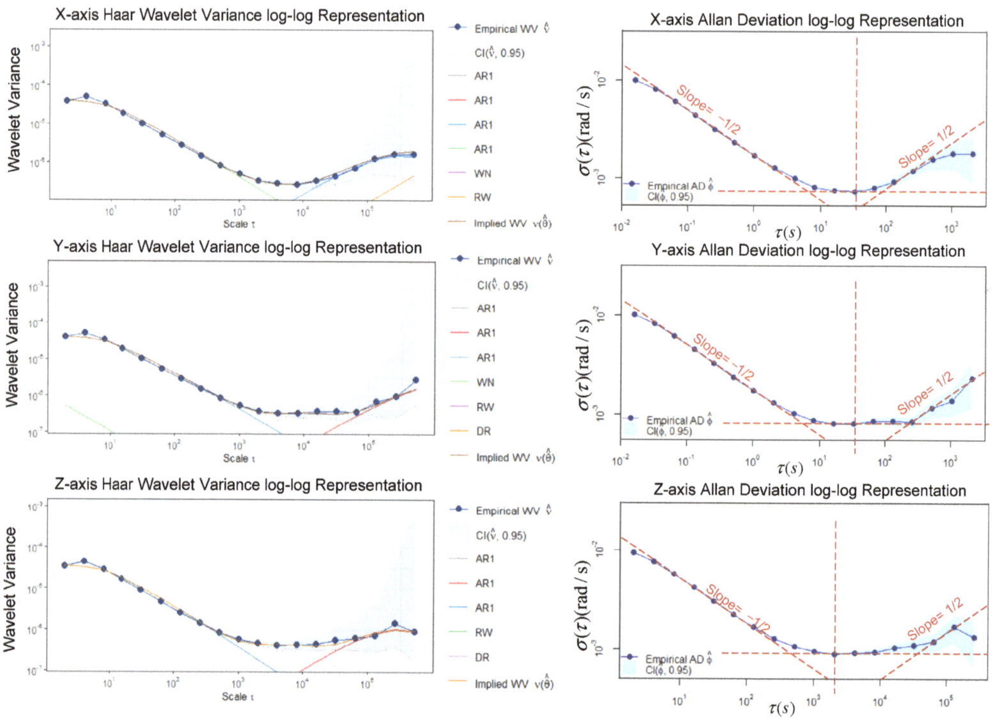

Figure 5. GMWM log–log plot of the three-axis gyroscope (**left column**) and AV log–log plot of the three-axis gyroscope (**right column**). The red dashed lines are auxiliary lines to help identify linear regions.

3.1.2. Experiment Validation

The five gyro stochastic models to be validated are as follows:

Model 1: Loosely coupled navigation solution;

Model 2: AV-based model;

Model 3–5: GMWM-based model with three minimum GOF values (values decrease from 3 to 5).

The navigation drifts of the five models at the end of three GNSS outages are shown in Figure 6, and the maximum navigation errors of the five models at the end of three GNSS outages are shown in Table 1. Figure 6 indicates that during the three GNSS outages, the GMWM-based model generally outperforms the other two models, with the AV-based model in the middle of the two models. Among the three models generated by the GMWM method, the best model with minimum GOF values gives the best results with 206 m, 153 m, and 182 m during three outages, respectively. Compared with the loosely coupled navigation solution, the best GMWM model has an accuracy increase of 26.8%, 37.2%, and 38.3% during three outages, respectively.

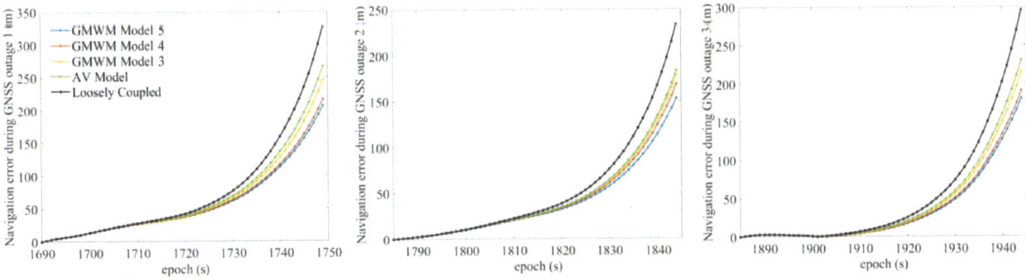

Figure 6. The navigation drift of the five models at the end of three GNSS outages.

Table 1. The maximum navigation error of the five models at the end of three GNSS outages.

Model		Navigation Drift at the End of GNSS Outage (m)		
		Outage 1	**Outage 2**	**Outage 3**
Loosely coupled		326	243	295
AV		264	186	227
GMWM	Model 3	254	178	215
	Model 4	216	164	191
	Model 5	206	153	182

3.2. GNSS Available Experiment

As mentioned above, human judgment in the identification of five noise terms will absolutely cause deviation in estimation parameters when performing Allan variance. Aiming to solve this problem, reference [27] provided a method to automate this process by maximizing the likelihood function of the assumed state-space models of interest using a constrained version of the expectation maximization (EM) algorithm [28]. Hence, we compared the algorithm for online stochastic error modeling proposed with the AV-based method, the EM-based method, and the traditional EKF solution to verify its efficiency.

In this experiment, the trajectory was produced by a car with SPAN NovAtel-CPT driving around an urban area of Xuzhou on 16 April 2018. The reference trajectory was provided by SPAN NovAtel-CPT under the post-processed solution, as shown in Figure 7. The whole trajectory lasts 50 min and the IMU data were sampled at 125 Hz. The IMU data accumulate during the static epochs detected by static state detecting with an adaptive threshold. Autonomous modeling prepares the best stochastic error model as the augmented vector in INS/GNSS integration with EKF.

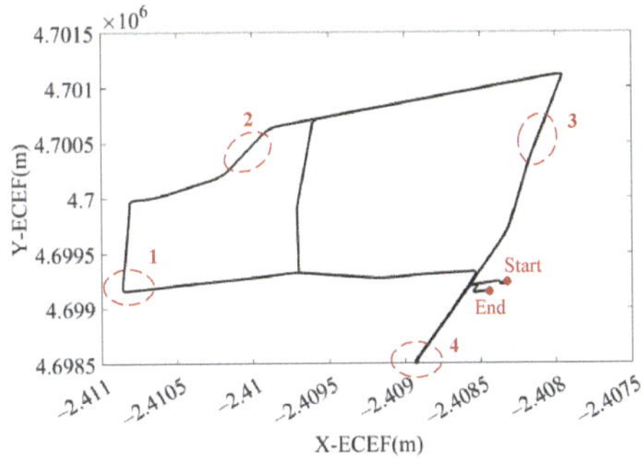

Figure 7. The reference trajectory of the GNSS available experiment in ECEF. The end and the start is marked in red point. Four sections are given detailed figures for further analysis (in dashed red circle).

3.2.1. Static State Detection

The result of static state detection with adaptive threshold for the X axis during the whole trajectory is shown in Figure 8. Comparing the curve of test statistic and car velocity, it is clear that this method successfully detects almost all of the static states, including six relatively long stops and other temporary stops. The six long stops are marked by the light red area for further experimental analysis. For the static states marked by the red stars, the standard deviation statistics of the fixed window have obvious statistical characteristics, i.e., they blow the adaptive threshold, which can effectively identify the static epochs. In particular, in the right column are two zoomed-in figures of the detected results. It shows how the threshold (dark red lines) adjust to the actual circumstances of the detection. Since the previous epochs are identified as static states, the adaptive threshold increases to extend this inertia. However, when the test statistic fluctuates sharply, the adaptive threshold shows an immediate decrease.

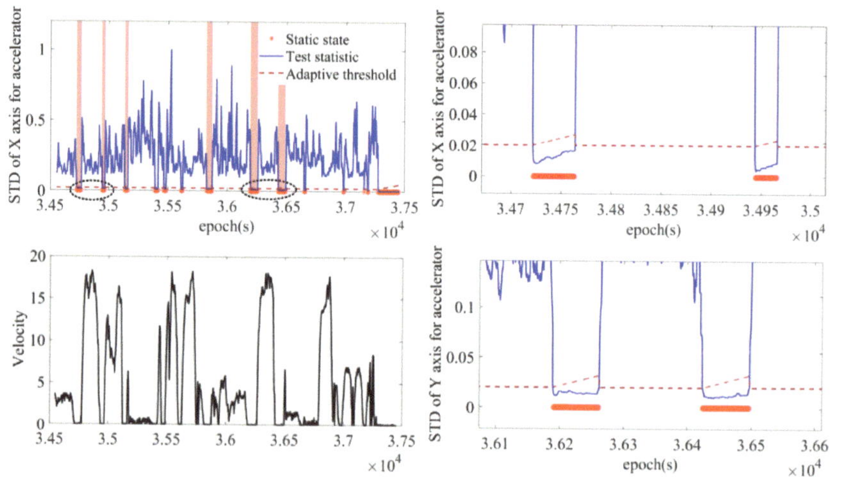

Figure 8. The result of static state detection with adaptive threshold compared with the velocity curve (**left column**) and two detailed figures of static states (**right column**). Two sections are given detailed figures for further analysis (in dashed black circle).

3.2.2. Stochastic Error Modeling

Based on the detection results of static states, raw IMU data accumulate during the static epochs, reaching a duration of 318 s. The accumulation of static data was fed to the GMWM-based model, the AV-based model, and the EM-based model. For a clear view of the estimation results of these two methods, Figure 9 simply gives the estimation results based on all the static data accumulated throughout the whole trajectory. The parameter estimation results of the AV-based method, the EM-based method, and the GMWM-based method are shown in Tables 2–4, respectively.

It is worth mentioning that AV plot-plot of X axis and Y axis does not present a typical U- or V-shaped curve. Only WN and RW can be identified from the X axis and only WN and BI can be identified from the Y axis. The amount of data accumulation is quite small; therefore, it requires a much larger amount of data to present the specific characteristics of other processes. Moreover, the human identification in each process is probably unreliable, then the parameters estimated by performing linear regression of these visually identified linear regions may have significant deviations. The EM-based approach is very sensitive to the initial values of parameters. When the initial values are "far" from the true values, the EM-based approach is likely to converge to a local minima. Hence, the initial values are set to the results estimated by AV.

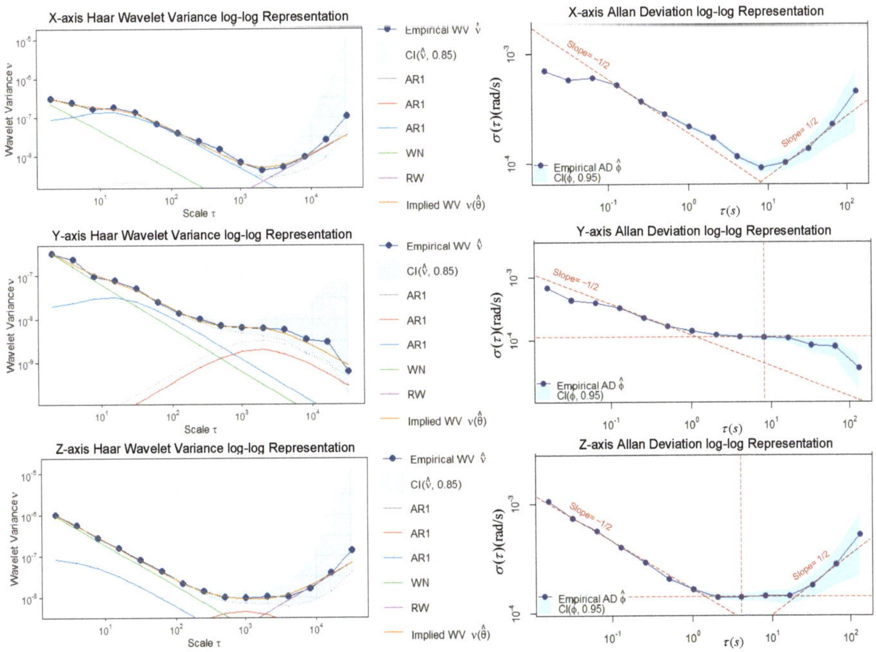

Figure 9. A GMWM log–log plot of the three-axis gyroscope (**left column**) and an AV log–log plot of the three-axis gyroscope (**right column**). The red dashed lines are auxiliary lines to help identify linear regions.

Table 2. The parameter estimation results of AV-based method.

Parameter	X	Y	Z
σ_{WN}	1.763×10^{-3}	1.347×10^{-2}	1.309×10^{-2}
σ_{RW}	7.171×10^{-3}		6.334×10^{-3}
σ_{BI}		1.762×10^{-4}	1.407×10^{-4}
T_{BI}		4.096	4.096

Table 3. The parameter estimation results of EM-based method.

Parameter	X	Y	Z
σ_{WN}	1.822×10^{-3}	1.441×10^{-2}	1.358×10^{-2}
σ_{RW}	7.167×10^{-3}		6.401×10^{-3}
σ_{BI}		1.759×10^{-4}	1.413×10^{-4}
T_{BI}		4.092	4.093

Table 4. The parameter estimation results of GMWM-based method.

Parameter		X	Y	Z
		Model = 3 × AR + WN + RW	**Model = 3 × AR + WN + RW**	**Model = 3 × AR + WN + RW**
AR	σ_1	3.009×10^{-9}	7.156×10^{-11}	4.461×10^{-10}
	β_1	8.574×10^{-1}	9.979×10^{-1}	9.298×10^{-1}
	σ_2	1.723×10^{-9}	4.498×10^{-11}	1.871×10^{-10}
	β_2	8.852×10^{-1}	9.978×10^{-1}	9.961×10^{-1}
	σ_3	3.064×10^{-7}	6.96×10^{-8}	2.340×10^{-7}
	β_3	7.512×10^{-1}	7.611×10^{-1}	3.866×10^{-1}
WN	σ_{WN}	4.551×10^{-7}	6.264×10^{-7}	1.878×10^{-6}
RW	σ_{RW}	1.276×10^{-11}	4.171×10^{-15}	2.506×10^{-11}

3.2.3. Experiment Validation

Based on the stochastic error model generated above, three models were compared as follows:

Model 1: INS/GNSS integration within EKF;

Model 2: EKF with AV-based model;

Model 3: EKF with EM based model;

Model 4: EKF with an online GMWM-based model.

The reference trajectory is shown in the left column in Figure 10 and the four detailed figures of trajectories generated by the four models are shown on the right. It is clear that the algorithm proposed is the closest to the reference trajectory regarding both turns and straight lines. Figure 11 indicates the navigation error of the three models in X, Y and Z of the ECEF coordinate system. The online GMWM model reflects a more accurate positioning result than the other two models. Table 5 shows the navigation error root mean square error (RMS) of the three models. It clarifies that the online GMWM model has the smallest navigation errors of 1.3265 m, 1.4384 m, and 1.6629 m in X, Y, and Z, respectively. Compared with the AV-based method, it has an improvement of 16.6%, 14.3%, and 14.7% in the navigation accuracy of X, Y, and Z, respectively. Compared with the EM-based method, it has an improvement of 10.8%, 8.2%, and 7.1% in the navigation accuracy of X, Y, and Z, respectively. It significantly validates the effect of the online algorithm.

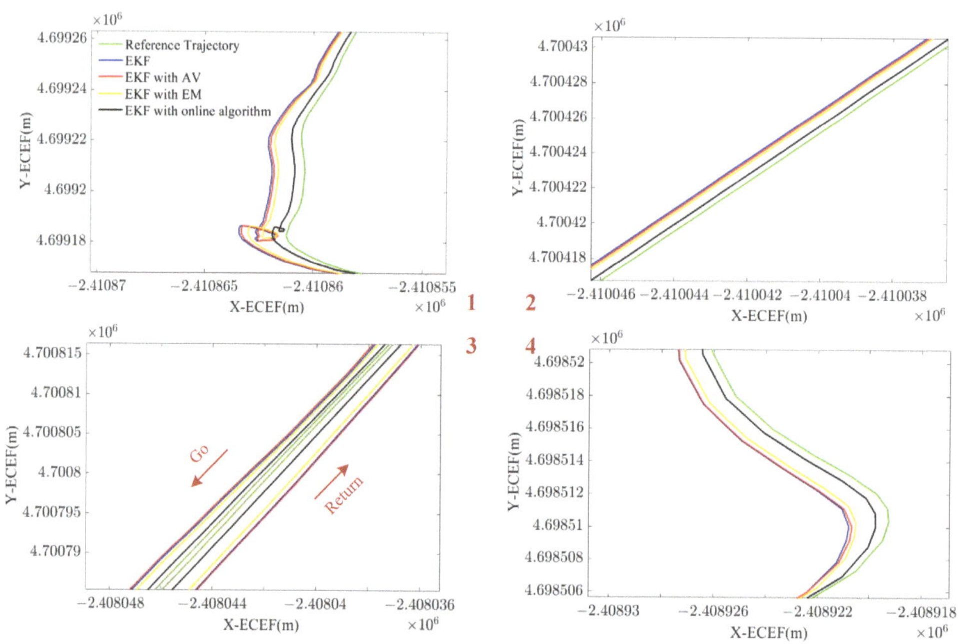

Figure 10. Four detailed figures of trajectories generated by the four models.

Moreover, the stochastic error model become more and more accurate with the accumulation of the static data. Table 6 shows the navigation error RMS of the online algorithm for the trajectory sections between two adjacent long stops. The positioning accuracy increases in X, Y, and Y as the car drives, which further confirms the ability of the online GMWM algorithm to enhance the navigation accuracy.

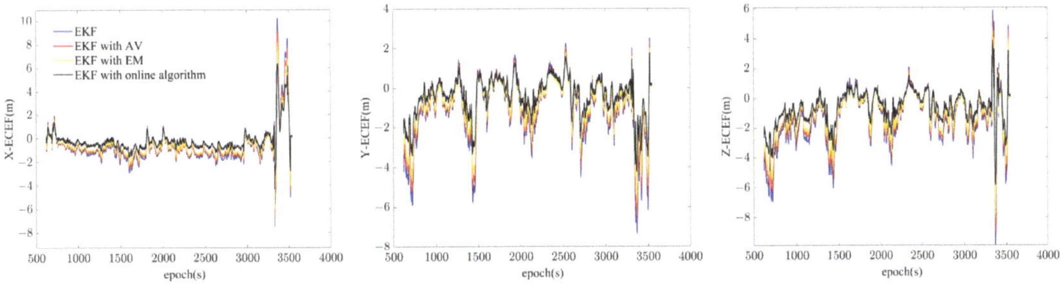

Figure 11. The navigation error of the three models in X, Y, and Z of ECEF.

Table 5. The navigation error RMS of the three models.

Model	Navigation Error RMS (m)		
	X	Y	Z
Model 1	1.6675	1.7170	2.0956
Model 2	1.5916	1.6802	1.9492
Model 3	1.4877	1.5665	1.7902
Model 4	1.3265	1.4384	1.6629

Table 6. The navigation error RMS of the online GMWM algorithm between the two adjacent long stops and the time-taken after long stops.

Stops	Navigation Error RMS (m)			Time-Taken
	X	Y	Z	
1–2	1.5379	1.6442	1.8184	1.4 min
2–3	1.5267	1.6279	1.7950	1.877 s
3–4	1.4860	1.6118	1.7692	1.996 s
4–5	1.4369	1.5543	1.6974	2.184 s
5–6	1.3872	1.4265	1.6483	2.338 s
6–end	1.2969	1.4098	1.6318	2.434 s

Particular attention should be given to the time-taken for the online algorithm. All the static epochs detected take up 318 s in the whole trajectory. Due to the relatively small amount of data accumulation, calculations with the online algorithm take little time. After the first relatively long stop, identifying the model structure and estimating the parameters takes 1.4 min. After that, only the estimation of the fixed model structure will be performed, which takes much less time, i.e., it is finished within 3 s. Table 6 also gives the time-taken after six long stops in the right column. After the sixth long stop, the static data accumulated reach the largest amount in the whole trajectory. It means that the stochastic error model can work as an augmented vector within EKF solution, with a delay of 2.434 s. Therefore, the algorithm has an excellent performance in online modeling of stochastic errors.

4. Conclusions

This paper proposes a new method for online modeling of stochastic errors of inertial sensors, which combines static state detection with an adaptive threshold and the autonomous stochastic error model based on GMWM. Firstly, the limitations of other widely used stochastic modeling methods including AV and PSD are analyzed. Then, two experiments are designed to compare the online GMWM algorithm with the AV-based method and the EM-based method. The GNSS denial experiment proves the feasibility of the proposed autonomous stochastic error model based on GMWM. Meanwhile, it reveals the capability of the online GMWM algorithm to estimate the stochastic error and limit the navigation drift. The GNSS available experiment demonstrates that the online GMWM can significantly improve the navigation accuracy as the vehicle runs. Moreover, it validates the online performance as well.

This paper mainly focuses on exploring the stochastic error of inertial sensors, which is related to the natural characteristics of the sensor itself. Meanwhile, for vehicle-mounted sensors, the dynamic environment of the vehicle may affect the error behavior of the sensor as well. Given the conclusions of our research, a stochastic error model in a static environment can work instead of a model at low speed. Therefore, further study is warranted on how to construct the error models and estimate the parameters based on dynamic characteristics. We will focus on how the inertial error model changes with vehicle maneuvers at high speed. Future work aims to set up and conduct experiments which enable the construction and analysis of error signals acquired in dynamic environments. Therefore the observability of some processes and the justification of employing complex stochastic models for MEMS inertial sensors can only then be fully verified.

Author Contributions: Conceptualization, methodology, software, validation, formal analysis, investigation, resources, data curation, writing—original draft preparation, visualization: L.Z. (Luodi Zhao) writing—review and editing, supervision, project administration, funding acquisition: L.Z. (Long Zhao). All authors have read and agreed to the published version of the manuscript.

Funding: This research was funded by the National Science Foundation of China, grant numbers 42274037 and 41874034, the Beijing Natural Science Foundation, grant number 4202041, and the National key research and development program of China, grant number 2020YFB0505804.

Institutional Review Board Statement: Not applicable.

Informed Consent Statement: Not applicable.

Data Availability Statement: Not applicable.

Acknowledgments: Thanks to China University of Mining and Technology for the data collected in Xuzhou on 16 April 2018.

Conflicts of Interest: The authors declare no conflict of interest.

References

1. Berman, Z. Efficient Error Model Construction. In *Itzhack Y. Bar-Itzhack Memorial Symposium on Estimation, Navigation, and Spacecraft Control*; Springer: Berlin/Heidelberg, Germany, 2012.
2. Fong, W.T.; Ong, S.K.; Nee, A.Y.C. Methods for in-field user calibration of an inertial measurement unit without external equipment. *Meas. Sci. Technol.* **2008**, *19*, 085202. [CrossRef]
3. Titterton, D.; Weston, J. Strapdown inertial navigation technology, 2nd edition. *IEEE Aerosp. Electron. Syst. Mag.* **2005**, *20*, 33–34. [CrossRef]
4. Allan, D.W. Statistics of Atomic Frequency Standards. *Proc. IEEE* **1966**, *54*, 221–230. [CrossRef]
5. Guerrier, S. Integration of Skew-Redundant MEMS-IMU with GPS for Improved Navigation Performance. 2008. Available online: https://www.academia.edu/15379282/Integration_of_Skew_Redundant_MEMS_IMU_with_GPS_for_Improved_Navigation_Performance (accessed on 22 November 2022).
6. Guerrier, S. Improving Accuracy with Multiple Sensors: Study of Redundant MEMS-IMU/GPS Configurations. In *Ion Gnss*; Institute of Navigation: Savannah, GA, USA, 2009.
7. Hou and Haiying, Modeling Inertial Sensors Errors Using Allan Variance [Microform]. Available online: https://prism.ucalgary.ca/handle/1880/41609 (accessed on 22 November 2022).
8. El-Sheimy, N.; Hou, H.; Niu, X. Analysis and Modeling of Inertial Sensors Using Allan Variance. *IEEE Trans. Instrum. Meas.* **2008**, *57*, 140–149. [CrossRef]
9. Xing, Z.; Gebre-Egziabher, D. Modeling and bounding low cost inertial sensor errors. In Proceedings of the Position, Location & Navigation Symposium, Twin Cities, MN, USA, 5–8 May 2008.
10. Vaccaro, R.J.; Zaki, A.S. Statistical modeling of rate gyros. *IEEE Trans. Instrum. Meas.* **2012**, *61*, 673–684. [CrossRef]
11. Greenhall, C.A. Spectral ambiguity of Allan variance. *IEEE Trans. Instrum. Meas.* **1998**, *47*, 623–627. [CrossRef]
12. Percival, D.B.; Walden, A.T. *Wavelet Methods for Time Series Analysis*; Cambridge Series in Statistical and Probabilistic Mathematics; IEEE: Monterey, CA, USA, 2000.
13. Percival, D.P. On estimation of the wavelet variance. *Biometrika* **1995**, *82*, 619–631. [CrossRef]

14. Guerrier, S.; Skaloud, J.; Stebler, Y.; Victoria-Feser, M.-P. Wavelet-variance-based estimation for composite stochastic processes. *JASA J. Am. Stat. Assoc.* **2013**, *108*, 1021–1030. [CrossRef] [PubMed]
15. Stebler, Y.; Guerrier, S.; Skaloud, J.; Victoria-Feser, M.-P. Generalized method of wavelet moments for inertial navigation filter design. *IEEE Trans. Aerosp. Electron. Syst.* **2014**, *50*, 2269–2283. [CrossRef]
16. Narasimhappa, M.; Mahindrakar, A.D.; Guizilini, V.C.; Terra, M.H.; Sabat, S.L. MEMS-Based IMU Drift Minimization: Sage Husa Adaptive Robust Kalman Filtering. *IEEE Sens. J.* **2020**, *20*, 250–260. [CrossRef]
17. Jin, K.; Chai, H.; SU, C.; Xiang, M. Adaptive Kalman filter method with colored noise for fiber optic gyroscope random drift. *Acta Geod. Cartogr. Sin.* **2022**, *51*, 80–86.
18. Harris, D.; Mátyás, L. Introduction to the Generalized Method of Moments Estimation. In *Generalized Method of Moments Estimation*; Themes in Modern Econometrics; Mátyás, L., Ed.; Cambridge University Press: Cambridge, UK, 1999; pp. 3–30.
19. Savage, P.G. Analytical Modeling of Sensor Quantization in Strapdown Inertial Navigation Error Equations. *J. Guid. Control Dyn.* **2002**, *25*, 833–842. [CrossRef]
20. *Std 1293-2018*; IEEE Standard Specification Format Guide and Test Procedure for Linear Single-Axis, Nongyroscopic Accelerometers. IEEE: Piscataway, NHJ, USA, 2019; pp. 1–271. [CrossRef]
21. Abdel-Hamid, W. *Accuracy Enhancement of Integrated MEMS-IMU/GPS Systems for Land Vehicular Navigation Applications*; Library and Archives Canada: Calgary, CA, USA, 2005.
22. Nassar, S. *Improving the Inertial Navigation System (INS) Error Model for INS and INS/DGPS Applications*; National Library of Canada: Calgary, CA, Canada, 2003.
23. Guerrier, S.; Molinari, R.; Stebler, Y. Theoretical Limitations of Allan Variance-based Regression for Time Series Model Estimation. *IEEE Signal Process. Lett.* **2016**, *23*, 597–601. [CrossRef]
24. Martí, E.D.; Martín, D.; García, J.; de la Escalera, A.; Molina, J.M.; Armingol, J.M. Context-Aided Sensor Fusion for Enhanced Urban Navigation. *Sensors* **2012**, *12*, 16802–16837. [CrossRef] [PubMed]
25. González-de-la-Rosa, J.; Sierra-Fernández, J.; Florencias-Oliveros, O.; Espinosa-Gavira, M.; Palomares-Salas, J.; Agüera-Pérez, A. Improving frequency characterization for power systems using the Allan variance and a GPS-controlled reference: Measurement procedure, test and validation. *Measurement* **2022**, *198*, 111395. [CrossRef]
26. Ri, Y.G.; Sin, C.M.; Kang, J.G. Statistical modelling of rate gyros based on fully overlapping Allan variance. *IET Sci. Meas. Technol.* **2022**, *16*, 69–77. [CrossRef]
27. Stebler, Y.; Guerrier, S.; Skaloud, J.; Victoria-Feser, M. Constrained expectation-maximization algorithm for stochastic inertial error modeling: Study of feasibility. *Meas. Sci. Technol.* **2011**, *22*, 085204. [CrossRef]
28. Chen, L.; Chen, Z.; Ji, Z. Expectation–Maximization-Based Simultaneous Localization and Mapping for Millimeter-Wave Communication Systems. *Sensors* **2022**, *22*, 6941. [CrossRef] [PubMed]

sensors

MDPI

Article

Provably Secure Mutual Authentication and Key Agreement Scheme Using PUF in Internet of Drones Deployments

Yohan Park [1], Daeun Ryu [1], Deokkyu Kwon [2,*] and Youngho Park [2,*]

[1] School of Computer Engineering, Keimyung University, Daegu 42601, Republic of Korea;
[2] School of Electronics Engineering, Kyungpook National University, Daegu 41566, Republic of Korea
* Correspondence: kdk145@knu.ac.kr (D.K.); parkyh@knu.ac.kr (Y.P.); Tel.: +82-53-950-7842 (Y.P.)

Abstract: Internet of Drones (IoD), designed to coordinate the access of unmanned aerial vehicles (UAVs), is a specific application of the Internet of Things (IoT). Drones are used to control airspace and offer services such as rescue, traffic surveillance, environmental monitoring, delivery and so on. However, IoD continues to suffer from privacy and security issues. Firstly, messages are transmitted over public channels in IoD environments, which compromises data security. Further, sensitive data can also be extracted from stolen mobile devices of remote users. Moreover, drones are susceptible to physical capture and manipulation by adversaries, which are called drone capture attacks. Thus, the development of a secure and lightweight authentication scheme is essential to overcoming these security vulnerabilities, even on resource-constrained drones. In 2021, Akram et al. proposed a secure and lightweight user–drone authentication scheme for drone networks. However, we discovered that Akram et al.'s scheme is susceptible to user and drone impersonation, verification table leakage, and denial of service (DoS) attacks. Furthermore, their scheme cannot provide perfect forward secrecy. To overcome the aforementioned security vulnerabilities, we propose a secure mutual authentication and key agreement scheme between user and drone pairs. The proposed scheme utilizes physical unclonable function (PUF) to give drones uniqueness and resistance against drone stolen attacks. Moreover, the proposed scheme uses a fuzzy extractor to utilize the biometrics of users as secret parameters. We analyze the security of the proposed scheme using informal security analysis, Burrows–Abadi–Needham (BAN) logic, a Real-or-Random (RoR) model, and Automated Verification of Internet Security Protocols and Applications (AVISPA) simulation. We also compared the security features and performance of the proposed scheme and the existing related schemes. Therefore, we demonstrate that the proposed scheme is suitable for IoD environments that can provide users with secure and convenient wireless communications.

Keywords: AVISPA; BAN logic; Internet of Drones; mutual authentication; PUF

Citation: Park, Y.; Ryu, D.; Kwon, D.; Park, Y. Provably Secure Mutual Authentication and Key Agreement Scheme Using PUF in Internet of Drones Deployments. *Sensors* **2023**, *23*, 2034. https://doi.org/10.3390/s23042034

Academic Editor: Constantin Caruntu and Ciprian Romeo Comşa

Received: 15 December 2022
Revised: 7 February 2023
Accepted: 8 February 2023
Published: 10 February 2023

1. Introduction

Internet of Drones (IoD) [1], which is often referred to as an unmanned aerial vehicles (UAVs) network, is a layered network control architecture designed to coordinate the access of drones. Drones in IoD environments can perform various flight tasks by embedding various sensors, actuators, recorders, batteries, computations, and communication modules. Figure 1 shows the basic structure of a drone in IoD environments. With these modules, drones are used to control the airspace and offer services such as rescue, healthcare, traffic surveillance, environmental monitoring, delivery, and search to users [2]. The IoD architecture generally comprises remote users, a control server, and drones. Remote users query the information of drones to receive useful services. The control server is centrally located in the wireless communication flow, mediating and providing a seamless data exchange process between remote users and drones. Drones, located in their own flying zone, collect surrounding environment information and send it to users through the control center.

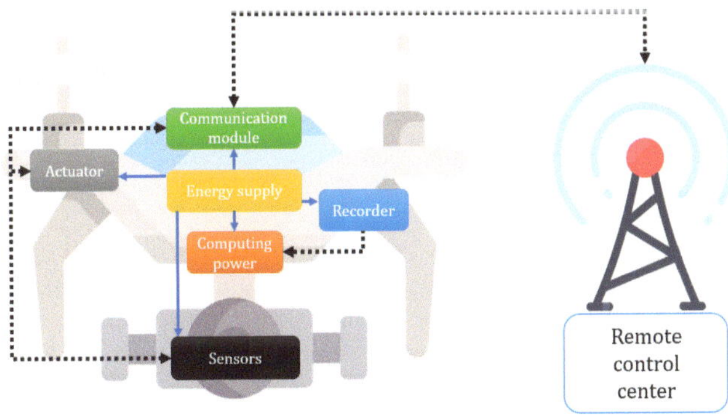

Figure 1. Basic structure of the drone in IoD environments.

Although IoD environments offer useful services to users, they can suffer from several privacy and security issues [3]. Firstly, IoD environments can be vulnerable to various security attacks, such as eavesdropping, deleting, and intercepting, because all messages are transmitted via a public channel. Moreover, the mobile devices of remote users can be stolen/lost, and the sensitive stored data of these devices can threaten the whole IoD environment. Additionally, drones can be physically captured by malicious adversaries who can try to impersonate them using secret information extracted from drones using power analysis attacks. Finally, drones in IoD environments are designed to use restricted power, computation, and storage sources because the entire energy source is preferentially devoted to flying tasks. Thus, a secure and lightweight authentication scheme is necessary, considering the above security vulnerabilities and specific features of IoD environments.

In 2021, Akram et al. [4] proposed a user–drone access scheme designed to be secure and lightweight for drone networks. The authors claimed that the scheme resists user, control center, and drone impersonation attacks and provides anonymity and untraceability. However, we find that Akram et al.'s scheme is vulnerable to drone impersonation, verification table leakage, and denial of service (DoS) attacks. In addition, their scheme cannot ensure perfect forward secrecy and fails to guarantee correctness. To improve these vulnerabilities, we propose a mutual authentication and key agreement (MAKA) scheme that can provide convenient services to users with high security and efficiency for IoD environments. In the proposed scheme, we utilize biometrics [5] to resist various security attacks, such as offline guessing attacks on user devices. Moreover, we apply physical unclonable function (PUF) [6] technology to prevent cloning and physical attacks of drones using power analysis attacks. Considering real-time communication in IoD environments and the limited computation resources of user devices and drones, we only utilize hash functions and exclusive-OR operators, which are reliable in terms of computation and communication overheads.

1.1. Research Contributions

- We review and perform a security analysis of Akram et al.'s scheme. Then, we propose a MAKA scheme designed to ensure high security using biometrics and PUF. Hash functions and exclusive-OR operations are used for lightweight architecture, making the proposed scheme suitable for drone networks. Moreover, a fuzzy extractor and PUF are applied in the proposed scheme to enhance the security level.
- We prove the security robustness of the proposed scheme using the Automated Verification of Internet Security Protocols and Applications (AVISPA) simulation tool [7,8], Real-or-Random (RoR) model [9], and Burrows–Abadi–Needham (BAN) logic [10].
- We perform an informal analysis to ensure that the proposed scheme can provide security against various attacks, including offline password guessing, session key

disclosure, verification table leakage, impersonation, and DoS attacks. Additionally we show that the proposed scheme can achieve mutual authentication, perfect forward secrecy, untraceability, and anonymity.

- We evaluate and compare the security features, communication, and computation costs of the proposed scheme with existing authentication schemes, including Akram et al.'s scheme.

1.2. Organization

In Section 2, we introduce existing studies on IoD environments. We provide a system model as well as an adversary model, fuzzy extractor, and PUF used in the proposed scheme in Section 3. Then, we show Akram et al.'s scheme in Section 4. Section 5 describes security vulnerabilities discovered in Akram et al.'s scheme. The proposed scheme is introduced in Section 6. Security analyses, i.e., BAN logic, RoR model, AVISPA, are shown in Section 7 and performance analyses, i.e., security features, communication, computation costs, are shown in Section 8. In Section 9, we conclude our paper and describe future works.

2. Related Works

Since the basic concept of IoD environments was introduced by Gharibi et al. [1] various authentication schemes have been proposed over the past few years. In 2018 Wazid et al. [11] proposed an authentication scheme to provide remote users with drone services based on three-factor technology. To apply lightweight communication services Wazid et al. utilize hash function and exclusive-OR operators. However, their scheme cannot prevent privileged insider and impersonation attacks. In 2019, Teng et al. [12] analyzed security vulnerabilities, named "attacker mode", which can happen in IoD environments. Thus, they proposed an authentication scheme utilizing the elliptic curve digital signature algorithm (ECDSA) to verify the legitimacy of identity signatures on drones. However, Teng et al.'s scheme was designed as an authentication scheme involving two-way authentication between drones based on ECC, which incurs a large computational overhead. Srinivas et al. [13] proposed a temporal credential-based authentication for IoD networks. Srinivas et al. argued that security and efficiency are the main requirements for the IoD environment, and a lightweight authentication protocol is essential to satisfy these requirements. In their scheme, the authors claimed that it can resist various security attacks such as a stolen mobile device, replay, MITM, ephemeral secret leakage (ESL), impersonation, password and/or biometric update, and remote drone capture attacks. In 2020, Ali et al. [14] pointed out that Srinivas et al.'s scheme [13] does not provide untraceability and resists stolen verifier attacks. To overcome that, Ali et al. suggested a lightweight authentication scheme for drones using symmetric key primitives and temporal credentials. Ever [15] suggested a framework for mobile sinks used in drones using bilinear pairing and ECC, which has a large computational cost. However, Ever's protocol cannot provide user anonymity and untraceability [16]. In 2022, Wu et al. [17] proposed a drone communication scheme for 5G networks. They argued that several existing IoD protocols have high computation overheads because of using a public key infrastructure (PKI) mechanism. Therefore, they only utilized hash functions and exclusive-OR operators. In the same year, Tanveer et al. [18] proposed an authentication mechanism for IoD environments. They used an AES-CBC-256 cipher and ECC to ensure the anonymity of users. Although the above schemes [11–15,17,18] provide useful services such as healthcare, rescue, and traffic surveillance, they can suffer from physical attacks because each drone cannot protect security parameters from power analysis attacks.

To strengthen the authentication process and access control of drones, various PUF-based authentication schemes have been proposed. Alladi et al. [19] proposed a two-stage authentication protocol that divided drone hierarchies for smart drone networks. In Alladi et al.'s scheme, each drone equipped with PUF communicates with a ground station through a leader drone, reducing network overhead. Thus, the authors claimed their scheme does not require the storage of secret keys in drones, protecting it from impersonation, drone tampering, and MITM attacks. In the same years, Pu et al. [20] proposed an authentication protocol for drone environments using PUF and chaotic systems. The authors used the challenge–response pair of the PUF as the seed value of the chaotic system to jumble the message randomly. In 2021, Zhang et al. [21] suggested a three-party authentication scheme for IoD environments. In Zhang et al.'s scheme, the head drone manages member drones and mediates the communication between the ground station and member drones. The entire process of their scheme only uses hash functions and XOR operations. Moreover, the authors introduced PUF systems to prevent physical capture attacks.

In 2021, Akram et al. [4] suggested a scheme for secure and efficient drone access in IoD networks. The authors demonstrated that various security attacks, e.g., user, control center, and drone impersonation attacks, can be prevented in their scheme. However, our security analysis indicates that their scheme is vulnerable to DoS, session key disclosure, stolen-verifier, and drone impersonation attacks and cannot provide perfect forward secrecy.

We summarize the cryptographic techniques and the advantages and limitations of the existing related schemes [4,11–15,17–21] in Table 1. Although previous authentication schemes can provide convenient services to users, they still have high computational and communication overhead and security drawback problems. Therefore, we propose a secure drone-access scheme to improve these security flaws considering lightweight communication characteristics of IoD environments. The proposed scheme can provide stolen mobile device and drone impersonation attacks using biometric and PUF technologies, respectively. Moreover, the proposed scheme can support efficient communications using only hash functions and exclusive-OR operators.

Table 1. Cryptographic technologies and properties of the related schemes for IoD environments.

Schemes	Cryptographic Technologies	Advantages and Limitations
Wazid et al. [11]	* Hash functions * Fuzzy extractor	* Presented IoD environments and utilized biometrics information to ensure the security of remote users * Vulnerable to privileged insider and impersonation attacks
Teng et al. [12]	* ECDSA	* Defined security threats in IoD environments named "attacker mode" * Requires large computation overheads
Srinivas et al. [13]	* Hash functions * Fuzzy extractor	* Used temporal credentials for mutual authentication * Vulnerable to untraceability and stolen verifier attacks
Ali et al. [14]	* Hash functions * Fuzzy extractor * Symmetric key primitives	* Anonymous and lightweight security solution using temporal credentials and symmetric key primitives * Vulnerable to ESL, physical and cloning attacks
Ever et al. [15]	* Bilinear pairings * ECC	* Analyzed studies utilized UAVs as mobile sinks * Require high computation overheads * Cannot provide anonymity and untraceability
Wu et al. [17]	* Hash functions * Fuzzy extractor	* Proposed a drone-to-user authentication scheme for 5G networks * Vulnerable to physical attacks due to the stored parameters in UAV
Tanveer et al. [18]	* Hash functions * Fuzzy extractor * ECC * Symmetric key primitives	* Provides anonymous communication to users using AES and ECC * Vulnerable to physical attacks due to the stored parameters in UAV
Alladi et al. [19]	* PUF * Message authentication code * Symmetric key primitives	* Classified drones by layer and proposed PUF-based two-stage authentication protocol * Vulnerable to replay, insider, server spoofing, DoS attacks
Pu et al. [20]	* PUF * Chaotic system	* Used PUF and chaotic map technologies to generate random key * Vulnerable to physical attacks because of a stored challenge value in the memory of UAV
Zhang et al. [21]	* Hash functions * Fuzzy extractor * FourQ * Symmetric key primitives	* Proposed authentication scheme using FourQ and BPV pre-computation technologies * Require high computation and communication overheads * Cannot provide user anonymity
Akram et al. [4]	* Hash functions * Fuzzy extractor * Symmetric key primitives	* Provide privacy of location information to remote users and drones * Vulnerable to drone impersonation, stolen verifier, and DoS attacks, and have correctness problem

3. Preliminaries

We present the system model and adversary model for IoD environments. Moreover we introduce some relevant preliminaries to understand this paper.

3.1. System Model

As shown in Figure 2, IoD environments consist of a control center, users and drones According to the IoD environment model, various drones collect the data in their particular zones in a target field and transmit the data to the server. External users are required to connect to the server to obtain data from the deployed drones. For access, secure authentication is necessary between the user and drone via the control center. Subsequently the user and drone pair share a session key and begin communication. The details of this process are as follows.

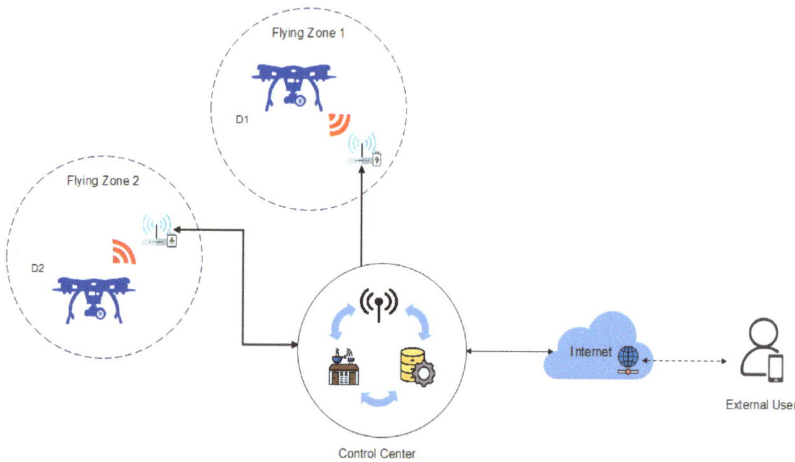

Figure 2. The general system model of IoD environments.

- Remote user (U_m): A remote user U_m owns a mobile device to receive IoD services To communicate with a drone D_n, U_m must register with the control center. U_m utilizes biometric technology in addition to identity and password to store sensitive information safely.
- Control center: The control center is a trusted third party with enough computation and storage capacities. Therefore, the control center perform a role as the system manager of IoD environments. Furthermore, the control center authenticates with both U_m and D_n information and helps U_m to access the D_n. The control center generates secret keys for U_m and D_n against their identities.
- Drone (D_n): A drone D_n collects the data in their particular flying zone and must be registered by the control center to communicate with U_m. Then, D_n sends the data to =U_m through the control center. Moreover, D_n has restricted computation and storage capacities.

3.2. Adversary Model

We follow the widely used adversary model, named the "Dolev–Yao (DY) adversary model" [22,23]. Under the DY model, the entities involved in the IoD environments, i.e., U_m and D_n, are not assumed to be trustworthy, and the communication of the channel is insecure. Therefore, an adversary $\mathcal{A}$ can modify or delete the transmitted messages and also can eavesdrop on the exchanged messages. Furthermore, drones move around in unattended hostile areas with collected sensor data. Thus, they are vulnerable to physical capture attacks [11,24], and the sensitive data stored in the drone can be extracted using the power analysis attacks.

3.3. Fuzzy Extractor

The fuzzy extractor [25] is widely accepted to verify the biometric authentication. A biometric key can be generated with a biometric template such as fingerprints, faces and irises. The fuzzy extractor is defined with the following two algorithms:

- $Gen(Bio_m) = (\alpha_m, \beta_m)$: It is a probabilistic algorithm to generate a secret key α_m. The user inputs biometric Bio_m, the output of this function is the secret parameter α_m, and the public reproduction parameter β_m.
- $Rep(Bio^*_m, \beta_m) = (\alpha_m)$: It is a deterministic algorithm to recreate the original α_m. The function accepts a noisy user biometric Bio^*_m and controls the noise using the public reproduction parameter β_m. Then, this algorithm reproduces the original biometric secret key α_m.

3.4. Physical Unclonable Function

PUF is a physical circuit that maps a bit-string pair called "challenge–response pair" [6]. When an input challenge value is entered into the PUF circuit, it produces a value that isan arbitrary string of bits. In this paper, we use PUF to generate secret values instead of stringing them in the memory of the drone and obtain a stable response good enough for security using fuzzy extractors. The property of PUF is as below.

- The PUF is a physical microstructure of the device.
- It is extremely difficult or impossible to clone the PUF circuit.
- An unpredictable response value must be output.
- It is possible to evaluate and implement a PUF circuit easily.

4. Revisit of Akram et al.'s Scheme

Akram et al. [4] suggested a drone-access authentication protocol for surveillance tasks in a smart city. Akram et al.'s scheme is composed of the following phases: (1) user registration; (2) drone registration; (3) authentication and key agreement (AKA) phases. Table 2 shows the whole notation and description in their scheme.

Table 2. Notations and descriptions.

Notation	Description
ID_m, ID_n	Identity of the user and drone
SID_c, SID_m, SID_n	Pseudonym of the control center, user and drone
Bio_m	Biometric of the user
k_m, k_n	Master private key of the user and drone
s, MSK	Secret keys of the control center
$Rep(.)$	Fuzzy biometric reproduction
$Gen(.)$	Fuzzy biometric generator
a_1, a_2, a_3	Random numbers
SK	Session key
$h(.)$	Hash function
$\|\|$	Concatenation operator
$\oplus$	Exclusive-OR operator

4.1. Registration Phase

4.1.1. Remote User Registration Phase

Step 1: The user inputs their own ID_m, PW_m and imprints Bio_m. Then, U_m calculates $Gen(Bio_m) = (\alpha_m, \beta_m)$ and sends ID_m to the control center.

Step 2: The control center calculates $SID_m = h(ID_m||s)$, $k_m = h(SID_m||MSK)$ and generates a random number a_m. After that, the control center computes $MID_m = Enc_{MSK}$ $(SID_m||\alpha_m)$ and sends $\{k_m, SID_m, SID_n\}$ to U_m.

Step 3: U_m computes $\gamma_m = h(ID_m||PW_m||\alpha_m) \oplus k_m$, $SID^u_m = h(ID_m||PW_m) \oplus SID_m$. Then, U_m stores $\{\gamma_m, SID^u_m, SID_n\}$.

4.1.2. Drone Registration Phase

Step 1: D_n selects ID_n and sends it to the control center.

Step 2: The control center computes $SID_n = h(ID_n||s)$, $k_n = h(SID_n||MSK)$ and stores $\{ID_n, k_n, SID_n\}$ in its database. Then, the control center sends $\{k_n, SID_n\}$ to D_n.

Step 3: When D_n receives $\{k_n, SID_n\}$, D_n saves them in the memory.

4.2. AKA Phase

Step 1: U_m inputs ID_m, PW_m and also imprints Bio_m. Then, U_m computes $\alpha_m = Rep(Bio_m, \beta_m)$, $SID_m = SID_m^u \oplus h(ID_m||PW_m)$, $k_m = \gamma_m \oplus h(ID_m||PW_m||\alpha_m)$. Afterward, U_m generates a_1 and computes $A_1 = h(SID_m||SID_c||k_m) \oplus a_1$, $A_2 = h(SID_m||SID_c||k_m||a_1) \oplus SID_n$ and $A_3 = h(SID_m||SID_n||SID_c||k_m||a_1)$. Finally, U_m sends $\{MID_m, A_1, A_2, A_3\}$ to the control center.

Step 2: The control center retrieves $(SID_m||\alpha_m) = Dec_{MSK}(MID_m)$. Then, the control center computes $k_m = h(SID_m||\ MSK)$, $a_1^* = A_1 \oplus h(SID_m^*||SIDc||k_m^*)$ and $SID_n^* = A_2 \oplus h(SID_m^*||SID_c||k_m^*||a_1^*)$, and verifies k_n against SID_n^*. Then, the control center computes $A_3^* = h(SID_m^*||SID_n^*||SID_c||k_m^*||a_1^*)$ and checks $A_3^* \stackrel{?}{=} A_3$. The control center generates a_2, a_m^{new} and computes $MID_m^{new} = Enc_{MSK}(SID_m||a_m^{new})$, $A_4 = h(SID_n^*||k_n) \oplus (a_1^*||a_2||MID_m^{new})$, $A_5 = h(SID_n^*||SID_c||k_n||a_1^*) \oplus SID_m^*$ and $A_6 = h(SID_m^*||SID_n^*||SID_c||k_n||a_1^*||a_2)$. Finally, the control center sends $\{A_4, A_5, A_6\}$ to the drone D_n.

Step 3: D_n computes $(a_1^{**}||a_2^*||MID_m^{new}) = A_4 \oplus h(SID_n||k_n)$, $SID_m^{**} = A_5 \oplus h(SID_n||SID_c||k_n||a_1^{**})$ and $A_6^* = h(SID_M^{**}||SID_n||\ SID_c||k_n||a_1^{**}||a_2^*)$. Then, D_n checks $A_6^* \stackrel{?}{=} A_6$ and generates a_3. After that, D_n computes $A_7 = h(SID_n||SID_m^{**}||a_1^{**}) \oplus (a_2||a_3^*||MID_m^{new})$, $A_8 = h(a_1^{**}||a_2||a_3^*)$, $SK_{nm} = h(SID_m^{**}||SID_n||SID_c||A_8)$ and $A_9 = h(SID_m^{**}||SID_n||\ SID_c||a_2||a_3^*||A_8)$. Finally, D_n sends $\{A_7, A_9\}$ to U_m.

Step 4: The U_m computes $(a_2^*||\ a_3^{**}||MID_m^{new}) = A_7 \oplus h(SID_n||SID_m||a_1)$, $A_8^* = h(a_1||a_2^*||a_3^{**})$ and $A_9^* = h(SID_m||SID_n||SID_c||\ a_2^*||a_3^{**}||A_8^*)$. Then, it validates $A_9^* \stackrel{?}{=} A_9$ and computes $SK_{nm} = h(SID_m^{**}||SID_n||\ SID_c||A_8^*)$.

5. Cryptanalysis of Akram et al.'s Scheme

According to Section 3.2, an adversary $\mathcal{A}$ can obtain a $\{\gamma_m, SID_m^u, SID_n\}$ from legitimate user's mobile device. Moreover, $\mathcal{A}$ can obtain $\{k_n, SID_n\}$ from a captured drone using a power analysis attack. With this information, various security attacks, i.e., session key disclosure, drone impersonation, stolen-verifier, DoS attacks, and perfect forward secrecy, can be executed by $\mathcal{A}$. The details are shown below.

5.1. Session Key Disclosure Attack

For $\mathcal{A}$ to generate a session key $SK_{nm} = h(SID_m||SID_n||SID_c||A_8)$, $\mathcal{A}$ has to obtain SID_m, SID_n and $A_8 = h(a_1||a_2||a_3)$. The procedures are as follows.

Step 1: $\mathcal{A}$ computes $(a_1||a_2||MID_m^{new}) = A_4 \oplus h(SID_n||\ k_n)$, $SID_m = A_5 \oplus h(SID_n||SID_c||k_n||a_1)$, and $(a_2||a_3||MID_m^{new}) = A_7 \oplus h(SID_n||SID_m||a_1)$.

Step 2: $\mathcal{A}$ calculates $SK_{nm} = h(SID_m||SID_n||SID_c||A_8)$.

Thus, Akram et al.'s scheme is insecure against session key disclosure attacks.

5.2. Drone Impersonation Attack

In this attack, we assume that $\mathcal{A}$ can capture drones D_n physically and obtain the value $\{SID_n, k_n\}$ stored in the memory of D_n. In order to be able to forward message $\{A_7, A_9\}$ on behalf of legal D_n, then $\mathcal{A}$ has to calculate the value of $A_7 = h(SID_n||SID_m||a_1) \oplus (a_2||a_3||MID_m^{new})$, $A_9 = h(SID_m||SID_n||SID_c||a_2||a_3||A_8)$. $\mathcal{A}$ can compute the A_7 and A_9 through the following below:

Step 1: The adversary $\mathcal{A}$ first intercepts $\{A_4, A_5, A_6\}$ transmitted by the public channel.

Step 2: $\mathcal{A}$ can obtain a_1, a_2, MID_m^{new} by computing $(a_1||a_2||MID_m^{new}) = A_4 \oplus h(SID_n||k_n)$.

Step 3: $\mathcal{A}$ can compute SID_m through $SID_m = A_5 \oplus h(SID_n||SID_c||k_n||a_1)$.

Step 4: $\mathcal{A}$ generates random a_3^* and computes $A_8^* = h(a_1||a_2||a_3^*)$.

Step 5: $\mathcal{A}$ can successfully compute $A_7^* = h(SID_n||SID_m|| a_1) \oplus (a_2||a_3^*||MID_m^{new})$, $A_9^* = h(SID_m||SID_n||SID_c|| a_2||a_3^*||A_8^*)$.

Therefore, Akram et al.'s scheme cannot resist drone impersonation attacks.

5.3. Stolen-Verifier Attack

When $\mathcal{A}$ obtains the table information $\{k_n, SID_n\}$ of the control center, $\mathcal{A}$ can calculate $SK_{nm} = h(SID_m||SID_n||SID_c||A_8)$. The steps are the same as Section 5.1. Therefore, Akram et al.'s scheme is vulnerable to stolen-verifier attacks.

5.4. Perfect Forward Secrecy

Let us suppose that the control center's long-term secret key MSK is compromised by the adversary $\mathcal{A}$, and $\mathcal{A}$ has captured all the previously transmitted messages MID_m, A_1, A_2 and A_4 through the public channel. $\mathcal{A}$ can retrieve SID_m through $(SID_m||a_m) = Dec_{MSK}(MID_m)$, compute $k_m = h(SID_m||MSK)$, $a_1 = A_1 \oplus h(SID_m||SID_c||k_m)$, $SID_n = A_2 \oplus h(SID_m||SID_c||k_m||a_1)$, and $k_n = h(SID_n||MSK)$. Furthermore, $\mathcal{A}$ can retrieve a_1 and a_2 through $(a_1||a_2||MID_m^{new}) = A_4 \oplus h(SID_n||k_n)$ and compute $A_8 = h(a_1||a_2||a_3)$. Finally, $\mathcal{A}$ computes the session key $SK_{nm} = h(SID_m||SID_n||SID_c||A_8)$. Thus, Akram et al.'s scheme does not provide perfect forward secrecy.

5.5. DoS Attack

In the AKA phase, the login process is not executed normally in the remote user (U_m) side. Afterward, the inputs ID_m, PW_m, and Bio_m, U_m compute α_m, SID_m, and k_m. Then, U_m immediately generates a random nonce and computes an authentication request message $\{MID_m, A_1, A_3\}$. Therefore, the adversary $\mathcal{A}$ can send unlimited amounts of login authentication request messages to the control center if $\mathcal{A}$ obtains a stolen/lost mobile device of U_m and inputs a randomly selected identity, password, and biometrics. These messages can threaten the load on the control center. Thus, Akram et al.'s scheme is vulnerable to DoS attacks.

5.6. Correctness

In the user registration phase, the control center calculates the value of MID_m. After that, the MID_m is not transmitted to U_m, and U_m cannot compute it because the MID_m is masked with MSK, which is the control center's secret key. However, in the AKA phase, U_m sends the MID_m to the control center as the first transmitted message. Thus, Akram et al.'s scheme has a correctness problem.

6. Proposed Scheme

The proposed scheme consists of the following phases: (1) initialization; (2) user registration; (3) drone registration; (4) MAKA. We show the flowchart of the proposed scheme in Figure 3. The proposed scheme is lightweight as it uses only the cryptographic one-way hash function and exclusive-OR operations, apart from the fuzzy extractor and PUF technique that is needed for verification at the user side and drone side, respectively.

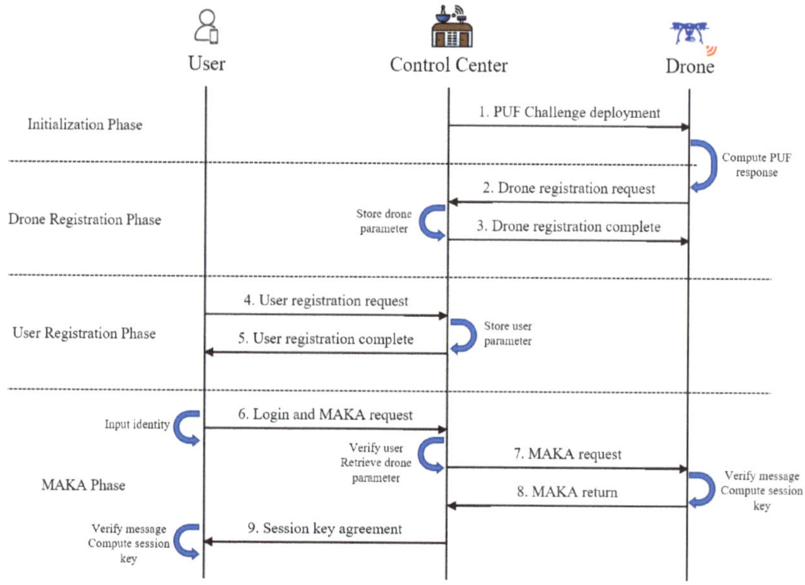

Figure 3. The overall flowchart of the proposed scheme.

6.1. Initialization Phase

This phase describes that the control center selects an identity and a challenge for the drone D_n before the registration phase. Detailed steps are illustrated in Figure 4 Additionally, this phase is performed via a secure channel.

Control Center		**Drone D_n**
Selects ID_n Generates a challenge CH_n		
	$\{ID_n, CH_n\}$ ⟶	
		Stores $\{ID_n, CH_n\}$ in the memory

Figure 4. Initialization phase of the proposed scheme.

Step 1: The control center selects an identity ID_n and a challenge CH_n and sends $\{ID_n, CH_n$ to the drone D_n.

Step 2: The drone stores $\{ID_n, CH_n\}$ in the memory.

6.2. Drone Registration Phase

In this phase, a drone D_n is registered at the control center to its deployment in the IoD environments through a secure channel. Detailed steps are illustrated in Figure 5.

Step 1: The drone D_n retrieves the challenge CH_n stored in the memory and computes $RE_n = PUF(CH_n)$, and $Gen(RE_n) = (\alpha_n, \beta_n)$. After that, the D_n sends $\{ID_n, CH_n\}$ to the control center.

Step 2: The control center generates a random number a_n and computes $SID_n = h(ID_n||s)$, $k_n = h(SID_n||s||a_n)$, and saves $\{ID_n, SID_n, a_n, CH_n\}$ in the database. Then, the control center sends $\{SID_n, k_n\}$ to the D_n.

Step 3: Finally, the D_n deletes the CH_n and computes $\gamma_n = h(ID_n||\alpha_n) \oplus k_n$, $SID_n^D = h(ID_n||\alpha_n||k_n) \oplus SID_n$, and stores $\{\gamma_n\}$ in its memory.

Drone D_n		Control Center D_n
Computes $RE_n = PUF(CH_n)$ $Gen(RE_n) = (\alpha_n, \beta_n)$		
	$\{ID_n, CH_n\}$ →	
		Random number a_n Computes $SID_n = h(ID_n\|\|s)$ $k_n = h(SID_n\|\|s\|\|a_n)$ Saves $\{ID_n, SID_n, a_n, CH_n\}$ in database
	← $\{SID_n, k_n\}$	
Delete CH_n from the memory $\gamma_n = h(ID_n\|\|\alpha_n) \oplus k_n$ $SID_n^D = h(ID_n\|\|\alpha_n\|\|k_n) \oplus SID_n$ Stores $\{\gamma_n\}$		

Figure 5. Drone registration phase of the proposed scheme.

6.3. User Registration Phase

In the user registration phase, a remote user U_m has to register at the control center to access the real-time information from an accessed drone D_n in IoD environments. This procure performs via a secure channel with the following steps. Figure 6 shows the details.

User U_m		Control Cetner
Selects ID_m, PW_m Imprint Bio_m Calculate $Gen(Bio_m) = (\alpha_m, \beta_m)$		
	$\{ID_m\}$ →	
		Random number a_m Computes $SID_m = h(ID_m\|\|s)$ $k_m = h(SID_m\|\|s\|\|a_m)$ $SID_m^* = SID_m \oplus h(s\|\|a_m)$ $MID_m = h(SID_m\|\|a_m)$ Stores $\{MID_m, SID_m^*, a_m\}$ in database
	← $\{k_m, SID_m, SID_n, MID_m\}$	
Computes $\gamma_m = h(ID_m\|\|PW_m\|\|\alpha_m) \oplus k_m$ $\delta_m = h(\alpha_m\|\|k_m\|\|SID_m)$ $SID_m^u = h(ID_m\|\|PW_m) \oplus SID_m$ $SID_n^u = h(PW_m\|\|\alpha_m) \oplus SID_n$ Stores $\{\gamma_m, \delta_m, SID_m^u, SID_n^u, MID_m\}$		

Figure 6. User registration phase of the proposed scheme.

Step 1: The user U_m selects an identity ID_m, a password PW_m, and a biometric template Bio_m. After that, the mobile device calculates $Gen(Bio_m) = (\alpha_m, \beta_m)$. The U_m sends $\{ID_m\}$ to the control center.

Step 2: The control center generates random number a_m and computes $SID_m = h(ID_m\,||s)$, $k_m = h(SID_m||s||a_m)$, $SID_m^* = SID_m \oplus h(s||a_m)$ and $MID_m = h(SID_m||a_m)$. Then, the control center stores $\{MID_m, SID_m^*, a_m\}$ in the database, and sends $\{k_m, SID_m, SID_n, MID_m\}$ to the U_m.

Step 3: The U_m computes $\gamma_m = h(ID_m||PW_m||\alpha_m) \oplus k_m$, $\delta_m = h(\alpha_m||k_m||SID_m)$, $SID_m^u = h(ID_m||PW_m) \oplus SID_m$, and $SID_n^u = h(PW_m||\alpha_m) \oplus SID_n$, and stores $\{\gamma_m, \delta_m, SID_m^u, SID_n^u, MID_m\}$ in the memory.

6.4. MAKA Phase

The following steps are performed among the U_m, the control center, and an accessed drone D_n through a public channel. To establish a session key for secure communication among them, they need to perform the MAKA processes. Details are illustrated in Figure 7.

User U_m	**Control Center**	**Drone D_n**

User U_m:

Inputs ID_m, PW_m
Imprints Bio_m
$\alpha_m = Rep(Bio_m, \beta_m)$
$SID_m = h(ID_m||PW_m) \oplus SID_m^u$
$SID_n = h(PW_m||\alpha_m) \oplus SID_n^u$
$k_m = h(ID_m||PW_m||\alpha_m) \oplus \gamma_m$
$\delta_m^* = h(\alpha_m||k_m||SID_m)$
Checks if $\delta_m^* \stackrel{?}{=} \delta_m$
Selects a_1
$A_1 = h(SID_m||SID_c||k_m) \oplus a_1$
$A_2 = h(SID_m||SID_c) \oplus SID_n$
$V_1 = h(SID_m||SID_n||SID_c||k_m||a_1)$

$\{MID_m, A_1, A_2, V_1\}$ → (to Control Center)

Control Center:

Checks whether
$MID_m = MID_m^{old}$ or $MID_m = MID_m^{new}$
If $(MID_m == MID_m^{old})$
{Retrieves $\{SID_m^*, a_m\}$ against MID_m^{old}}
If $(MID_m == MID_m^{new})$
{Retrieves $\{SID_m^*, a_m\}$ against MID_m^{new}}
$SID_m = SID_m^* \oplus h(s||a_m)$
$k_m = h(SID_m||s||a_m)$
$a_1 = A_1 \oplus h(SID_m||SID_c||k_m)$
$SID_n = A_2 \oplus h(SID_m||SID_c)$
$V_1^* = h(SID_m||SID_n||SID_c||k_m||a_1)$
Checks if $V_1^* \stackrel{?}{=} V_1$
$MID_m^{new} = h(SID_m||a_1)$
Updates MID_m^{new}
Checks for ID_n, a_n, CH_n against SID_n from its database
$k_n = h(SID_n||s||a_n)$
Selects a_2
$A_3 = h(SID_n||k_n) \oplus (a_1||a_2)$
$A_4 = h(SID_n||k_n||a_1) \oplus SID_m$
$A_5 = h(SID_c||ID_n) \oplus CH_n$
$V_2 = h(SID_m||SID_n||SID_c||k_n||a_1||a_2)$

$\{A_3, A_4, A_5, V_2\}$ → (to Drone D_n)

Drone D_n:

Computes $CH_n = A_5 \oplus h(SID_c||ID_n)$
$RE_n = PUF(CH_n)$
$\alpha_n = Rep(RE_n, \beta_n)$
$k_n = \gamma_n \oplus h(ID_n||\alpha_n)$
$SID_n = SID_n^D \oplus h(ID_n||\alpha_n||k_n)$
$(a_1||a_2) = A_3 \oplus h(SID_n||k_n)$
$SID_m = A_4 \oplus h(SID_n||k_n||a_1)$
$V_2^* = h(SID_m||SID_n||SID_c||k_n||a_1||a_2)$
Checks if $V_2^* \stackrel{?}{=} V_2$
Selects a_3
$A_6 = h(SID_m||SID_n||a_1) \oplus (a_2||a_3)$
$A_7 = h(SID_m||SID_n||SID_c)$
$SK = h(A_7||a_1||a_2||a_3)$
$V_3 = h(A_7||a_1||a_3||SK)$

← $\{A_6, V_3\}$ (to User U_m)

User U_m:

$(a_2||a_3) = A_6 \oplus h(SID_m||SID_n||a_1)$
$A_7 = h(SID_m||SID_n||SID_c)$
$SK = h(A_7||a_1||a_2||a_3)$
$V_3^* = h(A_7||a_1||a_3||SK)$
Checks if $V_3^* \stackrel{?}{=} V_3$
Updates MID_m^{new}

Figure 7. MAKA phase of the proposed scheme.

Step 1: The U_m inputs ID_m and PW_m, and imprints Bio_m. After that, U_m computes $\alpha_m = Rep(Bio_m, \beta_m)$, $SID_m = h(ID_m||PW_m) \oplus SID_m^u$, $SID_n = h(PW_m||\alpha_m) \oplus SID_n^u$, $k_m = h(ID_m||PW_m||\alpha_m) \oplus \gamma_m$, and $\delta_m^* = h(\alpha_m||k_m||SID_m)$, and checks $\delta_m^* \stackrel{?}{=} \delta_m$. Then, the U_m generates a random nonce a_1 and calculates $A_1 = h(SID_m||SID_c||k_m) \oplus a_1$, $A_2 = h(SID_m||SID_c) \oplus SID_n$, and $V_1 = h(SID_m||SID_n||SID_c||k_m||a_1)$. The U_m sends $\{MID_m, A_1, A_2, V_1\}$ to the control center.

Step 2: The control center checks whether $MID_m = MID_m^{old}$ or $MID_m = MID_m^{new}$. If $(MID_m == MID_m^{old})$ then, retrieves $\{SID_m^*, a_m\}$ against MID_m^{old}, and if $(MID_m == MID_m^{new})$, retrieves $\{SID_m^*, a_m\}$ against MID_m^{new}. After that, the control center computes $SID_m = SID_m^* \oplus h(s||a_m)$, $k_m = h(SID_m||s||a_m)$, $a_1 = A_1 \oplus h(SID_m||SID_c||k_m)$, $SID_n = A_2 \oplus h(SID_m||SID_c)$, and $V_1^* = h(SID_m||SID_n||SID_c||k_m||a_1)$. If

$V_1^* \stackrel{?}{=} V_1$ is correct, the control center computes $MID_m^{new} = h(SID_m||a_1)$ and updates MID_m^{new}. Then, the control center checks for ID_n, a_n, CH_n against SID_n from its database and computes $k_n = h(SID_n||s||a_n)$. The control center calculates $A_3 = h(SID_n||k_n) \oplus (a_1||a_2)$, $A_4 = h(SID_n||k_n||a_1) \oplus SID_m$, $A_5 = h(SID_c||ID_n) \oplus CH_n$, and $V_2 = h(SID_m||SID_n||SID_c||k_n||a_1||a_2)$ and sends $\{A_3, A_4, A_5, V_2\}$ to the drone.

Step 3: The drone D_n computes $CH_n = A_5 \oplus h(SID_c||ID_n)$, $RE_n = PUF(CH_n)$, $\alpha_n = Rep(RE_n, \beta_n)$, $k_n = \gamma_n \oplus h(ID_n||\alpha_n)$, $SID_n = SID_n^D \oplus h(ID_n||\alpha_n||k_n)$, $(a_1||a_2) = A_3 \oplus h(SID_n||k_n)$, $SID_m = A_4 \oplus h(SID_n||k_n||a_1)$, and $V_2^* = h(SID_m||SID_n||SID_c ||k_n||a_1||a_2)$. If $V_2^* \stackrel{?}{=} V_2$ is correct, the D_n generates a random nonce a_3, and calculates $A_6 = h(SID_m||SID_n||a_1) \oplus (a_2||a_3)$, $A_7 = h(SID_m||SID_n||SID_c)$, $SK = h(A_7||a_1||a_2||a_3)$, and $V_3 = h(A_7||a_1||a_3||SK)$. Then, the D_n sends $\{A_6, V_3\}$ to the U_m.

Step 4: The U_m computes $(a_2||a_3) = A_6 \oplus h(SID_m||SID_n ||a_1)$, $A_7 = h(SID_m||SID_n|| SID_c)$, $SK = h(A_7||a_1||a_2||a_3)$, and $V_3^* = h(A_7||a_1||a_3||SK)$ and checks $V_3^* \stackrel{?}{=} V_3$. Then, the U_m updates MID_m^{new}.

7. Security Analysis

To prove the security robustness of the proposed scheme, BAN logic, RoR model, and AVISPA simulation are used in this section. Using informal security analysis, we analyze the theoretical security of the proposed scheme.

7.1. BAN Logic

BAN logic [10] is a widely known formal proof used by many researchers to show mutual authentication of protocols [26–28]. Therefore, we apply the proposed scheme to BAN logic proof and verify mutual authentication. We introduce notations and descriptions for BAN logic in Table 3.

Table 3. Basic notations in BAN logic.

Notation	Description
$\mathcal{PR}_1, \mathcal{PR}_2$	Principals
MSG_1, MSG_2	Statements
SK	Session key
$\mathcal{PR}_1 \mid\equiv MSG_1$	$\mathcal{PR}_1$ **believes** MSG_1
$\mathcal{PR}_1 \mid\sim MSG_1$	$\mathcal{PR}_1$ once **said** MSG_1
$\mathcal{PR}_1 \Rightarrow MSG_1$	$\mathcal{PR}_1$ **controls** MSG_1
$\mathcal{PR}_1 \lhd MSG_1$	$\mathcal{PR}_1$ **receives** MSG_1
$\#MSG_1$	MSG_1 is **fresh**
$(MSG_1)_{KEY}$	MSG_1 is **encrypted** with KEY
$\mathcal{PR}_1 \stackrel{KEY}{\longleftrightarrow} \mathcal{PR}_2$	$\mathcal{PR}_1$ and $\mathcal{PR}_2$ have shared key KEY

7.1.1. Rules

In BAN logic, there are five logical rules: message meaning rule (MMR), nonce verification rule (NVR), jurisdiction rule (JR), belief rule (BR), and freshness rule (FR). Details are as follows.

1. MMR :

$$\frac{\mathcal{PR}_1 \Big| \equiv \mathcal{PR}_1 \stackrel{KEY}{\leftrightarrow} \mathcal{PR}_2, \quad \mathcal{PR}_1 \lhd (MSG_1)_{KEY}}{\mathcal{PR}_1| \equiv \mathcal{PR}_2| \sim MSG_1}$$

2. NVR :

$$\frac{\mathcal{PR}_1| \equiv \#(MSG_1), \quad \mathcal{PR}_1| \equiv \mathcal{PR}_2 \Big| \sim MSG_1}{\mathcal{PR}_1| \equiv \mathcal{PR}_2| \equiv MSG_1}$$

3. JR :

$$\frac{\mathcal{PR}_1| \equiv \mathcal{PR}_2 \Rightarrow MSG_1, \quad \mathcal{PR}_1| \equiv \mathcal{PR}_2| \equiv MSG_1}{\mathcal{PR}_1 \Big| \equiv MSG_1}$$

4. BR :

$$\frac{\mathcal{PR}_1 \Big| \equiv (MSG_1, MSG_2)}{\mathcal{PR}_1 \Big| \equiv MSG_1}$$

5. FR :

$$\frac{\mathcal{PR}_1 \Big| \equiv \#(MSG_1)}{\mathcal{PR}_1 \Big| \equiv \#(MSG_1, MSG_2)}$$

7.1.2. Goals

In the proposed scheme, there are four goals for the BAN logic. Let the user, control center, and drone be U_m, CC, and D_n, respectively.

Goal 1: $D_n| \equiv D_n \overset{SK}{\longleftrightarrow} U_m$

Goal 2: $D_n| \equiv U_m| \equiv D_n \overset{SK}{\longleftrightarrow} U_m$

Goal 3: $U_m| \equiv D_n \overset{SK}{\longleftrightarrow} U_m$

Goal 4: $U_m| \equiv D_n| \equiv D_n \overset{SK}{\longleftrightarrow} U_m$

7.1.3. Idealized Forms

Three messages, i.e., $\{MID_m, A_1, A_2, V_1\}$, $\{A_3, A_4, A_5, V_2\}$, and $\{A_6, V_3\}$, are transmitted via open channels in the proposed scheme. These messages are converted to idealized forms in BAN logic as below.

$Mes_1 : U_m \rightarrow CC : \{a_1, SID_n\}_{SID_m}$
$Mes_2 : CC \rightarrow D_n : \{a_1, a_2, SID_m\}_{k_n}$
$Mes_3 : D_n \rightarrow U_m : \{a_2, a_3\}_{SID_m}$

7.1.4. Assumptions

We show the assumptions using in BAN logic as follows.

AS_1: $CC| \equiv \#(a_1)$
AS_2: $D_n| \equiv \#(a_2)$
AS_3: $U_m| \equiv \#(a_3)$
AS_4: $D_n| \equiv U_m \Rightarrow (D_n \overset{SK}{\longleftrightarrow} U_m)$
AS_5: $U_m| \equiv D_n \Rightarrow (D_n \overset{SK}{\longleftrightarrow} U_m)$
AS_6: $CC| \equiv CC \overset{SID_m}{\longleftrightarrow} U_m$
AS_7: $D_n| \equiv CC \overset{k_n}{\longleftrightarrow} D_n$
AS_8: $U_m| \equiv D_n \overset{SID_m}{\longleftrightarrow} U_m$

7.1.5. BAN Logic Proof

Step 1: We can obtain RA_1 from the message Mes_1.

$$RA_1 : CC \triangleleft \{a_1, SID_n\}_{SID_m}$$

Step 2: We can obtain RA_2 from the rule MMR using RA_1 and AS_6.

$$RA_2 : CC| \equiv U_m| \sim (a_1, SID_n)$$

Step 3: We can obtain RA_3 from the rule FR using S_3 and AS_1.

$$RA_3 : CC| \equiv \#(a_1, SID_n)$$

Step 4: We can obtain RA_4 from the rule NVR using RA_2 and RA_3.

$$RA_4 : CC| \equiv U_m| \equiv (a_1, SID_n)$$

Step 5: We can obtain RA_5 from the message Mes_2.

$$RA_5 : D_n \triangleleft \{a_1, a_2, SID_m\}_{k_n}$$

Step 6: We can obtain RA_6 from the MMR using RA_5 and AS_7.

$$RA_6 : D_n| \equiv CC| \sim (a_1, a_2, SID_m)$$

Step 7: We can obtain RA_7 from the FR using RA_6 and AS_2.

$$RA_7 : D_n| \equiv \#(a_1, a_2, SID_m)$$

Step 8: We can obtain RA_8 from the NVR using RA_6 and RA_7.

$$RA_8 : D_n| \equiv CC| \equiv (a_1, a_2, SID_m)$$

Step 9: We can obtain RA_9 from the message Mes_3.

$$RA_9 : U_m \triangleleft \{a_2, a_3\}_{SID_m}$$

Step 10: We can obtain RA_{10} from the MMR using RA_9 and AS_8.

$$RA_{10} : U_m| \equiv D_n| \sim (a_2, a_3)$$

Step 11: We can obtain RA_{11} from the NVR using RA_{10} and AS_3.

$$S_{11} : U_m| \equiv D_n| \equiv (a_2, a_3)$$

Step 12: We can obtain RA_{12} and RA_{13} from RA_8 and RA_{11}. Therefore, U_m and D_n can compute the session key $SK = h(A_7||a_1||a_2||a_3)$, where $A_7 = h(SID_m||SID_n||SID_c)$.

$$RA_{12} : D_n| \equiv U_m| \equiv (D_n \overset{SK}{\longleftrightarrow} U_m) \quad \textbf{(Goal 2)}$$
$$RA_{13} : U_m| \equiv D_n| \equiv (D_n \overset{SK}{\longleftrightarrow} U_m) \quad \textbf{(Goal 4)}$$

Step 13: We can obtain RA_{14} and RA_{15} from the jurisdiction rule using RA_{12} and AS_4, and RA_{13} and AS_5, respectively.

$$RA_{14} : D_n| \equiv (D_n \overset{SK}{\longleftrightarrow} U_m) \quad \textbf{(Goal 1)}$$
$$RA_{15} : U_n| \equiv (D_n \overset{SK}{\longleftrightarrow} U_m) \quad \textbf{(Goal 3)}$$

7.2. RoR Model

The Real-or-Random model [9] is a formal proof analysis that proves the session key security of the protocol. Thus, we establish a premise for applying the proposed scheme to the RoR model. There are participants, adversaries and queries in our scheme. Participants are the entities that communicate with each other in the proposed scheme. Therefore, participants are as follows: PAR_U^i, PAR_C^j, and PAR_D^k, where i, j, and k are the instances of user, control center, and drone, respectively. The adversary in RoR model can modify, delete, and eavesdrop the exchanged messages. With this ability, the adversary can perform various queries such as *Execute*, *CorruptDevice*, *Send*, and *Test*. We describe the details of these queries as below.

- *Execute*$(PAR_U^i, PAR_C^j, PAR_D^k)$: In this query, the adversary eavesdrop messages are transmitted via an open channel. Therefore, the adversary can obtain messages generated from PAR_U^i, PAR_C^j, and PAR_D^k. This query is a passive attack.
- *CorruptDevice*(PAR_U^i): In this query, the adversary can obtain secret parameters from PAR_U^i using a power analysis attack. Therefore, the query *CorruptDevice* is an active attack.
- *Send*(PAR): In this query, the adversary can send messages to all participants PAR_U^i, PAR_C^j, and PAR_D^k. Furthermore, the adversary can obtain returned messages from these participants. Thus, this query is an active attack
- *Test*(PAR): Before starting the game, an unbiased coin UC is flipped in this query. The adversary obtains $UC = 1$ when the session key is fresh. The adversary can also obtain $UC = 0$ when the session key of the proposed scheme cannot guarantee freshness. If not, the adversary obtains a "null value" $\perp$. To achieve a secure session key agreement, the adversary cannot discriminate between the session key and the random number.

Security Proof

Theorem 1. *The adversary AD attempts to compute the session key $SK = h(A_7||a_1||a_2\ ||a_3)$ in polynomial time. Therefore, we define the possibility that AD breaks the security of the session key as $\mathcal{MA}_{AD}(P)$. Moreover, we define that HA and PU are the range space of the function $h(.)$ and $PUF(.)$, respectively. The number of HA, PU, and Send queries are qu_{ha}, qu_{pu}, and qu_{se}, respectively. We define the secret biometric bits as B_m. At last, we define the Zipf's parameter [29] as C' and s'.*

$$\mathcal{MA}_{AD}(P) \leq \frac{qu_{ha}^2}{|HA|} + \frac{qu_{pu}^2}{|PU|} + 2max\{C'qu_{se}^{s'}, \frac{qu_{se}}{2^{B_m}}\}$$

Proof. The security proof in the proposed scheme is composed of five games GA_n $(n = 0, 1, 2, 3, 4)$. Before starting the game, we define A_{GA_n} as the probability that AD wins the game and $AD[A_{GA_k}]$ as the advantage of A_{GA_k}. We follow the security proof according to [30–32].

GA_0 : In GA_0, the adversary selects a random bit r. Thus, we obtain the following equation.

$$\mathcal{MA}_{AD}(P) = |2AD[A_{GA_0}] - 1| \tag{1}$$

GA_1 : In GA_1, the adversary eavesdrops messages $\{MID_m, A_1, A_2, V_1\}$, $\{A_3, A_4, A_5, V_2\}$, and $\{A_6, V_3\}$ using *Execute* query. Then, the adversary performs the *Test* query to obtain the session key $SK = h(A_7||a_1||a_2||a_3)$. To compute SK, the adversary must obtain the random nonces a_1, a_2, and a_3. Moreover, A_7 is composed of SID_m, SID_n, and SID_c, where SID_m is the secret parameter of user. Therefore, the adversary cannot calculate SK. Therefore, we can obtain the following equation.

$$|AD[A_{GA_1}]| = |AD[A_{GA_0}]| \tag{2}$$

GA_2 : In GA_2, the adversary utilizes *Send* and HA to attack the network. However, all of the parameters are masked in a cryptographic hash function that can prevent the hash collision problem. For this reason, the adversary cannot obtain the session key SK. According to the birthday paradox [33], we can obtain the following inequation.

$$|AD[A_{GA_2}] - AD[A_{GA_1}]| \leq \frac{qu_{ha}^2}{|HA|} \tag{3}$$

$GA_3:$ Similar to GA_2, the adversary utilizes queries *Send* and *PU* in this game. According to Section 3.4, the PUF is extremely difficult or impossible to clone. This means the adversary has no advantage in GA_3.

$$|AD[A_{GA_3}] - AD[A_{GA_2}]| \leq \frac{qu_{pu}^2}{|PU|} \tag{4}$$

$GA_4:$ This game is the final game in which the adversary extracts secret parameters $\{\gamma_m, \delta_m, SID_m^u, SID_n^u, MID_m\}$ from the device of the user using the query *CorruptDevice*. The adversary attempts to calculate *SK* from these parameters. However, each parameter consists of a password and the biometrics of a user, and this means that the adversary must guess the password and biometrics at the same time. Since this task is computationally infeasible, the adversary cannot compute *SK*. Therefore, we can obtain the following inequation using Zipf's law [29].

$$|AD[A_{GA_4}] - AD[A_{GA_2}]| \leq max\{C'qu_{se}^{s'}, \frac{qu_{se}}{2^{B_m}}\} \tag{5}$$

After the game, the adversary guesses the result bits r, and we can make the following equation.

$$AD[A_{GA_4}] = \frac{1}{2} \tag{6}$$

We can calculate and obtain Equation (7) using (1) and (2).

$$\frac{1}{2}\mathcal{MA}_{AD}(P) = |AD[A_{GA_0}] - \frac{1}{2}| = |AD[A_{GA_1}] - \frac{1}{2}| \tag{7}$$

Then, we can calculate and obtain Equation (8) from (6) and (7).

$$\frac{1}{2}\mathcal{MA}_{AD}(P) = |AD[A_{GA_1}] - AD[A_{GA_4}]| \tag{8}$$

The result (9) can be obtained using the triangular inequality.

$$\begin{aligned}
\frac{1}{2}\mathcal{MA}_{AD}(P) &= |AD[A_{GA_1}] - AD[A_{GA_4}]| \\
&\leq |AD[A_{GA_1}] - AD[A_{GA_3}]| \\
&+|AD[A_{GA_3}] - AD[A_{GA_4}]| \\
&\leq |AD[A_{GA_1}] - AD[A_{GA_2}]| \\
&+|AD[A_{GA_2}] - AD[A_{GA_3}]| \\
&+|AD[A_{GA_3}] - AD[A_{GA_4}]| \\
&\leq \frac{qu_{ha}^2}{2|HA|} + \frac{qu_{pu}^2}{2|PU|} + max\{C'qu_{se}^{s'}, \frac{qu_{se}}{2^{B_m}}\}
\end{aligned} \tag{9}$$

After multiplying (9) by 2, we can obtain the required result inequation.

$$\mathcal{MA}_{AD}(P) \leq \frac{qu_{ha}^2}{|HA|} + \frac{qu_{pu}^2}{|PU|} + 2max\{C'qu_{se}^{s'}, \frac{qu_{se}}{2^{B_m}}\}$$

Therefore, we can demonstrate that the proposed scheme can ensure the session key security by proving the Theorem 1. □

7.3. AVISPA Simulation

AVISPA [7,8] is a simulation tool that proves the security robustness of the proposed scheme against replay and MITM attacks. Therefore, various security protocols [23,34,35] are proved by using AVISPA. In this section, we explain the main data flow of AVISPA and show the simulation result.

Firstly, we need to write the proposed scheme as a programming language named "High-Level Protocol Specification Language (HLPSL)" in AVISPA. After writing in HLPSL code, the proposed scheme is converted to "Intermediate Format (IF)". Then, the translator in AVISPA starts analyzing the IF through the four backends: "On-the-Fly Model Checker (OFMC)", "Three Automata based on Automatic Approximations for Analysis of Security Protocol (TA4SP)", "SAT-based Model Checker (SATMC)", and "Constraint Logic-based Attack Searcher (CL-AtSe)". Because OFMC and CL-AtSe only support an exclusive-OR operator, the proposed scheme is executed in these backends. The analyzed result is recorded and summarized in the "Output Format (OF)". If there is a result of "SAFE" in OF, we can demonstrate that the proposed scheme can prevent replay and MITM attacks.

In AVISPA, we define roles to be suitable for the proposed scheme. Therefore, there are three roles in the proposed scheme: the user US, control center CC, and drone DR. Moreover, we show the session and environment roles in Figure 8.

```
role session(DR, CC, US : agent, SKusdr, SKccdr, SKccus : symmetric_key, PUF,H : hash_func)
def=
local SN1, SN2, SN3, RV1, RV2, RV3 : channel(dy)
composition
user(US, CC, DR, SKusdr, SKccus, SKccdr, PUF, H, SN1, RV1)
/\ controlcenter(US, CC, DR, SKusdr, SKccus, SKccdr, PUF, H, SN2, RV2)
/\ drone(US, CC, DR, SKusdr, SKccus, SKccdr, PUF, H, SN3, RV3)
end role
%%%%%%%%%%%%%%%%%%%%%%%%%%%%%%%%%%%%%%%%
role environment()
def=
const dr, cc, us : agent,
            puf, h : hash_func,
            skusdr, skccdr, skccus : symmetric_key,
            us_cc_aa1, us_dr_aa1, cc_dr_aa2, dr_us_aa3 : protocol_id,
            sp1, sp2, sp3, sp4, sp5, sp6 : protocol_id,
            idi : text
intruder_knowledge = {h, idi}
composition
session(us, cc, dr, skusdr, skccus, skccdr, puf, h)
/\session(i, cc, dr, skusdr, skccus, skccdr, puf, h)
/\session(us, i, dr, skusdr, skccus, skccdr, puf, h)
/\session(us, cc, i, skusdr, skccus, skccdr, puf, h)
end role
%%%%%%%%%%%%%%%%%%%%%%%%%%%%%%%%%%%%%%%%
goal
secrecy_of sp1,sp2
authentication_on us_cc_aa1
authentication_on us_dr_aa1
authentication_on cc_dr_aa2
authentication_on dr_us_aa3
end goal
environment()
```

Figure 8. Session and environment roles written in HLPSL.

Figure 9 shows the role of user US written in HLPSL code. State 1 is the user registration phase that US sends $\{ID_m\}$ to the CC through a secure channel. After receiving return message $\{k_m, SID_m, SID_n, MID_m\}$ from CC, US computes and stores γ_m, δ_m, SID^u_m, and SID^u_n in state 2. Then, US computes a login request message $\{MID_m, A_1, A_2, V_1\}$ to the CC. Note that $witness(US, CC, us\_cc\_aa1, Aa1')$ and $witness(US, DR, us\_dr\_aa1, Aa1')$ are functions to prove the freshness of random nonce a_1. Finally, US receives $\{A_6, V_3\}$ from DR and computes the session key $SK = h(A_7||a_1||a_2||\ a_3)$. The code $request(DR, US, dr\_us\_aa3, Aa3')$ means the acceptance of freshness for a_3.

```
%%%AVISPA Simulation
role user(DR, CC, US : agent, SKusdr, SKccus, SKccdr : symmetric_key, PUF,H :
hash_func, SND,RCV : channel(dy))

played_by US
def=
local State : nat,
                IDn, CHn, REn, SIDn, Kn, An, IDm, SIDm, Km, MIDm, Am, S, PWm,
BIOm, Gamma, Delta, SIDum, SIDun : text,
                A1, A2, A3, A4, A5, A6, A7, V1, V2, V3, Aa1, Aa2, Aa3, SIDc, SK:
text
                const sp1, us_cc_aa1, us_dr_aa1, cc_dr_aa2, dr_us_aa3 : protocol_id

init State := 0
transition
%%%%%%%User registration phase
1. State = 0 /\ RCV(start) =|>
State' := 1
/\ SND({IDm}_SKccus)

2. State = 1 /\ RCV({H(H(IDm. S). S. Am). H(IDm. S). H(IDn. S)}_SKccus) =|>
State' := 2
/\ Gamma' := xor(H(IDm. PWm. BIOm), H(H(IDm. S). S. Am))
/\ Delta' := H(BIOm. H(H(IDm. S). S. Am). H(IDm. S))
/\ SIDum' := xor(H(IDm. PWm), H(IDm. S))
/\ SIDun' := xor(H(PWm. BIOm), H(IDn. S))
%login and authentication phase
/\ Aa1' := new()
/\ A1' := xor(H(H(IDm. S). SIDc. H(H(IDm. S). S. Am)), Aa1')
/\ A2' := xor(H(H(IDm. S). SIDc), H(IDn. S))
/\ V1' := H(H(IDm. S). H(IDn. S). SIDc. H(H(IDm. S). S. Am). Aa1')
/\ SND(H(H(IDm. S). Am). A1'. A2'. V1')
/\ witness(US,CC,us_cc_aa1,Aa1')
/\ witness(US,DR,us_dr_aa1,Aa1')

3. State = 2 /\ RCV(xor(xor(H(H(IDm. S). H(IDn. S). Aa1'), Aa2'), Aa3'). H(H(H(IDm. S).
H(IDn. S). SIDc). Aa1'. Aa3'. H(H(H(IDm. S). H(IDn. S). SIDc). Aa1'. Aa2'. Aa3'))) =|>
State' := 3
/\SK' := H(H(H(IDm. S). H(IDn. S). SIDc). Aa1'. Aa2'. Aa3')
/\request(DR,US,dr_us_aa3,Aa3')

end role
```

Figure 9. User role written in HLPSL.

The AVISPA result is shown in Figure 10. As we mentioned before, we execute the proposed scheme in OFMC and CL-AtSe backends, and the summary of the result is "SAFE". Therefore, we prove that the proposed scheme can prevent replay and MITM attacks.

```
% OFMC
% Version of 2006/02/13
SUMMARY
 SAFE
DETAILS
 BOUNDED_NUMBER_OF_SESSIONS
PROTOCOL
 /home/span/span/testsuite/results/DAPSCS.if
GOAL
 as_specified
BACKEND
 OFMC
COMMENTS
STATISTICS
 parseTime: 0.00s
 searchTime: 7.69s
 visitedNodes: 1608 nodes
 depth: 12 plies
```

```
SUMMARY
 SAFE

DETAILS
 BOUNDED_NUMBER_OF_SESSIONS
 TYPED_MODEL

PROTOCOL
 /home/span/span/testsuite/results/DAPSCS.if

GOAL
 As Specified

BACKEND
 CL-AtSe

STATISTICS

 Analysed   : 3 states
 Reachable  : 0 states
 Translation: 0.09 seconds
 Computation: 0.00 seconds
```

Figure 10. AVISPA result.

7.4. Informal Security Analysis

We conduct an informal analysis of the proposed scheme to demonstrate the theoretical security robustness. Details are as below.

7.4.1. Stolen/lost Mobile Device Attack

If an adversary $\mathcal{A}$ obtains a lost mobile device of U_m, it can extract secret parameters $\{\gamma_m, \delta_m, SID_m^u, SID_n^u, MID_m\}$ using power analysis attacks. However, all of secret parameters are masked in the identity ID_m, password PW_m, and biometrics Bio_m information. Therefore, $\mathcal{A}$ must guess ID_m, PW_m, and Bio_m at the same time and this process is not practical. Thus, the proposed scheme is secure against stolen/lost mobile device attacks.

7.4.2. Offline Password-Guessing Attack

An adversary $\mathcal{A}$ can attempt an offline guessing attack using $\{MID_m, A_1, A_2, V_1\}$ $\{A_3, A_4, A_5, V_2\}$ and $\{A_6, V_3\}$, and the extracted values $\{\gamma_m, \delta_m, SID_m^u, SID_n^u, MID_m\}$ $\{\gamma_n\}$ from mobile device and drone, respectively. Using a password dictionary, $\mathcal{A}$ can guess $PW_{\mathcal{A}}^*$. However, $\mathcal{A}$ cannot know that $PW_{\mathcal{A}}^*$ is valid or not. It is because δ_m is masked with biometric secret key α_m. Therefore, the proposed scheme prevents offline password guessing attacks.

7.4.3. Impersonation Attack

(1) User impersonation attack: In this attack, an adversary $\mathcal{A}$ tries to disguise a legitimate user U_m. $\mathcal{A}$ has to make a valid login request message $\{MID_m, A_1, A_2, V_1\}$. $\mathcal{A}$ can obtain MID_m from the mobile device. However, without having the credentials SID_m, SID_n, and k_m, it is a difficult task for $\mathcal{A}$ to calculate MID_m, A_1, A_2, V_1. Thus, $\mathcal{A}$ cannot generate a valid login request message on behalf of U_m. Hence, the proposed scheme provides protection against user impersonation attacks.

(2) Control center impersonation attack: For this attack, let us suppose that $\mathcal{A}$ tries to send the message $\{A_3, A_4, A_5, V_2\}$ to the D_n on behalf of the CC. However, without having the credentials SID_m, SID_n, k_n, ID_n, and random nonce a_1, it is computationally hard for $\mathcal{A}$ to make a valid message. Therefore, the proposed scheme is resilient against the CC impersonation attack.

(3) Drone impersonation attack: This attack is a disguise attack in which a malicious adversary $\mathcal{A}$ conceals its identity information and attempts to behave as D_n. To do this, $\mathcal{A}$ computes $CH_{\mathcal{A}}^* = A_3 \oplus h(ID_n||\gamma_n)$. Since $PUF(.)$ is a physical unclonable circuit, $\mathcal{A}$ cannot compute RE_n. Therefore, it is impossible to compute $\alpha_n = Rep(RE_n, \beta_n)$, $SID_n = h(ID_n||\alpha_n)$, $k_n = \gamma_n \oplus SID_n$, $(SID_m||a_1||a_2) = A_2 \oplus h(SID_n||SID_c||k_n)$ to calculate $A_4 = h(SID_m||SID_n||a_1) \oplus (a_2||a_3)$. Thus, the proposed scheme can prevent drone impersonation attacks.

7.4.4. Replay and MITM Attacks

In the proposed scheme, all messages are masked in random nonce a_1, a_2, and a_3 to maintain the freshness. Moreover, each participant, e.g., remote user, control center, drone, checks the validity of the message by calculating and checking V_1^*, V_2^*, and V_3^*. Therefore, the proposed scheme can prevent replay and MITM attacks.

7.4.5. Physical and Cloning Attacks

For this attack, an adversary $\mathcal{A}$ intercepts a drone D_n and extracts the secret parameters $\{\gamma_n\}$ from the memory. However, $\mathcal{A}$ cannot compute the session key $SK = h(A_7||a_1||a_2||a_3)$ because each parameter in the message $\{A_3, A_4, A_5, V_2\}$ is masked in the PUF technology, which has an unclonable property. Thus, $\mathcal{A}$ cannot obtain any advantages from D_n, and this means that the proposed scheme is secure against physical or cloning attacks.

7.4.6. Privileged Insider Attack

In this attack, an adversary $\mathcal{A}$ is a privileged insider of the proposed system. Thus, $\mathcal{A}$ can obtain the registration request message $\{ID_m\}$ and secret parameters $\{\gamma_m, \delta_m, SID^u_m, SID^u_n, MID_m\}$ from the remote user U_m. However, without having PW_m and biometric secret key α_m of U_m, deriving secret credentials $SID_m = h(ID_m||PW_m) \oplus SID^u_m$ and $k_m = h(ID_m||PW_m||\alpha_m) \oplus \gamma_m$ is computationally infeasible. Thus, the proposed scheme prevents privileged insider attacks.

7.4.7. Ephemeral Security Leakage Attack

To prevent this security attack, the proposed scheme must maintain security even if random numbers are leaked. Thus, $\mathcal{A}$ obtains a_1, a_2, a_3, which are used during the AKA phase. However, $\mathcal{A}$ cannot calculate SID_m, k_m, and k_n without knowing the secret key s to the control center. Additionally, $\mathcal{A}$ cannot obtain any advantages to impersonate as a legitimate user U_m. Thus, the proposed scheme prevents ephemeral secret leakage (ESL) attacks.

7.4.8. Stolen-Verifier Attack

We can assume that an adversary $\mathcal{A}$ obtains table data $\{ID_n, SID_n, a_n, CH_n\}$ and $\{MID_m, SID^*_m, a_m\}$ from the database of the control center and attempts to calculate the session key $SK = h(A_7||a_1||a_2||a_3)$ or impersonate the control center. However, $\mathcal{A}$ cannot calculate the secret parameter SID_m, k_m and k_n without the secret keys of the control center and also cannot obtain random number a_1, a_2, a_3. Thus, $\mathcal{A}$ cannot compute SK or impersonate the control center. This means that the proposed scheme is resilient to stolen-verifier attacks.

7.4.9. User Anonymity and Untraceability

An adversary $\mathcal{A}$ cannot reveal the real identity ID_m of a legitimate user because of a cryptographic one-way hash function $h(.)$ masks ID_m with the secret key of the control center. Therefore, the proposed scheme provides the user's anonymity.

7.4.10. Perfect Forward Secrecy

If the master key s of the control center is leaked to an adversary $\mathcal{A}$, it can attempt to compute SK to attack the previous session. However, $\mathcal{A}$ cannot obtain the SK because $SK = h(A_7||a_1||a_2||a_3)$ does not include s. Moreover, if master secret key s of the control center is compromised, $\mathcal{A}$ cannot obtain $SID_m, SID_n, a_1, a_2, a_3$ because $\mathcal{A}$ cannot compute $SID_m = h(ID_m||s)$ without the real identity of the U_m, $SID_n = h(ID_n||\alpha_n)$ and without the secret key α_n. Therefore, $\mathcal{A}$ does not obtain any advantages over SK. This means that the proposed scheme guarantees perfect forward secrecy.

7.4.11. Mutual Authentication

In the MAKA phase, there are three messages $\{MID_m, A_1, A_2, V_1\}$, $\{A_3, A_4, A_5, V_2\}$, $\{A_6, V_3\}$ transmitted via public channels. Thus, each participant checks the legitimacy of the other participants and messages using V_1, V_2, and V_3 in the proposed scheme. If this process is successful, we can ensure authentication. Thus, the proposed scheme guarantees mutual authentication.

7.4.12. DoS Attack

If an adversary $\mathcal{A}$ tries to transmit $\{MID_m, A_1, A_2, V_1\}$ to the control center as a replay message, $\mathcal{A}$ has to pass the login phase by verifying the values of $\delta_m = h(\alpha_m||k_m||SID_m)$. However, $\mathcal{A}$ cannot construct a valid δ_m because $\mathcal{A}$ cannot obtain α_m, k_m, SID_m. Therefore, the replay message would not be sent to the control center. Thus, this proposed scheme can resist DoS attacks.

7.4.13. Drone Capture Attack

If an adversary $\mathcal{A}$ captures a drone D_n and obtains $\{\gamma_n\}$, $\mathcal{A}$ can try to threaten another legitimate drone D_{n1}. However, all of the drones are secure in PUF technology according to Section 7.4.5, and $\gamma_n = h(ID_n||\alpha_n) \oplus k_n$ is an independent parameter. Therefore, the proposed scheme can prevent drone capture attacks.

7.4.14. Session Key Disclosure Attack

To compute the session key $SK = h(A_7||a_1||a_2||a_3)$, an adversary$\mathcal{A}$ has to obtain SID_m, SID_n, a_1, a_2 and a_3. However, $\mathcal{A}$ cannot obtain any of these values because SID_m and SID_n are masked with secret key s and a_1, a_2 and a_3 are random numbers that are temporarily used in a session. Therefore, the proposed scheme is secure against session key disclosure attacks.

8. Performance Analysis

We demonstrate the security features of the proposed scheme with a related sch eme [4,14,18,21,24] in terms of "security functionalities", "communication costs", and "computation costs".

8.1. Security Features Comparison

In order to provide visualized information, we offer comprehensive security properties of the proposed scheme and related schemes [4,14,17,18,21,24] in a table. As shown in Table 4, we consider various security functionalities and attacks, including "stolen smart card/mobile device", "offline password guessing ", "impersonation", "replay", "privileged insider", "physical and cloning", "ESL", "verification table leakage", "user anonymity", "perfect forward secrecy", "mutual authentication", "DoS", "untraceability", "device/drone capture", and "correctness". Thus, our scheme offers secure and functional features as compared to the related schemes [4,14,18,21,24].

8.2. Communication Costs Comparison

We demonstrate the comparison analysis for communication costs of the proposed scheme with the other related schemes [4,14,17,18,21,24]. We refer to [4] and assume that the bit lengths for the hash function, random number, identity, PUF challenge, ECC point, and enc-decryption are 256, random, 160, 32, 160, and 128 bits, respectively. Thus, during the MAKA process of our scheme, the exchanged messages $\{MID_m, A_1, A_2, V_1\}$ require $(256 + 256 + 256 + 256 = 1024$bits), the message $\{A_3, A_4, A_5, V_2\}$ requires $(256 + 256 + 256 + 256 = 1024$bits), and the message $\{A_6, V_3\}$ requires $(256 + 256 = 512$bits), respectively. Table 5 shows the total communication costs of the proposed scheme and the related schemes.

Table 4. Security and functionality features (SFF) comparison.

SFF	[14]	[17]	[18]	[21]	[24]	[4]	Proposed
$SP1$	✓	✓	✓	✓	✓	✓	✓
$SP2$	✓	✓	✓	✓	✓	✓	✓
$SP3$	✓	✓	✓	✓	✓	✓	✓
$SP4$	✓	✓	✓	✓	✓	✓	✓
$SP5$	✓	✓	✓	✓	×	✓	✓
$SP6$	×	×	×	×	×	×	✓
$SP7$	×	✓	✓	✓	✓	✓	✓
$SP8$	✓	✓	✓	✓	×	×	✓
$SP9$	✓	✓	✓	✓	✓	✓	✓
$SP10$	×	✓	✓	✓	✓	✓	✓
$SP11$	✓	✓	✓	✓	✓	✓	✓
$SP12$	✓	✓	✓	✓	✓	×	✓

Table 4. *Cont.*

SFF	[14]	[17]	[18]	[21]	[24]	[4]	Proposed
$SP13$	✓	✓	✓	✓	✓	✓	✓
$SP14$	✓	✓	✓	✓	✓	×	✓
$SP15$	✓	✓	✓	✓	✓	×	✓

Note: $SP1$: stolen smart card/mobile device attack; $SP2$: offline password guessing attack; $SP3$: impersonation attack; $SP4$: replay attack; $SP5$: privileged-insider attack; $SP6$: physical and cloning attack; $SP7$: ESL attack; $SP8$: stolen-verifier attack; $SP9$: user anonymity; $SP10$: perfect forward secrecy; $SP11$: mutual authentication; $SP12$: DoS attack; $SP13$: untraceability; $SP14$: device/drone capture attack; $SP15$: correctness; ✓: Provide or support SFF. ×: Do not provide or support SFF.

Table 5. Comparison study of communication costs.

Schemes	Total Costs	Number of Messages
Ali et al. [14]	1696 bits	3 messages
Wu et al. [17]	3360 bits	3 messages
Tanveer et al. [18]	2240 bits	3 messages
Zhang et al. [21]	5760 bits	4 messages
Tanveer et al. [24]	1856 bits	3 messages
Akram et al. [4]	2304 bits	3 messages
Proposed	2560 bits	3 messages

Although our scheme has slightly higher communication costs than Akram et al.'s scheme [4], we offer better security functionalities and efficient computation costs compared to the related schemes [14,17,18,21,24]. Figure 11 illustrates the total communication costs of the proposed scheme and the related schemes.

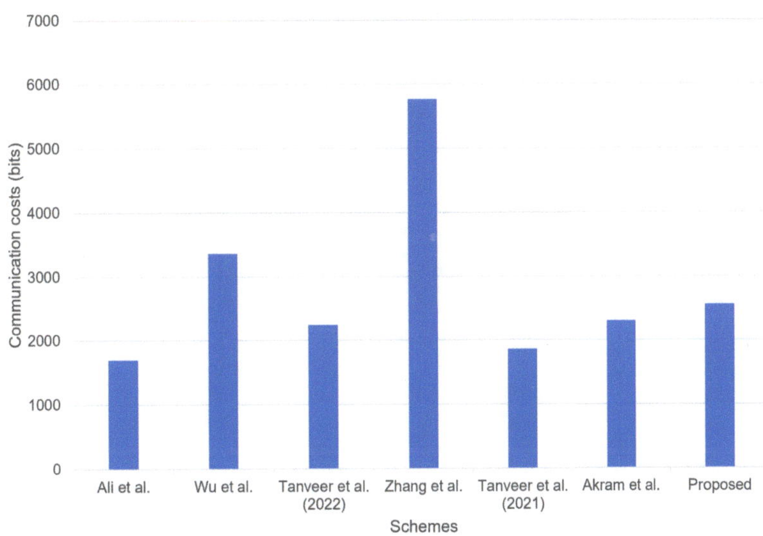

Figure 11. Communication costs comparison [4,14,17,18,21,24].

8.3. Computation Costs Comparison

We estimate the computation costs of the proposed scheme and [4,14,17,18,21,24] in the AKA phase. Referring to [18,21,24], we define that T_H, T_{ECC}, T_{ENC}, T_{FE}, T_{AC}, $T_{pm_{FourQ}}$, T_M, and T_O denote the hash function($\approx$0.029 ms), ECC multiplication($\approx$0.605 ms), enc-decryption time($\approx$0.036 ms), fuzzy extractor($\approx$0.605 ms), AEGIS($\approx$0.07 ms), FourQ point multiplication($\approx$1.199 ms), HMAC($\approx$0.053 ms), and BPV-online function($\approx$2.117 ms), respectively. Table 6 shows the total computation costs of the proposed scheme and the related schemes.

Table 6. Comparison study of computation costs.

Schemes	Remote User Side	Control Center Side	Drone Side	Total	Total Costs (s)
[14]	$10T_H + 1T_{FE}$	$7T_H$	$7T_H$	$24T_H + 1T_{FE}$	≈1.301 ms
[17]	$12T_H + 1T_{FE}$	$9T_H$	$8T_H$	$29T_H + 1T_{FE}$	≈1.446 ms
[18]	$9T_H + 4T_{ENC}$ $+3T_{ECC}$	$4T_H + 3T_{ENC} +$ $1T_{ECC}$	$7T_H + 2T_{ENC}$ $+2T_{ECC}$	$20T_H + 9T_{ENC} + 6T_{ECC}$	≈4.534 ms
[21]	$7T_H + 3T_{pmFourQ}+$ $1T_{ENC} + 1T_O + 1T_M$	$5T_H + 1T_{pmFourQ}$ $+2T_{ENC} + 1T_M$	$4T_H + 1T_{pmFourQ}$ $+1T_{ENC} + 1T_O$	$16T_H + 5T_{pmFourQ}$ $+4T_{ENC} + 2T_O + 2T_M$	≈10.943 ms
[24]	$6T_H + 3T_{AC}$ $+3T_{ECC} + 1T_{FE}$	$2T_H + 1T_{ECC} +$ $3T_{AC}$	$3T_H + 2T_{ECC} +$ $2T_{AC}$	$11T_H + 6T_{ECC}$ $+8T_{AC} + 1T_{FE}$	≈5.114 ms
[4]	$9T_H$	$7T_H + 2T_{ENC}$	$7T_H$	$23T_H + 2T_{ENC}$	≈0.739 ms
Ours	$11T_H + 1T_{FE}$	$11T_H$	$10T_H + 1T_{FE}$	$32T_H + 2T_{FE}$	≈2.138 ms

Compared with the proposed scheme and Akram et al.'s scheme, the proposed scheme consumes more computation costs. However, the proposed scheme utilizes the fuzzy extractor and PUF technologies and, therefore, provides much higher security to the entire IoD network systems than [4]. Figure 12 illustrates that the computational cost (delay) increases at the control center with an increasing number of users.

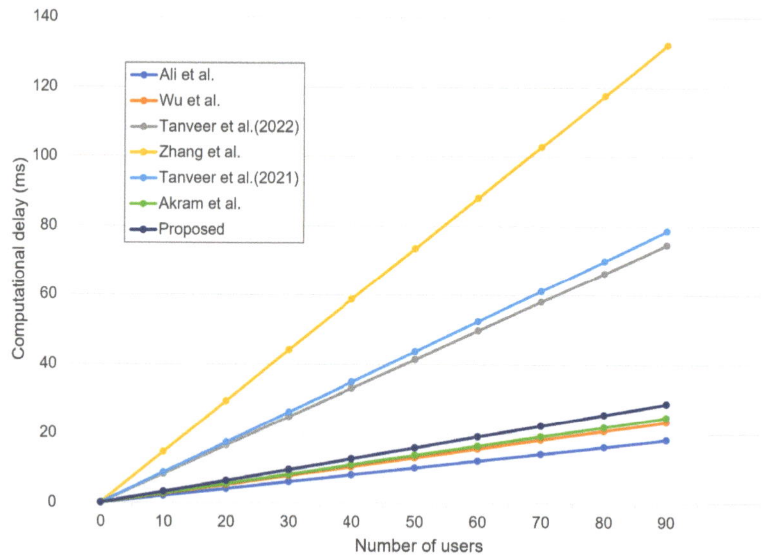

Figure 12. Computational delay at the control center with increasing the AKA requests [4,14,17,18,21,24].

9. Conclusions

In this study, we reviewed Akram et al.'s scheme, which was proposed for secure authentication between users and drones in IoD networks. In Akram et al.'s scheme, there are several security vulnerabilities, such as session key disclosure, drone impersonation, and stolen-verifier attacks. In addition, their scheme cannot ensure perfect forward secrecy and has correctness problems. To overcome the security flaws of their scheme and provide various functional features, we proposed a secure MAKA scheme using biometrics and PUF technologies. The proposed scheme can provide robustness to withstand various attacks, including session key disclosure, verification table leakage, impersonation, ESL, and privileged insider attacks. Moreover, the proposed scheme can achieve mutual authentication, perfect forward secrecy, and anonymity. To prove the session key security and mutual authentication, we analyzed the proposed scheme using an RoR model and BAN logic, respectively. Furthermore, we simulated the proposed scheme using AVISPA and showed that the proposed scheme is resilient against replay and MITM attacks. A comparative study of functionality features, efficiency, and security shows the effectiveness of the proposed scheme. Therefore, we can demonstrate that the proposed scheme has security robustness compared to existing user authentication protocols for IoD environments with reasonable computation and communication overheads. These characteristics show that the proposed scheme can provide users with high security reliability and high-speed communication in IoD environments. In future work, we intend to implement the proposed scheme in real environments using the mobile device as a user, a desktop as a server, and Raspberry PI 4 as a drone.

Author Contributions: Conceptualization, Y.P. (Yohan Park) and D.R.; Formal analysis, D.R. and D.K.; Methodology, Y.P. (Yohan Park) and D.K.; Software, D.K.; Validation, Y.P. (Yohan Park) and Y.P. (Youngho Park); Formal Proof, D.K.; Writing—original draft, and Y.P. (Yohan Park) and D.R.; Writing—review and editing, Y.P. (Yohan Park) and D.K.; Supervision, Y.P. (Yohan Park). All authors have read and agreed to the published version of the manuscript.

Funding: This work was supported by the Bisa Research Grant of Keimyung University in 2019.

Institutional Review Board Statement: Not applicable.

Informed Consent Statement: Not applicable.

Data Availability Statement: Not applicable.

Conflicts of Interest: The authors declare no conflict of interest.

References

1. Gharibi, M.; Boutaba, R.; Waslander, S.L. Internet of drones. *IEEE Access* **2016**, *4*, 1148–1162. [CrossRef]
2. Abualigah, L.; Diabat, A.; Sumari, P.; Gandomi, A.H. Applications, deployments, and integration of internet of drones (iod): A review. *IEEE Sens. J.* **2021**, *21*, 25532–25546. [CrossRef]
3. Lin, C.; He, D.; Kumar, N.; Choo, K.K.R.; Vinel, A.; Huang, X. Security and privacy for the internet of drones: Challenges and solutions. *IEEE Commun. Mag.* **2018**, *56*, 64–69. [CrossRef]
4. Akram, M.W.; Bashir, A.K.; Shamshad, S.; Saleem, M.A.; AlZubi, A.A.; Chaudhry, S.A.; Alzahrani, B.A.; Zikria, Y.B. A secure and lightweight drones-access protocol for smart city surveillance. *IEEE Trans. Intell. Transp. Syst.* **2021**, *23*, 19634–19643. [CrossRef]
5. Umar, M.; Islam, S.H.; Mahmood, K.; Ahmed, S.; Ghaffar, Z.; Saleem, M.A. Provable secure identity-based anonymous and privacy-preserving inter-vehicular authentication protocol for VANETS using PUF. *IEEE Trans. Veh. Technol.* **2021**, *70*, 12158–12167. [CrossRef]
6. Herder, C.; Yu, M.D.; Koushanfar, F.; Devadas, S. Physical unclonable functions and applications: A tutorial. *Proc. IEEE* **2014**, *102*, 1126–1141. [CrossRef]
7. AVISPA, T. Automated Validation of Internet Security Protocols and Applications. 2015. Available online: https://www.avispa-project.org/ (accessed on 10 February 2023).
8. Glouche, Y.; Genet, T.; Heen, O.; Courtay, O. A security protocol animator tool for AVISPA. In Proceedings of the ARTIST2 Workshop on Security Specification and Verification of Embedded Systems, Pisa, Italy, 18–20 May 2006; pp. 1–7.
9. Abdalla, M.; Fouque, P.A.; Pointcheval, D. Password-based authenticated key exchange in the three-party setting. In *Proceedings of the International Workshop on Public Key Cryptography*; Springer: Berlin/Heidelberg, Germany, 2005; pp. 65–84.
10. Burrows, M.; Abadi, M.; Needham, R. A logic of authentication. *ACM Trans. Comput. Syst. (TOCS)* **1990**, *8*, 18–36. [CrossRef]
11. Wazid, M.; Das, A.K.; Kumar, N.; Vasilakos, A.V.; Rodrigues, J.J. Design and analysis of secure lightweight remote user authentication and key agreement scheme in Internet of drones deployment. *IEEE Internet Things J.* **2018**, *6*, 3572–3584. [CrossRef]

12. Teng, L.; Jianfeng, M.; Pengbin, F.; Yue, M.; Xindi, M.; Jiawei, Z.; Gao, C.; Di, L. Lightweight security authentication mechanism towards UAV networks. In Proceedings of the 2019 International Conference on Networking and Network Applications (NaNA), Daegu City, Republic of Korea, 10–13 October 2019; pp. 379–384.
13. Srinivas, J.; Das, A.K.; Kumar, N.; Rodrigues, J.J. TCALAS: Temporal credential-based anonymous lightweight authentication scheme for Internet of drones environment. *IEEE Trans. Veh. Technol.* **2019**, *68*, 6903–6916. [CrossRef]
14. Ali, Z.; Chaudhry, S.A.; Ramzan, M.S.; Al-Turjman, F. Securing smart city surveillance: A lightweight authentication mechanism for unmanned vehicles. *IEEE Access* **2020**, *8*, 43711–43724. [CrossRef]
15. Ever, Y.K. A secure authentication scheme framework for mobile-sinks used in the internet of drones applications. *Comput. Commun.* **2020**, *155*, 143–149. [CrossRef]
16. Deebak, B.D.; Al-Turjman, F. A smart lightweight privacy preservation scheme for IoT-based UAV communication systems. *Comput. Commun.* **2020**, *162*, 102–117. [CrossRef]
17. Wu, T.; Guo, X.; Chen, Y.; Kumari, S.; Chen, C. Amassing the security: An enhanced authentication protocol for drone communications over 5G networks. *Drones* **2022**, *6*, 10–29. [CrossRef]
18. Tanveer, M.; Alkhayyat, A.; Naushad, A.; Kumar, N.; Alharbi, A.G.; et al. RUAM-IoD: A Robust User Authentication Mechanism for the Internet of Drones. *IEEE Access* **2022**, *10*, 19836–19851. [CrossRef]
19. Alladi, T.; Chamola, V.; Kumar, N.; et al. PARTH: A two-stage lightweight mutual authentication protocol for UAV surveillance networks. *Comput. Commun.* **2020**, *160*, 81–90. [CrossRef]
20. Pu, C.; Li, Y. Lightweight authentication protocol for unmanned aerial vehicles using physical unclonable function and chaotic system. In Proceedings of the 2020 IEEE International Symposium on Local and Metropolitan Area Networks (LANMAN), Orlando, FL, USA, 13–15 July 2020; pp. 1–6.
21. Zhang, N.; Jiang, Q.; Li, L.; Ma, X.; Ma, J. An efficient three-factor remote user authentication protocol based on BPV-FourQ for internet of drones. *Peer-to-Peer Netw. Appl.* **2021**, *14*, 3319–3332. [CrossRef]
22. Dolev, D.; Yao, A. On the security of public key protocols. *IEEE Trans. Inf. Theory* **1983**, *29*, 198–208. [CrossRef]
23. Chattaraj, D.; Bera, B.; Das, A.K.; Rodrigues, J.J.; Park, Y. Designing Fine-Grained Access Control for Software-Defined Networks Using Private Blockchain. *IEEE Internet Things J.* **2021**, *9*, 1542–1559. [CrossRef]
24. Tanveer, M.; Kumar, N.; Hassan, M.M.; et al. RAMP-IoD: A robust authenticated key management protocol for the Internet of Drones. *IEEE Internet Things J.* **2021**, *9*, 1339–1353. [CrossRef]
25. Dodis, Y.; Reyzin, L.; Smith, A. Fuzzy extractors: How to generate strong keys from biometrics and other noisy data. In *Proceedings of the International Conference on the Theory and Applications of Cryptographic Techniques*; Springer: Berlin/Heidelberg, Germany, 2004; pp. 523–540.
26. Kim, M.; Lee, J.; Park, K.; Park, Y.; Park, K.H.; Park, Y. Design of secure decentralized car-sharing system using blockchain. *IEEE Access* **2021**, *9*, 54796–54810. [CrossRef]
27. Kwon, D.K.; Yu, S.J.; Lee, J.Y.; Son, S.H.; Park, Y.H. WSN-SLAP: Secure and lightweight mutual authentication protocol for wireless sensor networks. *Sensors* **2021**, *21*, 936. [CrossRef]
28. Shashidhara, R.; Nayak, S.K.; Das, A.K.; Park, Y. On the design of lightweight and secure mutual authentication system for global roaming in resource-limited mobility networks. *IEEE Access* **2021**, *9*, 12879–12895. [CrossRef]
29. Wang, D.; Cheng, H.; Wang, P.; Huang, X.; Jian, G. Zipf's law in passwords. *IEEE Trans. Inf. Forensics Secur.* **2017**, *12*, 2776–2791. [CrossRef]
30. Bagga, P.; Das, A.K.; Wazid, M.; Rodrigues, J.J.; Choo, K.K.R.; Park, Y. On the design of mutual authentication and key agreement protocol in internet of vehicles-enabled intelligent transportation system. *IEEE Trans. Veh. Technol.* **2021**, *70*, 1736–1751. [CrossRef]
31. Son, S.; Lee, J.; Park, Y.; Park, Y.; Das, A.K. Design of blockchain-based lightweight V2I handover authentication protocol for VANET. *IEEE Trans. Netw. Sci. Eng.* **2022**, *9*, 1346–1358. [CrossRef]
32. Wazid, M.; Bagga, P.; Das, A.K.; Shetty, S.; Rodrigues, J.J.; Park, Y. AKM-IoV: Authenticated key management protocol in fog computing-based Internet of vehicles deployment. *IEEE Internet Things J.* **2019**, *6*, 8804–8817. [CrossRef]
33. Boyko, V.; MacKenzie, P.; Patel, S. Provably secure password-authenticated key exchange using Diffie-Hellman. In *Proceedings of the International Conference on the Theory and Applications of Cryptographic Techniques*; Springer: Berlin/Heidelberg, Germany, 2000; pp. 156–171.
34. Kwon, D.; Son, S.; Park, Y.; Kim, H.; Park, Y.; Lee, S.; Jeon, Y. Design of Secure Handover Authentication Scheme for Urban Air Mobility Environments. *IEEE Access* **2022**, *10*, 42529–42541. [CrossRef]
35. Ryu, J.; Oh, J.; Kwon, D.; Son, S.; Lee, J.; Park, Y.; Park, Y. Secure ECC-based three-factor mutual authentication protocol for telecare medical information system. *IEEE Access* **2022**, *10*, 11511–11526. [CrossRef]

sensors

MDPI

Article

Experimental Study on Longitudinal Acceleration of Urban Buses and Coaches in Different Road Maneuvers

Damian Frej, Paweł Grabski, Rafał S. Jurecki and Emilia M. Szumska *

Faculty of Mechatronics and Mechanical Engineering, Kielce University of Technology, al. Tysiąclecia Państwa Polskiego 7, 25-314 Kielce, Poland
* Correspondence: eszumska@tu.kielce.pl

Abstract: A vehicle's longitudinal acceleration is a parameter often used for determining vehicle motion dynamics. This parameter can also be used to evaluate driver behavior and passenger comfort analysis. The paper presents the results of longitudinal acceleration tests of city buses and coaches recorded during rapid acceleration and braking maneuvers. The presented test results demonstrate that longitudinal acceleration is significantly affected by road conditions and surface type. In addition, the paper presents the values of longitudinal accelerations of city buses and coaches during their regular operation. These results were obtained on the basis of registration of vehicle traffic parameters in a continuous and long-term manner. The test results showed that the maximum deceleration values recorded during the tests of city buses and coaches in real traffic conditions were much lower than the maximum deceleration values found during sudden braking maneuvers. This proves that the tested drivers in real conditions did not have to use sudden braking. The maximum positive acceleration values recorded in acceleration maneuvers were slightly higher than the acceleration values logged during the rapid acceleration tests on the track.

Keywords: vehicle dynamics and stability; longitudinal acceleration; vehicle testing

Citation: Frej, D.; Grabski, P.; Jurecki, R.S.; Szumska, E.M. Experimental Study on Longitudinal Acceleration of Urban Buses and Coaches in Different Road Maneuvers. *Sensors* **2023**, *23*, 3125. https://doi.org/10.3390/s23063125

Academic Editors: Constantin Caruntu and Ciprian Romeo Comşa

Received: 10 February 2023
Revised: 3 March 2023
Accepted: 10 March 2023
Published: 15 March 2023

1. Introduction

Urban buses and coaches are some of the most important collective transport means. An urban bus is a vehicle planned to transport numerous people over small distances. In addition to seating, city buses have room for standing passengers, spacious and comfortable aisles among seats, and spacious entrance doors. An urban bus is adapted to move in urban conditions characterized by frequent accelerations and braking, and thereby its maximum speed usually is not very high. Passengers in these vehicles do not have seat belts at their disposal and may therefore be at risk of injury in driving situations that require dynamic driver responses.

Coaches are vehicles that usually carry out passenger transport at long distances, including interurban and international routes. These vehicles usually do not use traditional stops and are most often not restricted by rigid time schedules. Coaches are only equipped with seats and have dedicated spaces for luggage. Coaches are also characterized by higher passenger comfort than urban buses. Passengers have at their disposal comfortable, adjustable seats with armrests and tables, air-conditioning, and monitors for watching movies, and even fridges and a bar with hot drinks. Coach passengers should use the seat belts installed at each seat while driving.

Urban buses and coaches differ in terms of dynamic driving properties. Acceleration is one of the parameters that enable the testing of differences between a vehicle's dynamic parameters. Longitudinal acceleration is directly related to acceleration and braking maneuvers. Acceleration and deceleration values depend on the maneuver's intensity applied by the driver and vehicle type. The recording of longitudinal acceleration during experimental testing can be done with the use of various devices, including a 3-axis accelerometer, acceleration sensors with a GPS module, and recorders that read data from the CAN bus using OBD connections. Lately, it is becoming increasingly popular to use smartphones with dedicated applications. However, they are characterized by various measurement accuracies.

Accelerometers are widely used in vehicle dynamics studies. In vehicle monitoring it is necessary to differentiate between acceleration, deceleration, and lateral acceleration because these data can be used to effectively identify and classify vehicle maneuvers. Three-axis accelerometers are used to measure vehicle acceleration relative to three perpendicular coordinate axles. Examples of using three-axis accelerometers for vehicle dynamics testing can be found in [1–3]. Currently, there are increasing numbers of studies in which accelerometers are mounted in wheel rims [4–6].

The method of collecting acceleration profiles using GPS sensors allows for automatic and continuous recording of acceleration values while driving without the need to modify the vehicle's design or to design the experiment in a specific way. Sensors with a GPS module provide information about the vehicle's instantaneous position, instantaneous speed, instantaneous acceleration values, distance traveled, and travel time. Examples of this method's application in recording acceleration values and other dynamic parameters of a vehicle while driving are presented, among others, in [7–10].

In many papers, the dynamic properties of a vehicle in real conditions were evaluated using data derived from a CAN (controller area network) bus. A CAN bus is a two-wire network with real-time data transmission. Each electrical module of the vehicle is monitored and controlled using one or more sensors that provide notification to and interoperate with the main control unit (MCU). Microcontrollers (that operate the sensors) communicate with the MCU and between one another using typical communication standards based on a bus such as the CAN bus. The data that can be obtained from the CAN bus include longitudinal and lateral acceleration, brake pedal use, accelerator pedal use, or engine speed. Examples of using data derived from a CAN bus to test the dynamic parameters of a vehicle can be found in [11–14].

Another popular method of collecting vehicle driving parameters is to use an OBD connection that plugs into a dedicated port on the vehicle. As mentioned earlier, the vehicle's electronic units communicate with one another using a network (CAN bus or similar) to monitor and transmit data. Vehicle dynamic parameters are derived from a microcontroller that uses the OBD protocol (on-board diagnostics). OBD is the vehicle's reporting and diagnostic function that allows the vehicle operator or technician to access the condition of engine subsystems. Vehicles usually have an external port that enables the downloading of values of selected parameters and reading errors, as well as identifying the error code source. Access to the monitored values is provided by using OBD scanners. OBD connections can be used to record lateral acceleration and longitudinal acceleration, engine speed, acceleration and braking pedal position, or fuel consumption. The method of collecting data using an OBD connection to obtain the vehicle's dynamic parameters is presented in studies [15–18].

Using a smartphone as a device to record acceleration profiles is relatively inexpensive; however, the data captured are limited to features allowed by the smartphone. Common sensors provided by a smartphone include accelerometers, gyroscopes, and global positioning systems (GPS). Based on the possibility to record the vehicle's acceleration and instantaneous position, dedicated applications were developed that constitute an element of the measurement system. For example, in papers [19–22], it is possible to find an application that determines the driver's profile using a statistical model based on the recorded longitudinal and lateral acceleration values while driving.

Acceleration profiles are used to analyze and assess many aspects related to vehicle operation and safety. Many papers present results of experimental testing aimed at determining the relation between acceleration momentum and speed [23–27]. In the literature, it is possible to find a comparison of longitudinal acceleration values recorded by several types of vehicles. Analyses conducted in works [28–30] show that accelerating a vehicle to a higher desired speed requires a longer acceleration time and a longer acceleration distance. Vehicles braking from a higher initial speed show longer deceleration times, longer deceleration distances, and lower deceleration values. This is important in terms of the active safety of buses.

Due to the specificity of urban traffic conditions and high traffic intensity, public transport can be much more often involved in accidents or collisions. It is estimated that several hundred accidents involving buses take place each year. For example, in 2020, 189 accidents involving urban buses, 6 fatalities, and 232 wounded were recorded in Poland. In the case of other buses (coaches), 34 accidents involving 4 fatalities and 73 wounded were recorded. Traffic incidents that take place on different road types and in different atmospheric conditions are subject to evaluations of road traffic experts aimed at determining the causes of the traffic accident. These evaluations require obtaining the limit values of acceleration recorded during sudden braking and acceleration, which can be applied as a result of the tests described in this paper. The necessity of updating the related data must also be noted. This is related to the continuous improvement of vehicle braking systems, wheel slip control systems during acceleration and braking (e.g., BAS, ABS, ASR, TCS, ATC, ESP), and the use of, e.g., electric drive systems.

Longitudinal acceleration studies and analyses obtained during testing conducted in the normal operation of these form of transport enable the detection of dangerous driver behavior when driving a bus. When analyzing the data in a longer time interval, it becomes possible to determine the probability of accidents and collisions caused by a driver's dangerous driving. The acceleration and deceleration characteristics can point to different driver danger behaviors, thereby making it possible to eliminate them. The results described in studies [31–33] established that each driver uniquely perceives the environment and reacts subjectively to changing road conditions, which is reflected in acceleration values. Several studies have been reported in the literature in which driving style and driver behavior are determined by longitudinal acceleration profiles. Better and more effective brakes in a vehicle enable faster deceleration to avoid an accident, on one hand, but on the other hand, higher deceleration can cause dangerous situations among the vehicle's passengers [34,35]. Considering the specificity of bus designs, e.g., possible standing places, such tests can provide an opportunity to determine situations in which passengers can feel discomfort during travel. Many studies and scientific papers confirm that longitudinal acceleration is one of the factors that determine the comfort of passengers in urban buses and coaches. Papers [36–39] present the results of a study with the evaluation of acceleration values considered to be uncomfortable by seated standing and moving passengers on the bus, registered at different speeds and traffic conditions.

A literature review allowed for the observation that there are not very many studies that demonstrate the real longitudinal acceleration of urban buses and coaches.

The objective of this study is to present methodology to determine the extreme values of longitudinal accelerations recorded in test conditions and then compare them with the acceleration values obtained in driving conditions in real traffic conditions. Extreme longitudinal acceleration values are expressed as acceleration and deceleration ranges. These values were determined for city buses and coaches during sudden braking and acceleration maneuvers. The testing conducted as part of this research project featured two stages. In the first stage, special test sections were used to conduct sudden braking maneuvers and intense acceleration maneuvers from a standing start. This made it possible to determine the maximum acceleration and deceleration values under laboratory conditions. In the second stage, longitudinal acceleration was analyzed while driving the test vehicles in real road traffic conditions. The acceleration values collected under real conditions could be compared to the maximum values set in the track tests. The presented results are part of a project aimed at developing an application to determine and assess the driving style of drivers of several types of vehicles.

Longitudinal acceleration is one of the parameters used to examine a driver's driving style. These parameters are important in the case of urban buses and coaches due to passenger comfort. Reliable deceleration and acceleration values of vehicles are important for car experts. Acceleration values differ depending on the vehicle type. Programs intended for complex evaluation of road incidents utilize sophisticated vehicle models aimed at conducting a space and time analysis that covers any type of moving object. Many of these programs require the specification of the vehicle's longitudinal acceleration. A literature review allowed for the observation that not many publications demonstrate updated results of longitudinal acceleration tests conducted on urban buses and coaches. Obtaining true and reliable longitudinal acceleration values requires appropriate laboratory tests using specialized equipment.

This paper shows the results of part of the research carried out as part of a project to determine the driving style of drivers of different types of vehicles (including urban buses and long-distance coaches) under different road conditions. The maximum and minimum values of longitudinal acceleration determined in the test conditions were the basis for the initial evaluation of the drivers. High acceleration values recorded by the driver during regular driving were considered potentially dangerous and risky behavior. The ultimate goal of the project, of which the presented results are part, was to use longitudinal acceleration as one of the factors determining driver behavior. Thanks to this, it was possible to identify drivers with dangerous behavior that could lead to dangerous road incidents in the future.

In this paper, Section 2 shows the methodology of the conducted research. The tests of accelerations were carried out on the track and in a real-world driving condition. The parameter recorded during the tests was longitudinal acceleration. Section 3 presents the acceleration values recorded during rapid acceleration and braking maneuvers as well as the acceleration values recorded while driving in real traffic conditions. Section 4 discusses the differences between the acceleration values obtained by the city bus and coach in track tests and the acceleration values collected during the normal operation of the vehicles. The obtained results were compared with the results of similar tests presented in the literature. Finally, Section 5 presents the main conclusions from the analyzes carried out in the study.

2. Research Methodology

In this paper, an experimental study of vehicles was performed. The parameter investigated was longitudinal acceleration, recorded using specialized measuring equipment. The experimental research was carried out in two stages. The first stage was carried out under specific measurement conditions on specially prepared measuring sections on the track. The second was carried out in real conditions during regular vehicle operation. The data collected during the experimental tests were then statistically analyzed using the Statistica program (StatSoft, version 13).

As mentioned above, buses differ in terms of dynamic properties during driving based on their intended purpose and design. It was therefore decided to conduct testing of longitudinal acceleration values for two specific types of buses, i.e., an urban bus and a coach. These vehicles differ, due to their purpose, in their dynamic performance. The testing was conducted in two stages.

In the first stage, measurements were made during acceleration and rapid braking maneuvers. In acceleration maneuvers, the driver had to smoothly reach a speed of about 50 km/h from the starting position as quickly as possible. In braking maneuvers, the driver had to stop the vehicle as quickly as possible from an initial speed of 50 km/h. Measurements were carried out on test tracks with various surfaces: dry asphalt, wet asphalt, dry concrete surface, and wet concrete surface.

In the first stage of testing, the vehicles were equipped with a measurement system that consisted of the following:

- An optoelectronic sensor (Corrsys Datron S-350® Aqua) that recorded the vehicle's motion components, including the instantaneous speed (Figure 1a);

- A multi-functional acceleration sensor (TAA®) that recorded the linear acceleration, a sensor that measured the linear and angular acceleration (TANS®) relative to the X, Y, and Z axes (Figure 1b);
- A data acquisition station (Datron uEEP12®) connected with a tablet and the software (ARMS®) (Figure 1c,d).

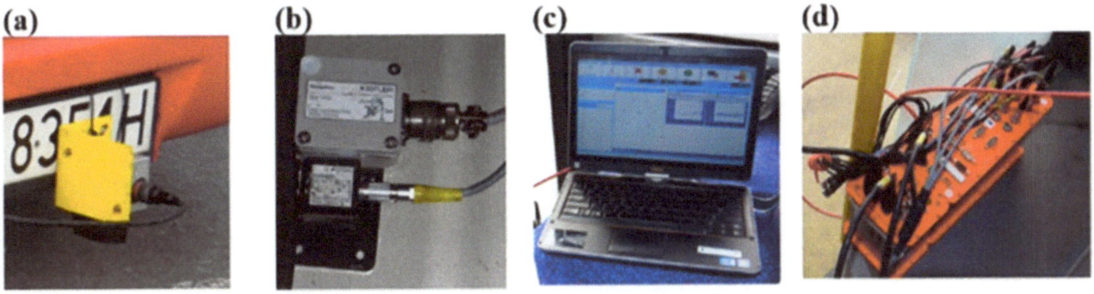

Figure 1. Measurement system during acceleration and braking maneuvers: (**a**) optoelectronic sensor, (**b**) acceleration sensors, (**c**) control tablet with software, (**d**) data acquisition station.

The measurement system enabled data recording with a frequency of 100 Hz, thereby allowing for detailed determination of the waveforms of the maneuvers' selected dynamic parameters.

The test vehicles underwent appropriate metering before the testing (Figure 2) to ensure adequate repeatability of the results. The parameters of the test vehicles used in the first stage of testing are shown in Table 1.

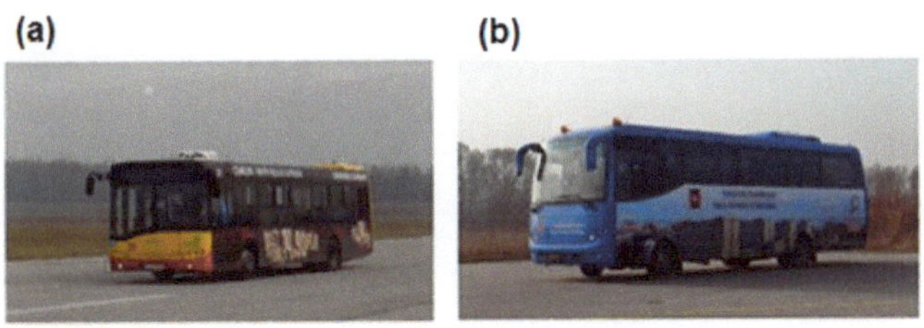

Figure 2. Test vehicles during acceleration and braking maneuvers: (**a**) urban bus, (**b**) coach.

Table 1. Test vehicle technical parameters.

Parameter	Urban bus (Figure 2a) Solaris Urbino 12	Coach (Figure 2b) Automex Apollo
Maximum output power (kW)	224	160
Maximum torque (Nm)	1200	810
Length (m)	12.00	8.70
Mass (m)	2.55	2.42
Height (m)	3.04	3.27
Wheelbase (m)	5.90	4.22
Curb mass (kg)	10,900	7645
Total mass (kg)	18,000	10,500
Tire	Continental 275/70 R22.5	Fulda RegioForce 245/70R17.5
Number of passengers		
standing	43	30
sitting	61	-

The aim of the first part of the study was to determine the range of limit values of longitudinal acceleration for the analyzed vehicles on various types of road surfaces. At this stage, the minimum values of the longitudinal acceleration during the rapid braking tests and the maximum acceleration values during the rapid acceleration tests were determined. The specified acceleration range can provide a reference for assessing driving style and detecting unsafe driver behavior.

The second stage of testing included a long-term recording data system installed in urban buses and coaches. The measurement system recorded vehicle velocity as a function of time and acceleration longitudinal related to time. In addition, a sensor with a GPS module was used to estimate the instantaneous position of the test vehicle. These vehicles were driven at different times of day and in various atmospheric conditions during traditional (planned) transport tasks.

Time profiles of longitudinal acceleration were recorded at this stage of testing. The measurement system used featured a sensor with a GPS module that recorded the vehicle's instantaneous position, velocity as a function of time, and acceleration related to time. The sensor's computer system saved the data on an SD card with a frequency of 25 Hz. It periodically archived the measurement data at a fixed time interval and designated maximum acceleration values in this interval. The data covering maximum (acceleration) and minimum (deceleration) values were used in further statistical analyses. The tests were conducted on 25 urban buses performing transport tasks for 7 consecutive days and on 8 coaches for 16 consecutive days.

3. Results

3.1. Longitudinal Acceleration Recorded during Sudden Braking

Figure 3 presents examples of deceleration profiles during braking from the initial speed of 50 km/h recorded on the track for an urban bus and a coach.

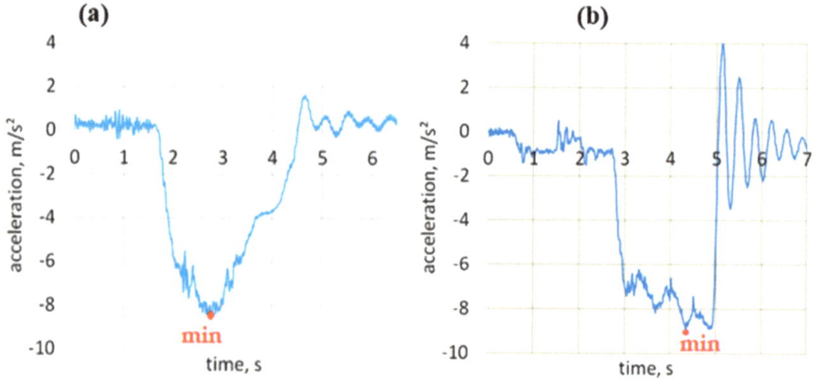

Figure 3. Acceleration profiles during braking for an urban bus (**a**) and a coach (**b**).

Multiple measurement cycles involved the recording of the vehicle's motion parameters, including longitudinal acceleration profiles as a function of time, which were then used to read the minimum values. The data obtained in this way were subjected to statistical analysis. Results are shown in Figure 4.

Figure 4 shows box plots of the minimum values of longitudinal accelerations noted for an urban bus on the sudden braking maneuvers from the initial speed of 50 km/h. These were designed based on 20 series measurements.

The average of minimal negative acceleration (calculated from the minimum values) of an urban bus recorded during extreme braking maneuvers on a dry asphalt surface was -7.96 m/s^2. The minimum value of acceleration noted during the sudden braking maneuvers was -8.54 m/s^2. The acceleration of the urban bus noted on a wet asphalt surface was -7.35 m/s^2. The lowest acceleration value noted during maneuvers on a wet asphalt surface was -7.88 m/s^2.

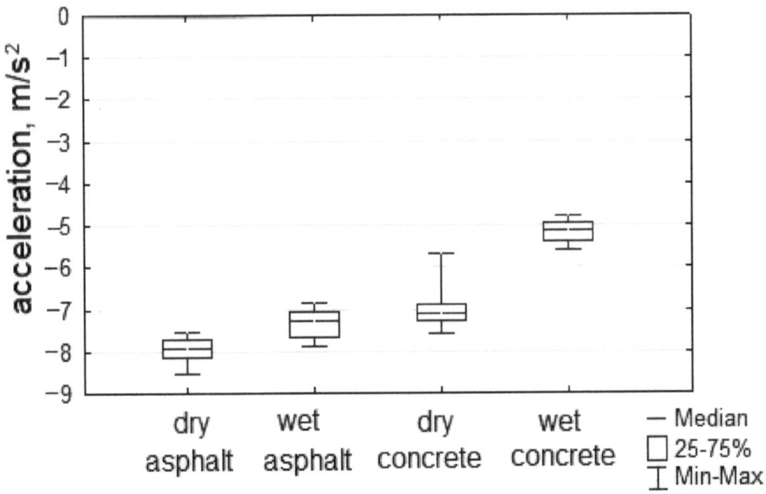

Figure 4. Longitudinal acceleration of urban bus during braking.

A concrete surface is characterized by a lower grip index than an asphalt surface. The average acceleration determined based on the maneuvers performed was −7.01 m/s^2, while the standard deviation was 0.49 m/s^2. The lowest acceleration noted during braking maneuvers of an urban bus on a dry concrete surface was −7.58 m/s^2. The average acceleration of an urban bus on a wet concrete surface was −5.15 m/s^2. The acceleration lower value noted in braking on a wet concrete surface was −5.58 m/s^2. As can be seen, the value of the standard deviation was small, which means that the obtained acceleration results were clustered around the average.

Table 2 presents selected statistical parameters of the urban bus deceleration values collected during braking maneuvers on different surfaces.

Table 2. Statistical characteristics of the acceleration value in the braking maneuvers of the urban bus.

Surface/Parameter	Max m/s^2	Min m/s^2	Average m/s^2	Standard Deviation m/s^2
Dry asphalt	−7.54	−8.54	−7.96	0.27
Wet asphalt	−6.85	−7.88	−7.35	0.32
Dry concrete	−5.68	−7.58	−7.01	0.49
Wet concrete	−4.76	−5.58	−5.15	0.26

When measuring the acceleration of an urban bus in sudden braking maneuvers, a higher value was recorded on a dry asphalt surface (−8.54 m/s^2) and the lowest on a wet concrete surface (−4.76 m/s^2). The average value of the minimum accelerations recorded during the tests on the dry asphalt surface and wet asphalt surface differed by 8%. The average of the minimum accelerations obtained from braking tests on dry concrete surfaces was 27% greater than the average acceleration obtained from braking maneuvers on wet concrete surfaces.

Figure 5 shows box plots of coach longitudinal acceleration values noted for a braking maneuver from the initial speed of 50 km/h.

The average value of acceleration determined for the series of measurements performed by the vehicle was −8.56 m/s^2. The minimum value of acceleration noted during braking maneuvers was −8.86 m/s^2. The average minimum acceleration value reached on the wet asphalt surface was −8.12 m/s^2. The lowest value of acceleration achieved during braking on the wet asphalt surface was −8.74 m/s^2.

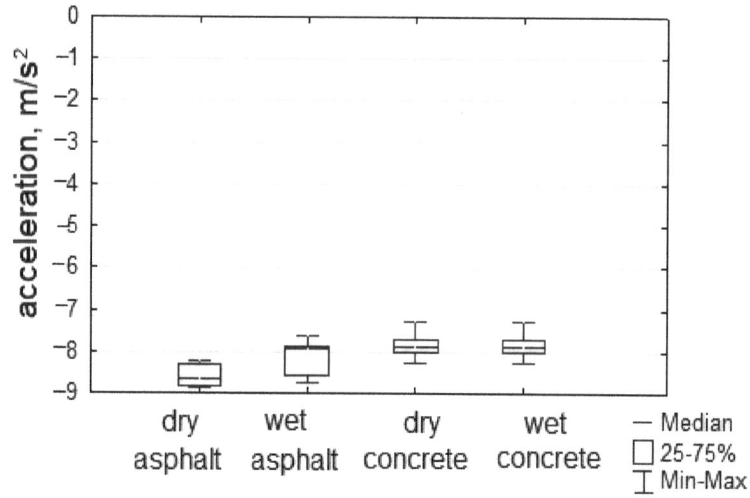

Figure 5. Longitudinal acceleration of coach during braking.

When analyzing the braking tests of the coach on a dry concrete surface, the average minimum acceleration value was -7.84 m/s^2 with a standard deviation of 0.29 m/s^2. The lowest acceleration value noted in sudden braking tests of the coach on the concrete dry surface was -8.25 m/s^2. On the concrete wet surface, the average value of minimum acceleration was -7.84 m/s^2, and the standard deviation was 0.30 m/s^2. The highest braking intensity on the concrete wet surface was -8.24 m/s^2. Table 3 presents selected statistical parameters of the coach acceleration value from braking maneuvers carried out on various surfaces.

Table 3. Statistical characteristics of the acceleration value in the braking maneuvers of the coach.

Surface/Parameter	Max m/s^2	Min m/s^2	Average m/s^2	Standard Deviation m/s^2
Dry asphalt	−8.22	−8.86	−8.56	0.26
Wet asphalt	−7.63	−8.74	−8.12	0.38
Dry concrete	−7.28	−8.25	−7.84	0.29
Wet concrete	−7.28	−8.24	−7.84	0.30

The highest value of acceleration was recorded during sudden braking maneuvers on a dry asphalt surface, which was -8.86 m/s^2. The average of the maximum deceleration values from all the rapid braking tests on the dry asphalt surface was 5% higher than the average of the minimum acceleration values from the wet asphalt surface tests. The lowest value of deceleration was recorded during rapid braking maneuvers on a concrete surface (-7.28 m/s^2).

3.2. Longitudinal Acceleration Recorded during Intense Acceleration

Figure 6 presents examples of acceleration profiles recorded during sudden acceleration from a standing start to 50 km/h. In this case, the recorded value was the acceleration over time, which was used to read the maximum acceleration.

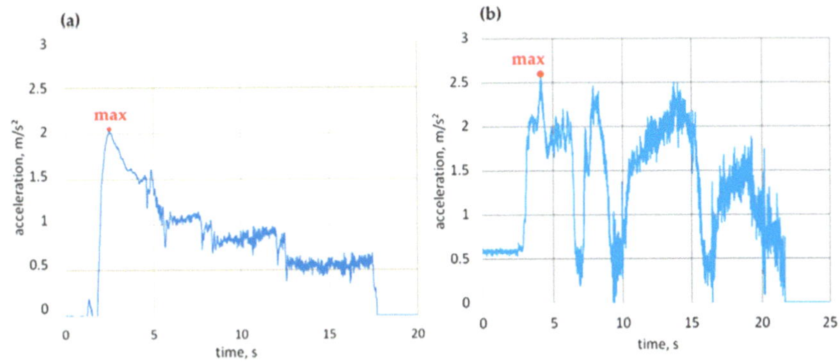

Figure 6. Profiles of acceleration during acceleration for an urban bus (**a**) and a coach (**b**).

A statistical analysis of the obtained maximum values is presented in Figure 7, which features box plots of urban bus acceleration recorded during sudden acceleration maneuvers.

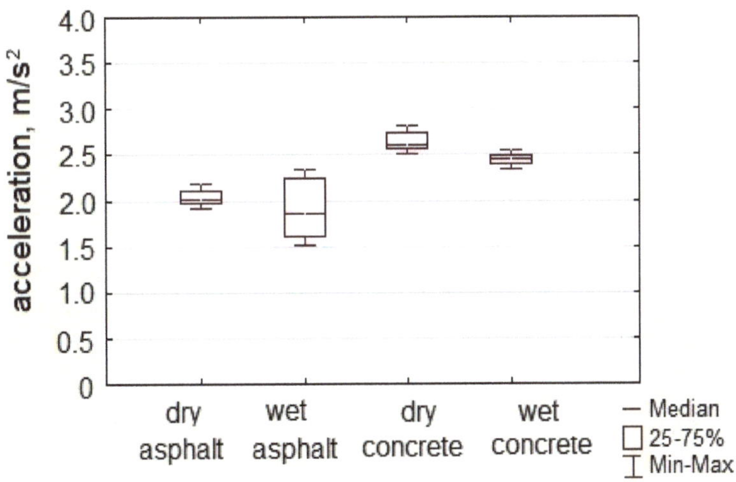

Figure 7. Longitudinal acceleration of urban buses during acceleration.

During sudden acceleration maneuvers of an urban bus on an asphalt dry surface, the average maximum acceleration value was 2.03 m/s^2. The maximum acceleration value achieved in maneuvers on dry asphalt pavement was 2.18 m/s^2.

During sudden acceleration maneuvers of the urban bus on a concrete dry surface, the average acceleration of the urban bus in these maneuvers was 2.64 m/s^2. The highest acceleration value noted in rapid acceleration maneuvers of an urban bus on a concrete dry surface was 2.81 m/s^2. On a concrete wet surface, the average value of urban bus acceleration was 2.43 m/s^2. The highest acceleration noted in rapid acceleration maneuvers on a concrete wet surface was 2.55 m/s^2.

Table 4 presents selected statistical parameters of the acceleration values of the urban bus obtained in the acceleration maneuvers carried out on various road surfaces.

Table 4. Statistical characteristics of the acceleration values in the acceleration maneuvers of the urban bus.

Parameter Surface	Max m/s^2	Min m/s^2	Average m/s^2	Standard Deviation m/s^2
Dry asphalt	2.18	1.92	2.03	0.08
Wet asphalt	2.33	1.52	1.91	0.31
Dry concrete	2.81	2.51	2.64	0.10
Wet concrete	2.55	2.21	2.43	0.09

Analyzing the maximum acceleration values achieved during the acceleration maneuver, it was noticed that the range of results for the urban bus on different surfaces was rather wide and ranged from 2.18 to 2.81 m/s^2. The average of acceleration maximum values recorded during acceleration maneuvers on asphalt dry surfaces and asphalt wet surfaces differed by 6%.

Figure 8 shows box plots of coach acceleration in sudden acceleration maneuvers up to 50 km/h.

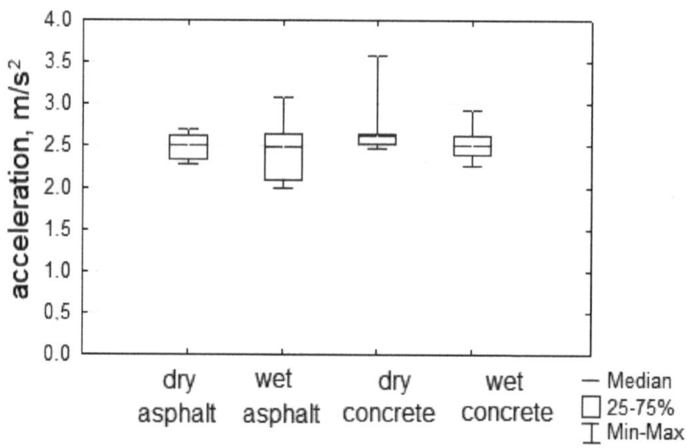

Figure 8. Longitudinal acceleration of coach during acceleration.

The coach's average maximum acceleration value during extreme acceleration maneuvers performed on an asphalt dry surface was 2.49 m/s^2. The maximum acceleration value noted in the maneuvers was 2.70 m/s^2. On an asphalt wet surface, the acceleration average value of the coach was 2.43 m/s^2. The highest acceleration value in maneuvers on an asphalt wet surface was 3.08 m/s^2.

During sudden acceleration maneuvers on asphalt dry pavement, the spread of acceleration values was 1.04 m/s^2, and the average acceleration was 2.69 m/s^2. The highest value of acceleration noted in sudden acceleration maneuvers on concrete dry pavement was 3.51 m/s^2. On concrete wet pavement, the average acceleration of the coach was 2.55 m/s^2, and the standard deviation was 0.21 m/s^2. The highest value of acceleration in the acceleration maneuvers of the coach on the concrete wet surface was 2.93 m/s^2.

Table 5 presents selected statistical parameters of the acceleration values obtained by the coach in the acceleration maneuvers carried out on different surfaces.

Table 5. Statistical characteristics of the acceleration value in the coach acceleration maneuvers.

Parameter Surface	Max m/s^2	Min m/s^2	Average m/s^2	Standard Deviation m/s^2
Dry asphalt	2.70	2.27	2.49	0.16
Wet asphalt	3.08	2.00	2.43	0.34
Dry concrete	3.51	2.47	2.69	0.29
Wet concrete	2.93	2.26	2.55	0.21

The maximum value of acceleration achieved by the coach during the acceleration maneuver on the analyzed surfaces was within the range of 2.70 to 3.51 m/s^2. Low values of the standard deviation confirmed the repeatability of the measurements. However, the average maximum acceleration value was within the range of 2.43 to 2.69 m/s^2. The highest on a dry concrete surface was 3.51 m/s^2.

3.3. Longitudinal Acceleration of Urban Bus and Coach in Real Road Traffic Conditions

The second stage of testing involved measurements in real traffic conditions specific to both vehicle types. For urban buses, these conditions involved regular travels on various routes performed during different days of the week and at different times. Transport by coaches was performed on short and longer sections.

The tests involved the recording of longitudinal acceleration in real road traffic conditions. Figure 9 presents examples of longitudinal acceleration profiles recorded for an urban bus and a coach.

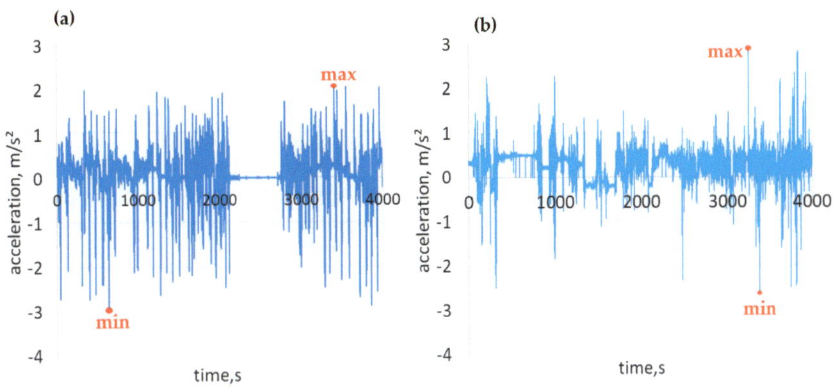

Figure 9. Longitudinal acceleration profiles during regular travels (**a**) for an urban bus and (**b**) for a coach.

With the use of the measuring equipment installed in the vehicles, the acceleration longitudinal values were continuously measured in real conditions during the regular operation of the vehicle. An analysis of the accelerations obtained provided insight into drivers' behaviors under different road conditions. In this study, only the values of longitudinal accelerations—maximum (during acceleration) and minimum (during braking)—were presented. Statistical analysis was performed on these parameters.

Positive values of the recorded acceleration corresponded to the acceleration maneuver, while negative values correspond to the braking maneuver and are identical to the deceleration presented in many publications. Figure 10 presents the ranges of negative acceleration (deceleration) recorded in real traffic conditions during braking for an urban bus and a coach.

As can be seen in real road conditions, the dominant share of maximum accelerations for the urban bus ranged from −3.0 to −2.5 m/s^2 (42%), and for the coach from −2.0 to −1.5 m/s^2 (44%).

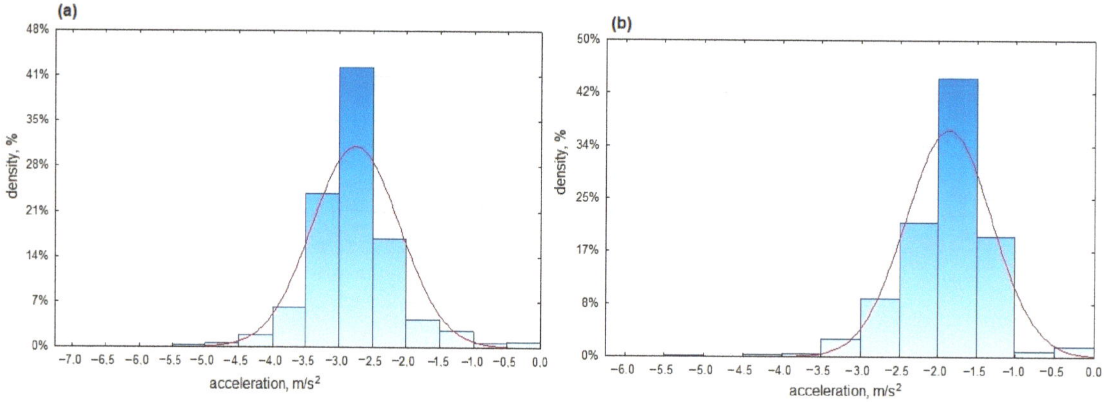

Figure 10. Acceleration recorded during braking in real traffic conditions (**a**) for an urban bus and (**b**) for a coach.

Table 6 presents selected statistical parameters of the values of acceleration accumulated in real conditions during the regular operation of the coach and urban bus.

Table 6. Statistical characteristics of the acceleration values accumulated under the regular driving conditions of the analyzed vehicles during braking.

Vehicle/Parameter	Min m/s^2	Average m/s^2	Standard Deviation m/s^2
Urban bus	−6.01	−2.78	2.78
Coach	−5.31	−1.92	0.50

The minimum acceleration values recorded in real road traffic conditions differed. The average acceleration of urban buses during braking in real traffic conditions amounted to about -2.8 m/s^2, while the minimum value recorded during the travel reached -6 m/s^2. This shows that drivers relatively rarely achieve high deceleration values when braking, which may be related to the care for passenger safety. The few higher deceleration values are usually caused by traffic situations. Additionally, when we analyzed the acceleration values obtained under real conditions, it can be seen that the standard deviation reached much higher values than in the track tests. This indicates a large dispersion of values around the average values. This is due to the peculiarities of urban traffic, where the driver is forced to perform many braking and acceleration maneuvers.

For coaches, the average acceleration during braking was slightly lower and amounted to approx. -2 m/s^2, while the standard deviation was only 0.50 m/s^2. The minimum value of acceleration noted in braking in real traffic conditions was -5.31 m/s^2.

Figure 11 shows the acceleration ranges of an urban bus and a coach during acceleration maneuvers.

As can be seen in real road conditions, the dominant shares of maximum accelerations for the urban bus ranged from 1.5 to 2.5 m/s^2 (82%) and for coach from 1.0 to 2.0 m/s^2 (82%).

Table 7 presents selected statistical parameters of acceleration values accumulated in real conditions during the regular operation of a coach and a city bus.

Table 7. Statistical characteristics of acceleration values collected in regular driving conditions of the analyzed vehicles during acceleration maneuvers.

Vehicle/Parameter	Max m/s^2	Average m/s^2	Standard Deviation m/s^2
Urban bus	4.91	2.10	2.78
Coach	3.99	1.66	0.50

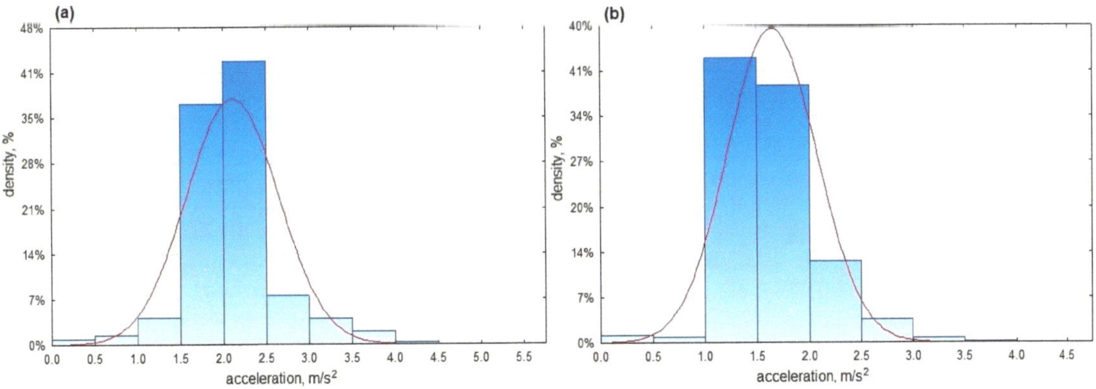

Figure 11. Acceleration recorded in real traffic conditions (**a**) for an urban bus and (**b**) for a coach.

The average maximum acceleration values achieved by the urban buses in real traffic conditions amounted to 2.1 m/s^2. The maximum of the acceleration values noted for an urban bus in real traffic conditions amounted to approximately 4.9 m/s^2. The temporary high values of accelerations in the urban bus indicated dynamic phenomena during normal operations related to frequent starting and braking, e.g., when moving in a traffic jam. The average acceleration of coaches in real traffic conditions was slightly lower and amounted to 1.66 m/s^2, and the standard deviation was 0.39 m/s^2. The maximum acceleration noted in real traffic conditions amounted to approximately 4 m/s^2.

4. Discussion

The presented study of longitudinal acceleration was separated into several parts. The first part featured sudden braking and acceleration maneuvers, and it can be observed that a change in the road surface condition and type affected the obtained acceleration values. The acceleration values obtained during braking for an urban bus demonstrate that the highest deceleration (negative value of longitudinal acceleration) was achieved on an asphalt dry surface. The maximum deceleration value recorded during the braking tests of the urban bus was 8.54 m/s^2. The lowest deceleration value was recorded during sudden braking maneuvers on wet concrete surfaces (5.58 m/s^2).

The deceleration achieved by a coach was substantially higher than the deceleration achieved by an urban bus during braking maneuvers on each of the analyzed surfaces. The differences are especially evident in deceleration values achieved on concrete surfaces. The maximum deceleration achieved by an urban bus on a concrete wet surface was 32% higher than the maximum deceleration achieved by a coach.

It is necessary to consider many factors in the acceleration maneuver analysis and the acceleration values achieved during this maneuver. Hydro-mechanical transmissions are usually used in urban buses, while traditional drive units are still often used in coaches. During rapid acceleration maneuvers, the differences in maximum acceleration values analyzed for each type of vehicle are no more than 0.8 m/s^2. The highest acceleration values of the urban bus and coach were observed on the concrete surface.

It should be noted that the obtained acceleration values reflect the dynamic properties of a specific type of bus, type of road surface, and environmental conditions. However, by determining the values of extreme longitudinal accelerations in experimental conditions on the track, it can be assumed that they constitute a specific framework for the acceleration of city buses and coaches. Therefore, taking into account the aim of the project, measurements carried out in real traffic conditions are of primary importance. Obtaining by vehicles of this type of deceleration values close to the extreme values determined in this study may be a signal of a dangerous situation on the road. Similar research was presented in [40]. The results of acceleration maneuvers of rapid braking of coaches and urban buses were analyzed. During the tests of rapid acceleration to a speed of 42 km/h, the highes acceleration value was 0.37 g (3.63 m/s^2). During braking from initial speeds of 30 km/h and 50 km/h, the highest recorded deceleration value was 0.85 g (8.34 m/s^2).

An analysis of the longitudinal acceleration obtained during measurements in rea traffic conditions demonstrated that the minimum negative acceleration values recorded for urban buses were 12% higher than the values recorded for coaches. The average value of negative acceleration noted by urban buses was 30% higher than the average value of the negative acceleration of coaches. The minimum negative acceleration recorded for urban buses during the regular course in real traffic conditions amounted to -6.01 m/s^2, while for coaches, it amounted to -5.31 m/s^2. The testing conducted on urban buses and coaches in real traffic conditions featured no values similar to the maximum negative acceleration values recorded during sudden braking maneuvers. This proves that during the tests there was no dangerous emergency requiring a violent reaction from the driver.

When analyzing the acceleration values obtained during tests in real traffic conditions maximum acceleration values obtained for both types of vehicles differed by approximately 19%. The maximum acceleration recorded for urban buses amounted to 4.91 m/s^2, while for coaches it was 3.99 m/s^2.

The range of maximum values obtained during an urban bus's acceleration maneuvers on tested road surfaces amounted to 2.81 m/s^2, while the maximum acceleration value recorded in real traffic conditions amounted to 4.91 m/s^2. As can be seen in Figure 11, high acceleration values were obtained very rarely. For a coach, the maximum acceleration value recorded in real traffic conditions amounted to 3.99 m/s^2.

The driving style and behavior of a driver, especially a professional driver, affect the safety of other road users [41–43]. Depending on the driver's driving style, traffic conditions, and other road conditions, drivers may experience different values of longitudinal acceleration. When driving in real traffic conditions, the average acceleration and deceleration values determined in the tests were 2.10 m/s^2 and 2.78 m/s^2, respectively The values of longitudinal accelerations obtained for city buses presented in the paper are comparable to the results of similar tests described in the literature. The results of the study of the urban bus dynamic parameters during normal driving are presented in [44]. It was noticed that during heavy braking before traffic lights, the value of longitudinal acceleration was -0.32 g (3.14 m/s^2). When accelerating from a traffic light, the acceleration value was 0.16 g (1.57 m/s^2). In [45], the values of longitudinal accelerations of various vehicles during acceleration and braking at intersections with traffic lights were analyzed. The average value of acceleration of a city bus when starting from a stop before traffic lights was 0.62 m/s^2, and the maximum value was 1.57 m/s^2. The average deceleration when braking before traffic lights was 0.58 m/s^2, and the maximum was 1.28 m/s^2.

The acceleration of city buses during regular driving on the streets of Amsterdam was presented in [46]. The most common acceleration values were in the range of 1 m/s^2 do 2 m/s^2. The results presented in [40] show that the accelerations recorded during normal bus driving were usually less than 2 m/s^2. During the tests, emergency situations were also recorded, and then during acceleration, acceleration values above 4 m/s^2 were logged, and during braking the deceleration was over 8 m/s^2.

5. Conclusions

The aim of the article was to determine the maximum (extreme) values of acceleration and deceleration of urban buses and coaches in rapid acceleration and braking maneuvers and the values of longitudinal acceleration during their regular driving. The presented research is part of a project aimed at creating an application for assessing the driving style of drivers and quantifying drivers using many parameters, including longitudinal accelerations. The study did not conduct comparisons of vehicles in terms of dynamics, but to learn about and present the acceleration values of these vehicles in selected maneuvers on the track and in ordinary road conditions.

Based on the results of experimental tests of vehicles on various types of road surfaces, the values show that the average maximum and minimum longitudinal accelerations obtained for the city bus are higher than for the coach. Without taking into account the type of surface, the minimum acceleration values were -7.4 m/s^2 and -8.5 m/s^2, respectively. The determined limit values of longitudinal acceleration during sudden acceleration and braking maneuvers carried out on the track indicate that the measurements for the analyzed vehicles were carried out correctly. This can be verified by analyzing statistical parameters, e.g., standard deviation. They reached small values not exceeding 0.5 m/s^2 for the braking maneuver and about 0.35 m/s^2 for acceleration. In turn, the average values determined for all types of surfaces indicated that the average maximum acceleration values during acceleration were lower for the city bus than for the coach. Without taking into account the type of road surface, the average values were 2.5 m/s^2 and 3 m/s^2, respectively.

This work determined the longitudinal acceleration during sudden acceleration and braking maneuvers during regular driving during normal operation of these vehicles. The values obtained in these conditions, due to the impossibility of separating the performed defensive maneuvers from normal driving, resulted in highly variable acceleration values. Taking into account the acceleration distributions obtained, one can indicate the values commonly used in road traffic and those approaching the limit values—characterizing dangerous maneuvers. One should be aware that especially in real traffic conditions, the obtained acceleration values may depend not only on the type of vehicle but also on the specificity of a given city, e.g., road infrastructure system, or road layout. Additionally, traffic regulations and the size of fines for offenses in different countries can significantly determine the tendency of drivers to drive dangerously.

The presented methodology was used in a research project, where having a sufficiently large, representative measurement base could be used to determine the driving style of drivers. In this way, it was possible to evaluate drivers on the basis of simple measurements of longitudinal acceleration values. The values obtained in measurements in real road conditions significantly exceeded the limit acceleration values set in the track tests for the braking maneuver, which may indicate the occurrence of a dangerous road event.

Drivers have their own driving styles. Additionally, the types of roads may determine the possibility of obtaining different acceleration values in various situations. The type of road surface can change quite often, so it was omitted during real traffic analyses. The recorded values of acceleration and deceleration during experimental tests on the track are treated as extreme values. If during the regular driving of the vehicle the recorded acceleration values are close to the extreme values, then it can be expected that an emergency situation may have occurred. The presented method, which used continuous measurements of accelerations, allowed the drivers of the minimum determination to achieve high accelerations. Thanks to this, it was possible to analyze the driving style of professional drivers. Appropriate actions eliminating such drivers from work or forcing them to drive safely will help to reduce dangerous situations to which passengers of these vehicles may be exposed in the future.

The presented results are only part of a project devoted to the assessment of drivers' driving styles. Depending on the specificity of traffic, layout, and type of roads, these values may vary. Therefore, in the assumptions of the project, the fleet owner, based on the given dynamic parameters of the vehicles, can set a reference base and then, considering a specific parameter, compare the driving style of the drivers. In this way, it is possible to select drivers whose values of a given parameter differ significantly. In this paper, longitudinal acceleration was analyzed. In further research, analyses of other quantitative and qualitative parameters are planned, e.g., longitudinal acceleration, speed, overspeeding. The combination of various parameters will allow for a deeper analysis of the driver's driving style.

Author Contributions: Conceptualization, methodology, software, validation, writing—original draft preparation: D.F., P.G., R.S.J., and E.M.S.; writing—review and editing: R.S.J.; visualization: E.M.S.; supervision: R.S.J. All authors have read and agreed to the published version of the manuscript.

Funding: The research was carried out as part of the Innovative System Research Project Supporting the Motor Vehicle Insurance Risk Assessment dedicated to UBI (Usage Based Insurance) No. POIR.04.01.04 00 0004/19 00, financed by the National Centre for Research and Development.

Data Availability Statement: Data are contained within the article.

Conflicts of Interest: The authors declare no conflict of interest.

References

1. Pusty, T. Determining the trajectory of the vehicle motion on the basis of the recorded physical quantities describing its dynamics during double lane change maneuvers. *Arch. Automot. Eng. Arch. Motoryz.* **2022**, *96*, 96–102. [CrossRef]
2. Zuska, A.; Kurczyński, D.; Jackowski, J.T. Study of Loads Acting on the Load during the Sudden Braking of a Vehicle. *Appl. Sci.* **2023**, *13*, 1559. [CrossRef]
3. Szumska, E.; Stańczyk, T.L.; Zuska, A.; Grabski, P.; Jaśkiewicz, M.; Jurecki, R.; Kurczyński, D.; Łagowski, P. Experimental testing of longitudinal acceleration in urban buses. *IOP Conf. Ser. Mater. Sci. Eng.* **2022**, *1247*, 012017. [CrossRef]
4. Borecki, M.; Rychlik, A.; Olejnik, A.; Prus, P.; Szmidt, J.; Korwin-Pawlowski, M.L. Application of Wireless Accelerometer Mounted on Wheel Rim for Parked Car Monitoring. *Sensors* **2020**, *20*, 6088. [CrossRef] [PubMed]
5. Borecki, M.; Rychlik, A.; Vrublevskyi, O.; .Olejnik, A.; Korwin-Pawlowski, M.L. Method of Non-Invasive determination of wheel rim technical condition using vibration measurement and artificial neural network. *Measurement* **2021**, *185*, 110050. [CrossRef]
6. Youssef, A.A.; Al-Subaie, N.; El-Sheimy, N.; Elhabiby, M. Accelerometer-Based Wheel Odometer for Kinematics Determination. *Sensors* **2021**, *21*, 1327. [CrossRef]
7. Szumska, E.M.; Jurecki, R. The Effect of Aggressive Driving on Vehicle Parameters. *Energies* **2020**, *13*, 6675. [CrossRef]
8. Szumska, E.; Stańczyk, T.L. Preliminary driving style classification of the professional drivers. *Arch. Automot. Eng. Arch. Motoryz.* **2022**, *98*, 25–39. [CrossRef]
9. Brambilla, M.; Mascetti, P.; Mauri, A. Comparison of different driving style analysis approaches based on trip segmentation over GPS information. In Proceedings of the 2017 IEEE International Conference on Big Data (Big Data), Boston, MA, USA, 11–14 December 2017; pp. 3784–3791. [CrossRef]
10. Jurecki, R.S.; Stańczyk, T.L. A Methodology for Evaluating Driving Styles in Various Road Conditions. *Energies* **2021**, *14*, 3570. [CrossRef]
11. Azzopardi, M.A.; Azzopardi, J.P.; Farrugia, M.; Farrugia, M. Vehicle Dynamics Analysis from a Production Vehicle's CAN Bus Data Augmented with Additional IMU's. In Proceedings of the 2019 IEEE International Conference on Mechatronics (ICM), Ilmenau, Germany, 18–20 March 2019; pp. 277–283. [CrossRef]
12. Fugiglando, U.; Massaro, E.; Santi, P.; Milardo, S.; Abida, K.; Stahlmann, R.; Netter, F.; Ratti, C. Driving Behavior Analysis through CAN Bus Data in an Uncontrolled Environment. *IEEE Trans. Intell. Transp. Syst.* **2019**, *20*, 737–748. [CrossRef]
13. Vaitkus, V.; Lengvenis, P.; Žylius, G. Driving style classification using long-term accelerometer information. In Proceedings of the 19th International Conference on Methods and Models in Automation and Robotics (MMAR), Miedzyzdroje, Poland, 2–5 September 2014; pp. 641–644. [CrossRef]
14. Girbés, V.; Hernández, D.; Armesto, L.; Dols, J.F.; Sala, A. Drive Force and Longitudinal Dynamics Estimation in Heavy-Duty Vehicles. *Sensors* **2019**, *19*, 3515. [CrossRef] [PubMed]
15. Zhang, C.; Patel, M.; Buthpitiya, S.; Lyons, K.; Harrison, B.; Abowd, G.D. Driver Classification Based on Driving Behaviors. In Proceedings of the 21st International Conference on Intelligent User Interfaces (IUI '16), New York, NY, USA, 7–10 March 2016; Association for Computing Machinery: New York, NY, USA, 2016; pp. 80–84. [CrossRef]
16. Chen, Y.; Hwang, S. Combining OBD technology with acceleration, sensor to analyze aggressive driving behavior. *Am. J. Eng. Res.* **2018**, *7*, 139–144.
17. Malik, M.; Nandal, R.; Dalal, S.; Jalglan, V.; Le, D. Driving Pattern Profiling and Classification Using Deep Learning. *Intell. Autom. Soft Comput.* **2021**, *28*, 887–906. [CrossRef]
18. Wu, F.; Stern, R.; Churchill, M.; Delle Monache, M.L.; Han, K.; Piccoli, B.; Work, D.B. Measuring trajectories and fuel consumption in oscillatory traffic: Experimental results. In Proceedings of the TRB 2017—Transportation Research Board 96th Annual Meeting (TRB 2017), Washington, DC, USA, 8–12 January 2017; p. 14.
19. Marchuk, R.; Marchuk, N.; Sakhno, V.; Poliakov, V. To determine the stability of the metrobus in unstable driving modes. *Arch. Automot. Eng. Arch. Motoryz.* **2021**, *91*, 63–79. [CrossRef]
20. Meseguer, J.E.; Calafate, C.T.; Cano, J.C.; Manzoni, P. Driving Styles: A smartphone application to assess driver behavior. In Proceedings of the IEEE Symposium on Computers and Communications (ISCC), Split, Croatia, 7–10 July 2013; pp. 000535–000540. [CrossRef]

21. Drobiszewski, J.; Lozia, Z.; Zdanowicz, P. Verification of assessment method of car braking process using a mobile phone. In *Paragraf na Drodze. Prawne i Kryminalistyczne Problemy Ruchu Drogowego*; Wydawnictwo Instytutu Ekspertyz Sądowych: Kraków, Poland, 2015; pp. 55–74.
22. Hernández Sánchez, S.; Fernández Pozo, R.; Hernández Gómez, L.A. Estimating Vehicle Movement Direction from Smartphone Accelerometers Using Deep Neural Networks. *Sensors* **2018**, *18*, 2624. [CrossRef]
23. Camacho-Torregrosa, F.J.; Pérez-Zuriaga, A.M.; Campoy-Ungría, J.M.; García-García, A. New geometric design consistency model based on operating speed profiles for road safety evaluation. *Accid. Anal. Prev.* **2013**, *61*, 33–42. [CrossRef]
24. Montella, A.; Pariota, L.; Galante, F.; Imbriani, L.L.; Mauriello, F. Prediction of Drivers' Speed Behavior on Rural Motorways Based on an Instrumented Vehicle Study. *Transp. Res. Rec. J. Transp. Res. Board* **2014**, *2434*, 52–62. [CrossRef]
25. Brooks, R.M. Acceleration Characteristics of Vehicles in Rural Pennsylvania. *IJRRAS* **2012**, *12*, 449–453.
26. Omar, N.; Prasetijo, J.; Daniel, B.D.; Abdullah, M.A.E.; Ismail, I. Study of Car Acceleration and Deceleration Characteristics at Dangerous Route FT050. *IOP Conf. Ser. Earth Environ. Sci.* **2018**, *140*, 012078. [CrossRef]
27. Liu, R.; Zhao, X.; Zhu, X.; Ma, J. Statistical characteristics of driver acceleration behaviour and its probability model. *Proc. Inst. Mech. Eng. Part D J. Automob. Eng.* **2021**, *236*, 1–17. [CrossRef]
28. Bokarea, P.S.; Maurya, A.K. Acceleration-Deceleration Behaviour of Various Vehicle Types. *Transp. Res. Procedia* **2017**, *25*, 4733–4749. [CrossRef]
29. Kim, B.; Baek, Y. Sensor-Based Extraction Approaches of In-Vehicle Information for Driver Behavior Analysis. *Sensors* **2020**, *20*, 5197. [CrossRef] [PubMed]
30. Bogdanović, V.; Ruškić, N.; Papić, Z.; Simeunović, M. The Research of Vehicle Acceleration at Signalized Intersections. *Promet. Traffic Transp.* **2013**, *25*, 33–42. [CrossRef]
31. Eboli, L.; Mazzulla, G.; Pungillo, G. Combining speed and acceleration to define car users' safe or unsafe driving behavior. *Transp. Res. Part C: Emerg. Technol.* **2016**, *68*, 113–125. [CrossRef]
32. Vaiana, R.; Iuele, T.; Astarita, V.; Caruso, M.V.; Tassitani, A.; Zaffino, C.; Giofrè, V.P. Driving behavior and traffic safety: An acceleration-based safety evaluation procedure for smartphones. *Mod. Appl. Sci.* **2014**, *8*, 88. [CrossRef]
33. He, Y.; Yan, X.; Wu, C.; Chu, D. Effects of Driver's Unsafe Acceleration Behaviors on Passengers' Comfort for Coach Buses. In Proceedings of the Second International Conference on Transportation Information and Safety, Wuhan, China, 29 June–2 July 2013.
34. Krašna, S.; Keller, A.; Linder, A.; Silvano, A.P.; Xu, J.-C.; Thomson, R.; Klug, C. Human Response to Longitudinal Perturbations of Standing Passengers on Public Transport During Regular Operation. Front. *Bioeng. Biotechnol.* **2021**, *9*, 680883. [CrossRef]
35. Beurier, G. Analysis of the Discomfort Feeling of Standing Bus Passengers on the TEOR T1 Rouen Bus Lane. *Procedia Soc. Behav. Sci.* **2012**, *48*, 425–434. [CrossRef]
36. Merkisz, J.; Tarkowski, S. Dynamic factors and their impact on subjective feeling of comfort in city buses. *Postępy Nauki 0Techniki* **2012**, *14*, 169–178.
37. Barabino, B.; Eboli, L.; Mazzulla, G.; Mozzoni, S.; Murru, R.; Pungillo, G. An innovative methodology to define the bus comfort level. *Transp. Res. Procedia* **2019**, *41*, 461–470. [CrossRef]
38. Łagowski, P.; Stańczyk, T.; Szumska, E.; Zuska, A.; Grabski, P.; Jaśkiewicz, M.; Jurecki, R.; Kurczyński, D. Influence of road conditions on the values of longitudinal acceleration during braking of a tourist coach. *IOP Conf. Ser. Mater. Sci. Eng.* **2022**, *1247*, 012004. [CrossRef]
39. Jurecki, R.S.; Stańczyk, T.L.; Ziubiński, M. Analysis of the Structure of Driver Maneuvers in Different Road Conditions. *Energies* **2022**, *15*, 7073. [CrossRef]
40. D'Souza, C.; Paquet, V.L.; Lenker, J.A.; Steinfeld, E. Self-reported difficulty and preferences of wheeled mobility device users for simulated low-floor bus boarding, interior circulation and disembarking. *Assist. Technol.* **2019**, *14*, 109–121. [CrossRef] [PubMed]
41. Sheykhfard, A.; Haghighi, F.; Papadimitriou, E.; Van Gelder, P. Review and assessment of different perspectives of vehicle-pedestrian conflicts and crashes: Passive and active analysis approaches. *J. Traffic Transp. Eng.* **2021**, *8*, 681–702. [CrossRef]
42. Frej, D.; Szumska, E. Analysis of the Length of Highways and the Number of Motor Vehicles Impact on the Intensity of Road Accidents in Selected European Countries in 2010-2020. *Commun. Sci. Lett. Univ. Zilina* **2023**, *25*, A40–A60. [CrossRef]
43. Beňuš, J.; Poliak, M.; Lazaroiu, G. Violations of social regulation and traffic accidents in road freight transport. *Arch. Automot. Eng. Arch. Motoryz.* **2022**, *97*, 51–59. [CrossRef]
44. Palacio, A.; Tamburro, G.; O'Neill, D.; Simms, C.K. Non-collision injuries in urban buses—Strategies for prevention. *Accid. Anal. Prev.* **2009**, *41*, 1–9. [CrossRef]
45. Lee, J. Acceleration and Deceleration Rates in Interrupted Flow Based on Empirical Digital Tachograph Data. *Sustainability* **2022**, *14*, 11165. [CrossRef]
46. Graaf, B.D.; Van Weperen, W. The Retention of Blance: An Exploratory Study into the Limits of Acceleration the Human Body Can Withstand without Losing Equilibrium. *Hum. Factors* **1999**, *39*, 111–118. [CrossRef]

sensors

MDPI

Article

Model-Predictive Control for Omnidirectional Mobile Robots in Logistic Environments Based on Object Detection Using CNNs

Stefan-Daniel Achirei [1,*], Razvan Mocanu [2], Alexandru-Tudor Popovici [1] and Constantin-Catalin Dosoftei [2]

1 Department of Computer Engineering, "Gheorghe Asachi" Technical University of Iasi, 700050 Iasi, Romania
2 Department of Automatic Control and Applied Informatics, "Gheorghe Asachi" Technical University of Iasi, 700050 Iasi, Romania
* Correspondence: stefan-daniel.achirei@academic.tuiasi.ro

Abstract: Object detection is an essential component of autonomous mobile robotic systems, enabling robots to understand and interact with the environment. Object detection and recognition have made significant progress using convolutional neural networks (CNNs). Widely used in autonomous mobile robot applications, CNNs can quickly identify complicated image patterns, such as objects in a logistic environment. Integration of environment perception algorithms and motion control algorithms is a topic subjected to significant research. On the one hand, this paper presents an object detector to better understand the robot environment and the newly acquired dataset. The model was optimized to run on the mobile platform already on the robot. On the other hand, the paper introduces a model-based predictive controller to guide an omnidirectional robot to a particular position in a logistic environment based on an object map obtained from a custom-trained CNN detector and LIDAR data. Object detection contributes to a safe, optimal, and efficient path for the omnidirectional mobile robot. In a practical scenario, we deploy a custom-trained and optimized CNN model to detect specific objects in the warehouse environment. Then we evaluate, through simulation, a predictive control approach based on the detected objects using CNNs. Results are obtained in object detection using a custom-trained CNN with an in-house acquired data set on a mobile platform and in the optimal control for the omnidirectional mobile robot.

Keywords: omnidirectional mobile robots; object detection; convolutional neural networks; depth sensing; computer vision; discretized-time model; predictive control algorithm; navigation

Citation: Achirei, S.-D.; Mocanu, R.; Popovici, A.-T.; Dosoftei, C.-C. Model-Predictive Control for Omnidirectional Mobile Robots in Logistic Environments Based on Object Detection Using CNNs. *Sensors* **2023**, *23*, 4992. https://doi.org/10.3390/s23114992

Academic Editors: Ikhlas Abdel-Qader and Enrico Meli

Received: 22 March 2023
Revised: 6 May 2023
Accepted: 19 May 2023
Published: 23 May 2023

1. Introduction

The mobile robots sector has seen a global rise over the past decade. Industrial mobile robots are becoming more advanced to achieve higher levels of autonomy and efficiency in various industries [1]. These robots are equipped with sophisticated sensors, such as Light Detection and Ranging (LiDAR), stereo cameras, Inertial Measurement Unit (IMU), and a global positioning system or indoor positioning system, to gather information about the work environment and make well-informed decisions [2]. This is made possible by using complex algorithms for path planning, obstacle avoidance, and task execution. Furthermore, autonomous mobile robots, grouped in fleets, are often integrated with cloud-based technologies for remote monitoring and control, allowing for greater flexibility and scalability in their deployment.

Path planning is a crucial aspect of mobile robotics navigation because of the need to perform a task by moving from one point to another while avoiding obstacles and satisfying more constraints, among which are time, the level of autonomy given by the energy available, and significantly, maintaining safety margins regarding human operators and transported cargo. Mobile robot navigation is still one of the most researched topics of today, addressing two main categories: classical and heuristic navigation. In the variety of classical approaches, the most well-known algorithms, characterized by limited intelligence in [3], are cell decomposition, roadmap approach, and artificial potential field (APF). Heuristic approaches are more intelligent, including but not limited to the main components of computational intelligence (i.e., fuzzy logic, neural networks, and genetic algorithms). Researchers investigate solutions based on the particle swarm optimization algorithm, the FireFly algorithm, and the artificial BEE colony algorithm [4]. Combining classical and heuristic approaches, known as hybrid algorithms, offers better performances, especially for navigation in complex, dynamic environments [3].

In dynamic operational environments, the increased flexibility of autonomous mobile robots compared to automated guided vehicles is an added advantage due to decreased infrastructure setup and maintenance costs. Supplementary, the omnidirectional mobile robots (OMR), compared to other traction and steering systems (e.g., differential drive and Ackerman), provide three independent degrees of freedom (longitudinal and lateral translation, together with in-place rotation), motions that can be combined within certain speed and acceleration limits without producing any excessive wear on the ground contact surfaces. On the other hand, to obtain precise motion for the OMR, certain constraints apply to their suspension system and the smoothness of the ground surface.

Considering the computational requirements criteria, the planning technology of a mobile robot is divided into offline planning and online planning [5]. In offline planning, the path for the robot is pre-computed and stored in the robot's memory. The robot then follows the pre-computed path to reach its destination. This approach is suitable for deterministic environments with a priori information. When the mobile robot navigates and performs tasks in a dynamic and uncertain environment, it is necessary to use the online planning approach. The robot computes its path in real time based on its current location and the information obtained from its perception module.

Independent of the type of path planning algorithm, the OMR structure is beneficial because it better resembles the material point model used for simplifying the modeling of robots in motion planning simulations. In [6–9], a four-wheel's dynamic and kinematic modeling, OMR was studied using the Lagrange framework. Sliding mode control allows robust control for OMRs employing mecanum-wheels and rejects disturbances caused by unmodeled dynamics [10–12]. The nonholonomic model of the wheel was used to develop the dynamic equation of an OMR with four mecanum wheels [13]. The kinematic model of a three-wheeled mobile robot was used to create a predictive control model and filtered Smith predictor for steering the robot along predetermined paths [14]. A reduced dynamic model of the robot is the basis for developing a nonlinear model-predictive controller for trajectory tracking of a mecanum wheeled OMR [15]. A constrained quadratic programming problem is formulated towards optimizing the trajectory of a four-wheel omnidirectional robot [16]. Dynamic obstacles are considered in the work of the authors [17], whereas the numerical implementation presented in [18] is based on a three-wheeled omnidirectional robot. Distributed predictive control on a cooperative paradigm is discussed for a coalition of robots [19]. A nonlinear predictive control strategy with a self-rotating prediction horizon for OMR in uncertain environments is discussed. The appropriate prediction horizon was selected by incorporating the effects of moving velocity and road curvature on the system [20]. Adaptive model-predictive control, with friction compensation and incremental input constraints, is presented for an omnidirectional mobile robot [21]. Wrench equivalent optimality is used in a model-predictive control formulation to control a cable-driven robot [22]. Authors discuss an optimal controller to control the robot's motion on a minimum energy trajectory [23]. Recently, potential field methods have been used mainly due to their naturally inspired logic. These methods are also widely used in omnidirectional mobile robots due to their simplicity and performance in obstacle avoidance [24,25]. Timed elastic-band approaches utilize a predictive control strategy to steer the robot in a dynamic environment to tackle real-time trajectory planning tasks [26]. Because the optimization is confined to local minima, the original timed elastic-band planner may cause a route through obstacles. Researchers proposed an improved strategy for producing alternate sub-optimal trajectory clusters based on unique topologies [27]. To overcome the mismatch problem between the optimization graph and grid-based map, the authors suggested an egocentric map representation for a timed elastic band in an unknown environment [28]. These path-planning methods are viable and pragmatic, and acquiring a desired path in various scenarios is generally possible. Yet, these approaches could have multiple drawbacks, such as a local minimum, a low convergence rate, a lack of robustness, substantial computation, and so on. Additionally, in logistic environments where OMR robots are equipped with conveyor belts to transport cargo, it is essential to guarantee low translational and rotational accelerations for the safety of the transported cargo. Therefore, we propose a nonlinear predictive control strategy on a reduced model where we can include maximum acceleration and velocities of the wheels within inequality constraints derived from the obstacle positions obtained from environment perception sensors (i.e., LiDAR and video camera). To tackle the problem of local minima, we propose a variable cost function based on the proximity of obstacles ahead to balance the global objectives.

Object Detection for Mobile Platforms

Deep neural networks specifically created to analyze organized arrays of data (i.e., images) are known as convolutional neural networks, often called CNNs or ConvNets. CNNs offered solutions to computer vision challenges that are difficult to handle using conventional methods. They quickly advance to the state-of-the-art in areas such as semantic segmentation, object detection, and image classification. They are widely used in computer vision because they can quickly identify image patterns (such as lines, gradients, or more complex objects such as eyes and faces). CNNs are convolutional-layered feed-forward neural networks. CNNs attempt to mimic the structure of the human visual cortex with these specific layers.

Localization of object instances in images is implied by object detection. Object recognition generally assigns a class to the identified objects from a previously learned class list. Although object detection operates at the bounding-box level, it has no notion of different classes. The phrase "object detection" now encompasses both activities, even though they were initially two distinct jobs. So, before continuing, let's be clear that object detection includes both object localization and object recognition.

Object detection and recognition is an essential field of study in the context of autonomous systems. The models can be broadly divided into one-stage and two-stage detectors. One-stage detectors are designed to detect objects in a single step, making them faster and more suitable for real-time applications, such as path planning based on object detection for a moving system. On the other hand, two-stage detectors use a two-step process, first proposing regions of interest and then looking for objects within those areas. This approach excludes irrelevant parts of the image, and the process is highly parallelizable. However, it comes at the cost of being slower than one-stage detectors.

To meet the constraints of the Nvidia Jetson mobile platforms considered for the OMR, lightweight neural networks were investigated for object detection and recognition. Among the models evaluated, YoloV5 [29], SSD-Mobilenet-v1 [30], SSD-Mobilenet-v2-lite [31] and SSD-VGG16 [32] were trained and tested. Earlier, the YoloV4 [33] model had already made significant improvements over the previous iteration by introducing a new backbone architecture and modifying the neck of the model, resulting in an improvement of mean average precision (mAP) by 10% and an increase in FPS by 12%. Additionally, the training process has been optimized for single GPU architectures, like the Nvidia Jetson family, commonly used in embedded and mobile systems.

A particular implementation is YoloV5 [29], which differs from other Yolo implementations using the PyTorch framework [34] rather than the original Yolo Darknet repository. This implementation offers a wide range of architectural complexity, with ten models available, starting from the YoloV5n (nano), which uses only 1.9M parameters, up to the YoloV5x6 (extra large), which uses 70 times as many parameters (140M). The lightest models are recommended for Nvidia Jetson platforms.

Recently, an increasing interest has been in developing mobile device object detection and recognition algorithms. A popular approach is using the Single Shot Detector (SSD) [35] neural network, a one-step algorithm. To further improve the efficiency of the SSD algorithm on mobile devices, researchers have proposed various modifications to the SSD architecture, such as combining it with other neural network architectures. One such modification is the use of the SSD-MobileNet and SSD-Inception architectures, which combine the SSD300 [35] neural network with various backbone architectures, such as MobileNet [30] or Inception [36]. These architectures, such as the Nvidia Jetson development platforms, are recognized for their real-time object detection capabilities on mobile devices.

These methods for object detection perform very well in general detection tasks. Yet, there must be more datasets and pretrained models for objects specific to the OMR environment, such as fixed or mobile conveyors, charging stations, other OMRs, etc. We have acquired our dataset and deployed domain-specific models for object detection in the OMR environment. We summarize the main contributions of this paper to the field of object detection and OMR control in logistic environments below:

- Acquisition of a data set for object detection in the OMR environment;
- Investigation of domain-specific models for object detection and providing a model to be used in an OMR environment;
- Deployed an image acquisition and object detection module fit for the real-time task of OMR control;
- Proposed a joint perception&control strategy based on a non-linear model-predictive control;
- Avoid local minima by using switched cost function weights to navigate around obstacles while still achieving the overall objective of decreasing travel distance;
- Guarantee maximum wheel speed and acceleration through the constrained non-linear MPC in order to ensure safe transportation of cargo;

The rest of the paper is organized as follows: Section 2 discusses object detection in the context of OMR's logistic environment. First, some equipment experiments were conducted on the image acquisition sensor and the processing unit. We also describe the object detection dataset creation and object mapping in 2D and 3D perspectives. Section 3 is dedicated to the modeling and control of the OMR. We introduce the mathematical model used for developing the control strategy, followed by formulating the optimization problem considering the environmental objects. In the last two sections, we discuss the object detection results and the simulation of the control algorithm, conclude, and emphasize future work goals.

2. Object Detection for Omnidirectional Mobile Robots

2.1. Image Acquisition and Processing Unit

For the image acquisition unit, we analyzed four depth cameras. Depth information is needed to accurately place the detected objects on the 2D and 3D maps of the environment. The predictive control task relies on object maps. The most important features considered for the experiments were the correctness of the depth information and the integration of the camera with the Nvidia Jetson platforms, which are already in use on the Omnidirectional Robot.

All Zed cameras perform well in indoor environments, but, as can be seen in Figure 1, the far-depth information provided by Zed 2i is significantly better. Depth information is completely missing after 10 m for Intel RealSense. The best depth information is given by Zed 2i; it also has the largest FoV. Based on the image acquisition experiments performed in the OMR environment, Zed 2i was chosen to be integrated into the robot.

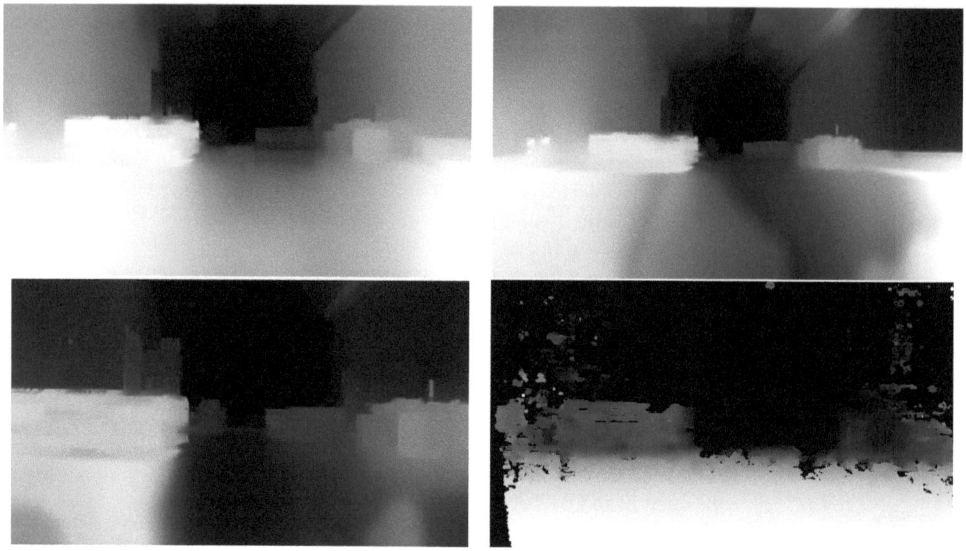

Figure 1. Depth information for ZED 1 (**top left**), ZED 2i (**top right**), ZED mini (**bottom left**), and Intel RealSense D435i (**bottom right**).

Nvidia Jetson system-on-chip platforms are already used on the OMR. Some experiments evaluated the computational capabilities, detection precision, and the dependency between inference time and resolution. Localization is very important in our defined use cases for the OMR environment. MS COCO dataset [37] was used for object detection evaluation across different lightweight neural networks such as Mobilenet [30,31] and Yolo [29,33] which are suitable for mobile platforms.

The neural networks used for the first experiment are optimized using TensorRT to run on Jetson mobile platforms. In Table 1, we can see the run-time measurements for the selected models from the SSD family. The same solution takes considerably more time to run on the Jetson Nano.

Table 1. Object detection evaluation of the SSD model family.

Architecture	FPS on Jetson Nano	FPS on Jetson Xavier AGX
SSD–Mobilenet–v1	10	83
SSD–Mobilenet–v2	7	61
SSD–Inception–v2	6	42

A second experiment aims to see how the processing time evolves depending on the image resolution. Table 2 presents the results in terms of FPS on a test subset from Cityscapes data set [38,39]. The results emphasize that the inference time depends on the size of the images provided at input. Thus, the higher the image resolution, the slower the model. Jetson Xavier AGX is 4 to 6 times faster than Jetson Nano, depending on the model and the input resolution.

Table 2. Inference Time vs. Image Resolution.

Architecture	Resolution	FPS on Jetson Nano	FPS on Jetson Xavier AGX
SSD–Mobilenet–v1	2048 × 1024 1024 × 512	15 fps 23 fps	91 fps 102 fps
SSD–Mobilenet–v2	2048 × 1024 1024 × 512	12 fps 18 fps	74 fps 76 fps
SSD–Inception–v2	2048 × 1024 1024 × 512	11 fps 15 fps	63 fps 63 fps

Following the analysis of hardware equipment and the experimental measurements, the Jetson architecture chosen to be integrated into the proposed solution for object detection in the OMR environment that meets the minimum requirements is the Nvidia Jetson Xavier AGX.

2.2. The Omnidirectional Robot Object Detection Dataset (OROD)

Enabling the efficient operation of autonomous robots is crucial for accurately detecting and recognizing objects specific to the OMR environment. Using the ZED 2i camera, we have acquired a new dataset for the object detection task that contains objects specific to the omnidirectional robot environment. The "Omnidirectional Robot Object Detection (OROD)" dataset includes charging stations, construction cones, mobile conveyors, and different types of fixed conveyors. The images in the dataset were captured using the camera mounted on an omnidirectional robot and were annotated with bounding boxes of objects. The dataset is intended to evaluate the performance of object detection algorithms in an omnidirectional robot environment.

The OROD dataset contains 1343 images, each labeled with the objects of interest in the scene. The images were collected in different environments, such as industrial warehouses and logistics centers, to reflect the various scenarios in which an omnidirectional robot operates. Additionally, the data set includes images with varying lighting conditions, occlusions, and different orientations of the objects to represent real-world challenges in object detection. The training subset was augmented for better results by applying flip, rotation, zoom, hue, saturation, blur, noise, etc. The original and augmented data sets were split according to the figures from Table 3. Examples of the augmented images can be visualized in Figure 2. The dataset augmentation process did not change the initial class distribution; it scaled by 3.

Table 3. OROD train-val-test split.

	Annotated Frames before/after Augmentation	Percentage before/after Augmentation
train-initial	940/2816	70/88
validation	269/269	20/8
test	134/134	10/4

The OROD dataset is the first to focus specifically on object detection in the context of an omnidirectional robot environment. It is intended to serve as a reference for evaluating the performance of object detection algorithms in this context and to promote research in this field.

The augmented data set and the raw dataset, both with YOLO annotations, are publicly available at https://universe.roboflow.com/gheorghe-asachi-technical-university-of-iasi/rmoa, accessed on 20 May 2023.

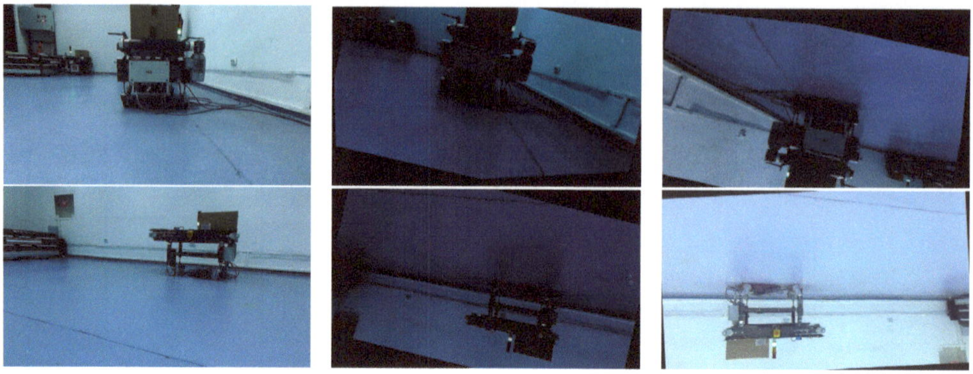

Figure 2. Acquired frame (**column 1**) and augmentation results (**columns 2 and 3**).

2.3. Detected Objects in 3D and Mapping

All objects detected by the custom-trained model, along with the distances to them are visible on the left side of Figure 3. Their 3D position is also exemplified on the right side. The distance between the scene object and the camera is measured from the back of the left eye of the camera and is given in meters.

In the context of an OMR moving through its environment, an important feature is to continuously be aware of its position and rotation relative to the starting point, the charging station, in our case.

Figure 3. Object detection and distance estimation in meters (**top**) and 3D point cloud mapping (**bottom**).

Examples of the OMR position and orientation are listed on the bottom of the frames in Figure 3. As a benefit of the IMU integration with Zed 2i, we can obtain the camera position, rotation, and quaternion orientation. In addition to the ZED 2i camera, the OMR is equipped with two LiDAR sensors for a 360-degree map. At this stage, the LiDAR data are empirically merged with the detected objects to obtain a bird's-eye view map of the entire environment. In Figure 4, we can see the obtained map of the environment with the detected objects shown in Figure 3.

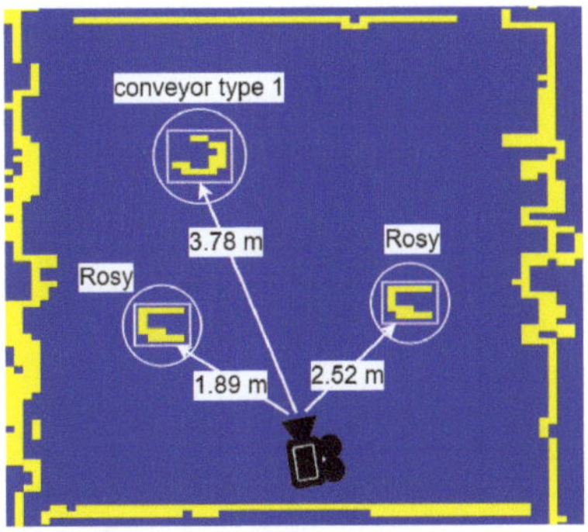

Figure 4. Bird's eye view mapping of detected objects.

3. Model-Predictive Motion Control of OMR

We derive the motion control strategy of the OMR based on a non-linear optimization algorithm as the core of the motion controller. We define in Section 3.1 the mathematical model used in the predictions step of the controller. The continuous time equations are discretized by the Euler method to realize the numerical implementation. Then, we define the physical constraints of the robot's actuators (i.e., omnidirectional wheel speed and acceleration) and the geometrical constraints of the objects (i.e., circumscribed circles of objects). We formulate the optimization problem considering the global objective of navigating on the shortest path, avoiding obstacles, and limiting the movement of the OMR within actuator limits.

3.1. Mathematical Model of 3D of Omnidirectional Robot

In this section, we define the discrete mathematical model used in the model-predictive controller to generate short-term paths and control the robot's movement along the predicted trajectory. Equation (1) depicts the inverse kinematics matrix representation:

$$\begin{bmatrix} \omega_1 \\ \omega_2 \\ \omega_3 \\ \omega_4 \end{bmatrix} = J \begin{bmatrix} v_x \\ v_y \\ \Omega \end{bmatrix} \tag{1}$$

where v_x and v_y are the longitudinal and lateral velocities of the OMR, respectively. Ω defines the angular speed along the normal axis, $\omega_j, j = \overline{1..4}$ are the individual wheels' angular velocities, while J is the inverse kinematic Jacobian matrix of the OMR defined in (2) [1]:

$$J = \frac{1}{R}\begin{bmatrix} 1 & 1 & -(l_x + l_y) \\ 1 & -1 & -(l_x + l_y) \\ 1 & 1 & (l_x + l_y) \\ 1 & -1 & (l_x + l_y) \end{bmatrix} \tag{2}$$

The forward kinematics of the 3DOF system are obtained from the lateral, longitudinal, and rotation velocities:

$$\begin{bmatrix} \frac{dx}{dt} \\ \frac{dy}{dt} \\ \frac{d\theta}{dt} \end{bmatrix} = \frac{R}{4}\begin{bmatrix} 1 & 1 & 1 & 1 \\ 1 & -1 & 1 & -1 \\ \frac{-1}{l_x+l_y} & \frac{-1}{l_x+l_y} & \frac{1}{l_x+l_y} & \frac{1}{l_x+l_y} \end{bmatrix} \begin{bmatrix} \omega_1 \\ \omega_2 \\ \omega_3 \\ \omega_4 \end{bmatrix} \tag{3}$$

where x, y, θ are plane coordinates and robot orientation, respectively. Moreover, R is the wheel radius, l_x defines the distance from the GC to the front axle, while l_y defines the half distance between the left and right wheels.

Pragmatically, it can be considered that deviations from the nominal kinematic model act on the system input. Therefore, we can design an input disturbance observer to compensate for unmodeled dynamics and disturbances. Let us define the disturbance acting on the system input as $F = [f_1, f_2, f_3, f_4]^t$, where the additive terms F act on the system inputs. The observer is designed considering the inverse kinematics of the process. An additional pole is added for the realizability of the observer. We define Q as a passive (i.e., unitary gain) first-order low-pass filter diagonal matrix. We define the estimated input disturbance as: $\hat{F} = [\hat{f}_1, \hat{f}_2, \hat{f}_3, \hat{f}_4]^t = -Q[\omega_1, \omega_2, \omega_3, \omega_4]^t + QJ[v_x, v_y, \dot{\theta}]^t$ Therefore the plant model becomes:

$$\begin{bmatrix} \frac{dx}{dt} \\ \frac{dy}{dt} \\ \frac{d\theta}{dt} \end{bmatrix} = \frac{R}{4} \begin{bmatrix} 1 & 1 & 1 & 1 \\ 1 & -1 & 1 & -1 \\ \frac{-1}{l_x+l_y} & \frac{-1}{l_x+l_y} & \frac{1}{l_x+l_y} & \frac{1}{l_x+l_y} \end{bmatrix} \begin{bmatrix} \omega_1 \\ \omega_2 \\ \omega_3 \\ \omega_4 \end{bmatrix} + \begin{bmatrix} f_1 \\ f_2 \\ f_3 \\ f_4 \end{bmatrix} \tag{4}$$

The discretized-time model of (3) is obtained by backward rectangle area approximation (i.e., Euler method). Therefore, the system Equation (3) can be re-written in the state space framework $\dot{X} = AX + (J_+)\omega$, where the state transition matrix is null, the state vector is $X = [x \;\; y \;\; \theta]^t$ while the input matrix J_+ is defined as $J_+ = (J^T J)^{-1} J^T$. Thus, we obtain the discretized-time model of the OMR in global coordinates:

$$X_{k+1} = I_3 X_k + (J_+) T_s \omega_k \tag{5}$$

where $I_3 \in \mathbb{R}^{3\times 3}$ unity matrix, $X_{k+1} = [x_k \; y_k \; \theta_k]^t$ is the state vector at iteration $k+1$, T_s is the sampling time and $\omega_k = [\omega_{1_k} \; \omega_{2_k} \; \omega_{3_k} \; \omega_{4_k}]^t$ is the input vector.

$$\begin{bmatrix} x_{k+1} \\ y_{k+1} \\ \theta_{k+1} \end{bmatrix} = \begin{bmatrix} x_k \\ y_k \\ \theta_k \end{bmatrix} + T_s \frac{R}{4} \begin{bmatrix} 1 & 1 & 1 & 1 \\ 1 & -1 & 1 & -1 \\ \frac{-1}{l_x+l_y} & \frac{-1}{l_x+l_y} & \frac{1}{l_x+l_y} & \frac{1}{l_x+l_y} \end{bmatrix} \begin{bmatrix} \omega_{1_k} \\ \omega_{2_k} \\ \omega_{3_k} \\ \omega_{4_k} \end{bmatrix} \tag{6}$$

To improve controller behavior w.r.t to deviations of the model and input perturbation, the extended discretized model can be used for states and output predictions within the MPC solver:

$$\begin{bmatrix} x_{k+1} \\ y_{k+1} \\ \theta_{k+1} \end{bmatrix} = \begin{bmatrix} x_k \\ y_k \\ \theta_k \end{bmatrix} + T_s \frac{R}{4} \begin{bmatrix} 1 & 1 & 1 & 1 \\ 1 & -1 & 1 & -1 \\ \frac{-1}{l_x+l_y} & \frac{-1}{l_x+l_y} & \frac{1}{l_x+l_y} & \frac{1}{l_x+l_y} \end{bmatrix} \begin{bmatrix} \omega_{1_k} \\ \omega_{2_k} \\ \omega_{3_k} \\ \omega_{4_k} \end{bmatrix} + T_s \begin{bmatrix} f_{1_k} \\ f_{2_k} \\ f_{3_k} \\ f_{4_k} \end{bmatrix} \tag{7}$$

Table 4 contains the parameters of the mobile robot and the sampling time considered for the time-discretization of the process.

Table 4. OMR parameters.

Parameter	Value
Wheel radius (R)	0.076 [m]
Distance from GC to front axle (l_x)	0.294 [m]
Half distance between left and right wheels (l_y)	0.2 [m]
Sampling time (T_s)	0.02 [s]

3.2. OMR Motion Optimization Problem

In the optimization problem, we aim to find the solution at time kT_s, comprised of actuator commands $\omega_j^{(i,k)}$ $\mathit{for}\;\; j = \overline{1..4}$, i$\in \{1 \ldots H\}$ satisfying actuator physical constraints, the geometric constraints, and to fulfill the global objective of traveling the shortest distance and avoiding the detected obstacles. Therefore, we formulate the optimization problem as follows. Find,

$$\begin{aligned} \min_{x(\cdot|k),\ y(\cdot|k),\ \theta(\cdot|k),\ d_p(\cdot|k)} \quad & J_k(X, X_r, \alpha) \\ \text{s.t.} \ -\omega_{UB} \leq \omega_{j, j=\overline{1..4}} & \leq \omega_{UB} \\ -a_{UB} \leq \dot{\omega}_{j, j=\overline{1..4}} & \leq a_{UB} \\ C_o & < 0 \end{aligned} \tag{8}$$

where J_k is the cost function defined in (9), $x(\cdot|k), y(\cdot|k), \theta(\cdot|k)$ are the solutions of the optimization problem; ω_{UB} and a_{UB} are the upper bounds of the angular velocity and acceleration of the wheels, respectively. C_o is the geometric constraints vector and is defined in (22).

The cost function J_k is defined by:

$$J_k(X, X_r, \alpha) = \frac{1}{2}\sum_{i=0}^{H-1}\left[w_x(\alpha)(x_r^{(i|k)} - x^{(i|k)})^2 + w_y(\alpha)(y_r^{(i|k)} - y^{(i|k)})^2 + w_\theta(\theta_r^{(i|k)} - \theta^{(i|k)})^2 + \right. \tag{9}$$

$$\left. + w_{Tx}(x_r^{(H-1|k)} - x^{(i|k)})^2 + w_{Ty}(y_r^{(H-1|k)} - y^{(i|k)})^2\right] + w_p(\alpha)\sum_{i=0}^{H-1}(d_p(X, X_0, X_f)^{(i|k)}) \tag{10}$$

where $X_r \in R^{H\times3}$ is the reference trajectory matrix of the OMR over the prediction horizon H:

$$X_r = \begin{bmatrix} x_r^{(0,k)} & y_r^{(0,k)} & \theta_r^{(0,k)} \\ x_r^{(1,k)} & y_r^{(1,k)} & \theta_r^{(1,k)} \\ \dots & \dots & \dots \\ x_r^{(H-1,k)} & y_r^{(H-1,k)} & \theta_r^{(H-1,k)} \end{bmatrix} \tag{11}$$

$d_p(X_0, X_f)$ is the length of the projection of the OMR geometric center over the ideal straight path connecting the starting (i.e., initial) node with the final node and is defined in (12):

$$d_p(X, X_0, X_f) = \sqrt{|L_1^2 - [(L_1^2 + L_3^2 - L_2^2)/(2L_3)]^2|} \tag{12}$$

with

$$L_1 = \sqrt{(x - x_0)^2 + (y - y_0)^2} \tag{13}$$

$$L_2 = \sqrt{(x - x_r^{(H-1)})^2 + (y - y_r^{(H-1)})^2} \tag{14}$$

$$L_3 = \sqrt{(x_0 - x_r^{(H-1)})^2 + (y_0 - y_r^{(H-1)})^2} \tag{15}$$

where L_1, L_2, and L_3 define the L2-norms between the OMR position, initial, and resting positions, while $X_0 = [x_0, y_0, \theta_0]^t$ and $X_f = [x_r^{(H-1)}, y_r^{(H-1)}, \theta_r^{(H-1)}]^t = [x_f, y_f, \theta_f]^t$ are the initial and final resting positions. In the cost function, we aim to penalize by weights $w_x(\alpha)$ and $w_y(\alpha)$ the deviation from the reference trajectory $x_r^{(i)}, y_r^{(i)}, i = \overline{0..H-1}$ defined by (25); by weight w_θ it is penalized the deviation from the desired orientation of the OMR. The set-point orientation $\theta_r^{(i)}, i = \overline{0..H-1}$ is such that the OMR remains with the frontal part facing the destination location. By w_{Tx}, we penalize the terminal cost of x_r and y_r to reduce the steady-state error. Therefore, $w_{Tx} > w_x$ and $w_{Ty} > w_y$; $w_p(\alpha)$ is a weight with two discrete states, and its value is a function of α which depends on the proximity (tolerance) of the closest object and is defined by (23).

The actuator constraints of the OMR are defined as:

$$-\omega_{UB} \leq \omega_{j, j=\overline{1..4}} \leq \omega_{UB} \tag{16}$$

$$-a_{UB} \leq \dot{\omega}_{j, j=\overline{1..4}} \leq a_{UB} \tag{17}$$

We define the physical-space constraints from the coordinates of the objects and their known sizes as:

$$-(x-x_o)^2-(y-y_o)^2+r_o^2\leq 0 \tag{18}$$

where $x_o=[x_{o_1}\dots x_{o_{no}}]^t$ and $y_o=[y_{o_1}\dots y_{o_{no}}]^t$ are the coordinates $r_o=[r_{o_1}\dots r_{o_{no}}]^t$ is the radius of the circle circumscribed about the polygon defining the object. Index *no* refers to *Number-of-Objects*, while, index *o* signifies the word *Objects*, also, index *a* refers to word *Actuators*. The global coordinates of the obstacles are obtained from the local coordinates of the OMR according to the equation below:

$$\begin{bmatrix}x_o\\y_o\end{bmatrix}=\begin{bmatrix}cos\phi & sin\phi\\-sin\phi & cos\phi\end{bmatrix}\left[\begin{bmatrix}x_l\\y_l\end{bmatrix}-\begin{bmatrix}x\\y\end{bmatrix}\right] \tag{19}$$

where ϕ is the angle from the global system's abscissa to the local system's abscissa. x_l and y_l are the local coordinates of the detected objects, and x and y are the global coordinates of the local system's origin. From inequality constraints (16) and (18), we obtain a concatenated vector of inequality constraints denoted by $C_k\in\mathbb{R}^{(n_a\cdot n_w\cdot H+n_o\cdot H)\times 1}\leq 0$:

$$C_k=[C_a^t,C_o^t]^t\in\mathbb{R}^{(n_a\cdot n_w\cdot H+n_o\cdot H)\times 1} \tag{20}$$

where $C_a\in R^{n_a\cdot n_w\cdot H\times 1}$, $C_o\in\mathbb{R}^{n_oH\times 1}$ are defined below:

$$C_a^{(k)}=\begin{bmatrix}|\omega_1^{(0|k)}|-\omega_{UB}\\|\omega_1^{(1|k)}|-\omega_{UB}\\\vdots\\|\omega_1^{(H-1|k)}|-\omega_{UB}\\|\omega_2^{(0|k)}|-\omega_{UB}\\\vdots\\|\omega_4^{(H-1|k)}|-\omega_{UB}\\|\dot{\omega}_1^{(0|k)}|-a_{UB}\\\vdots\\|\dot{\omega}_4^{(H-1|k)}|-a_{UB}\end{bmatrix}\leq 0 \tag{21}$$

$$C_o^{(k)}=\begin{bmatrix}-(x^{(0|k)}-x_{o_1})^2-(y^{(0|k)}-y_{o_1})^2+r_{o_1}^2\\-(x^{(1|k)}-x_{o_1})^2-(y^{(1|k)}-y_{o_1})^2+r_{o_1}^2\\\vdots\\-(x^{(H-1|k)}-x_{o_1})^2-(y^{(H-1|k)}-y_{o_1})^2+r_{o_1}^2\\-(x^{(0|k)}-x_{o_2})^2-(y^{(0|k)}-y_{o_2})^2+r_{o_2}^2\\\vdots\\-(x^{(H-1|k)}-x_{o_{no}})^2-(y^{(H-1|k)}-y_{o_{no}})^2+r_{o_{no}}^2\end{bmatrix}\leq 0 \tag{22}$$

In the previous equations, $n_a=2$ defines the number of constraints regarding actuators, and it is two because we included two types of actuator restrictions: angular speed and angular acceleration.

We approximate numerically $\dot{\omega}_j$ by $\dot{\omega}_j\approx\frac{(\omega_j^{(i)}-\omega_j^{(i-1)})}{T_s}$ where T_s is the sampling time.

In the cost function (9), we propose that $w_p(\alpha)$, $w_x(\alpha)$ and $w_y(\alpha)$ are switched between their two states based on the value of $\alpha = max_{1\leq j\leq n_o\cdot H}\, C_{o_j}$ which, practically, determines the minimum proximity to an obstacle from the object list. Therefore,

$$w_p(\alpha) = \begin{cases} w_{p_1}, & \text{if } \alpha > tol \\ w_{p_2}, & \text{if } \alpha \leq tol \end{cases}$$
$$w_x(\alpha) = w_y(\alpha) = \begin{cases} w_{xy_1}, & \text{if } \alpha > tol \\ w_{xy_2}, & \text{if } \alpha \leq tol \end{cases} \quad (23)$$

In the previous equation, *tol* defines the avoidance tolerance.

The set-point orientation over the control horizon H is defined as:

$$\theta_r^{(i)} = atan_2(y_r^{(H-1)} - y, x_r^{(H-1)} - x)\frac{360}{2\pi}, \text{for } i = \overline{0..H-1} \quad (24)$$

and the reference trajectory is given by a first-order static function where the slope λ and the bias ρ are given by:

$$\lambda = \begin{cases} \frac{y_f - y}{x_f - x} & , \text{if } x_f \neq x \\ 0, & \text{otherwise} \end{cases}$$

$$\rho = \begin{cases} y_f - \frac{y_f - y}{x_f - x} x_f & , \text{if } x_f \neq x \\ 0, & \text{otherwise} \end{cases}$$

$$\begin{aligned} x_r^{(i)} &= \frac{x_f - x}{H} i + x \ , i = \overline{0..H-2} \\ x_r^{(H-1)} &= x_f \\ y_r^{(i)} &= \lambda x_f^{(i)} + \rho \ , i = \overline{0..H-2} \\ y_r^{(H-1)} &= y_f \end{aligned} \quad (25)$$

3.3. Control Algorithm—One Step Optimization

The control algorithm core is the sequential quadratic optimizer with a constraint tolerance of 1.0×10^{-3} and an optimality tolerance of 1.0×10^{-4} deduced heuristically through multiple experiments. Under this parametrization, the behavior is fairly robust and predictable with respect to the initial robot position, final resting position, varying size obstacles, wheel speeds, and acceleration. The object lists consist of a matrix of object positions obtained from the perception module. In order to determine the radius of the obstacles, we use the Moor–Neighbour tracing algorithm with Jacob's stopping criteria, which provides the contour of the objects from LIDAR data. Beyond providing LIDAR data, CNN can provide estimates of object radius with higher precision based on the object class. In order to reduce the computation time, the optimization problem is reformulated at each sampling time, and we consider in the optimization only those objects within a maximum radius (d_{max}) relative to the OMR's geometric center. The avoidance radius for each object is determined from the actual object radius with an additional tolerance according to OMR's dimensions. The reference orientation θ_r, and reference trajectory (x_r, y_r) are determined at each sample time since the OMR position evolves from one pose to another, constantly changing the heading to the final resting position. The first computed command over the predicted horizon is applied to the process inputs. We summarize the control algorithm steps for one sampling time T_s in Algorithm 1.

Algorithm 1 Main control algorithm

Inputs:
Desired setpoint X_f *from mission planner,* $X_f \leftarrow [x_f \;\; y_f \;\; \theta_f]^t$;
Initial position X_0 *from perception module,* $X_0 \leftarrow [x_0 \;\; y_0 \;\; \theta_0]^t$;
Outputs:
Actuator commands over horizon H, $\omega_j^{(i,k)}, j = \overline{1..4}, i = \overline{0..H-1}$
Predicted path over horizon H, $x^{(i|k)}, y^{(i|k)}, \theta^{(i|k)}, i = \overline{0..H-1}$
Runtime
Acquire object list data: positions (x_o, y_o)*, radius* (r_o) *from perception module;*
Detect object boundaries from LiDAR data using Moore-Neighbor tracing algorithm with Jacob's stopping criteria [40]
[B,L] = bwboundaries(LiDAR data); (Matlab specific function)
l = 0;
for $k \in \{1 \ldots length(B)\}$ **do**
  Object boundary $\leftarrow B\{k\}$; *B is Matlab cell data-type, therefore brackets are '{}' for indexing*
  Ignore objects composed of a very small or very high number of pixels (usually are artifacts or room boundaries)
  if $Boundary_{Min} \leq numel(Object\ boundary)/2 \leq Boundary_{Max}$ **then**
    $l \leftarrow l + 1$;
    If the number of objects exceeds buffer size (MaxNoObjs), an error will be thrown, and optimization will not be started
    if $l > MaxNoObjs$ **then**
      $l \leftarrow -1$;
      break;
    $\overline{xy} \leftarrow mean(Object\ boundary) \in \mathrm{R}^{2\times 1}$ *Matlab specific function to determine mean value over each line of a matrix.*
    $x_o(l) \leftarrow \overline{xy}[2]$;
    $y_o(l) \leftarrow \overline{xy}[1]$;
    $r_o(l) \leftarrow max(|max(Objectboundary) - min(Objectboundary)|)$; *Matlab specific functions to determine min, max values of matrix rows; or* r_o *provided by CNN subsystem;*
noObjs $\leftarrow l$; *No. of all objects detected in the map;*
Determine relevant objects (within specified proximity d_{max}*);*
for $k \in \{1 \ldots noObjs\}$ *No. of all objects* **do**
  Calculate distance to each relevant object:
  $d_o \leftarrow \sqrt{(x_o[k] - x)^2 + (y_o[k] - y)^2}$;
  if $d_o \leq d_{max}$ **then**
    Update object radius to include tolerance w.r.t to OMR dimensions
    $r_o[k] \leftarrow r_o[k] + max(l_x, l_y)$;
    $n_o \leftarrow n_o + 1$;
Calculate reference trajectory $x_r^{(i)}, y_r^{(i)}$ *according to Equation (25);*
Calculate reference angle $\theta_r^{(i)}, i = \overline{0..H-1}$ *according to Equation (24);*
Input data to optimizer: Sampling time: T_s*; Object list:* x_o, y_o, r_o*; Number of objects* n_o*, Initial resting point* X_0*; Final resting point* X_f*, Run-time reference trajectory* X_r*; Previous optimized commands* ω_i
Optimize;
Save in buffer the optimized commands;
Provide to the process inputs the first (i.e., i = 0) command from the- control buffer;

Figure 5 depicts the control structure consisting of two main subsystems: Environment perception, Model-Predictive Controller, and the interconnection with the psychical process.

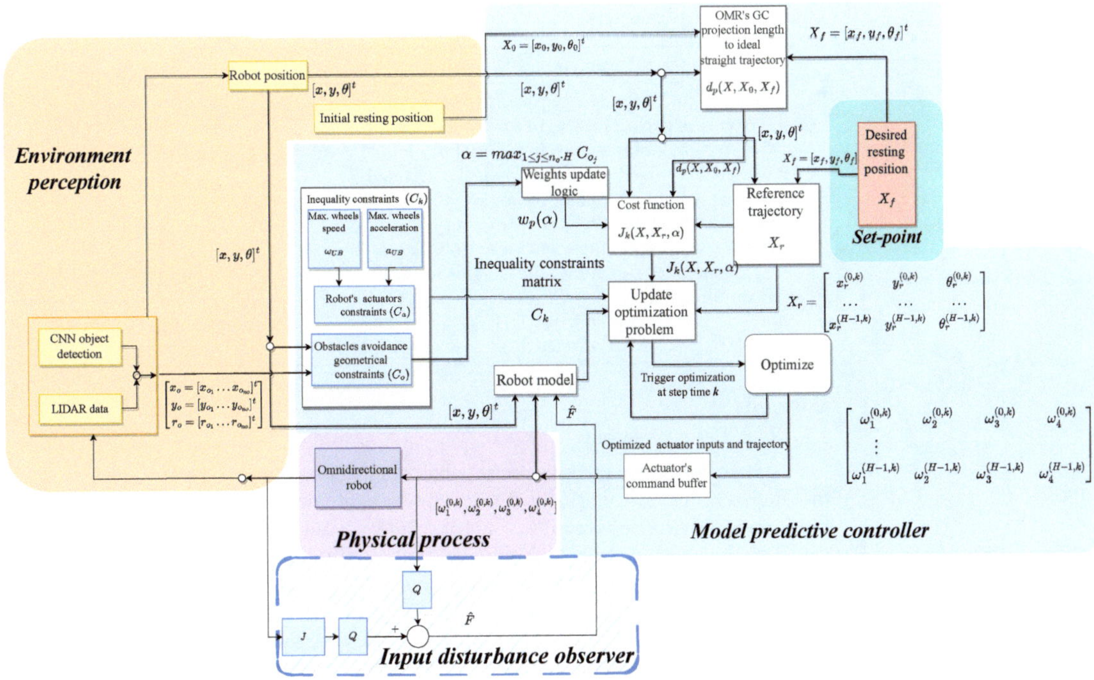

Figure 5. Illustrative block diagram of the control structure.

Figure 6 illustrates the main coordinates and notations used throughout the optimization problem. The projection d_p from the robot CG to the imaginary straight path connecting the initial X_0 and final X_f resting locations is noticeable. Moreover, the L2-norms used in calculating the cost function, L_1, L_2, and L_3 define the distances between the OMR, initial, and resting positions.

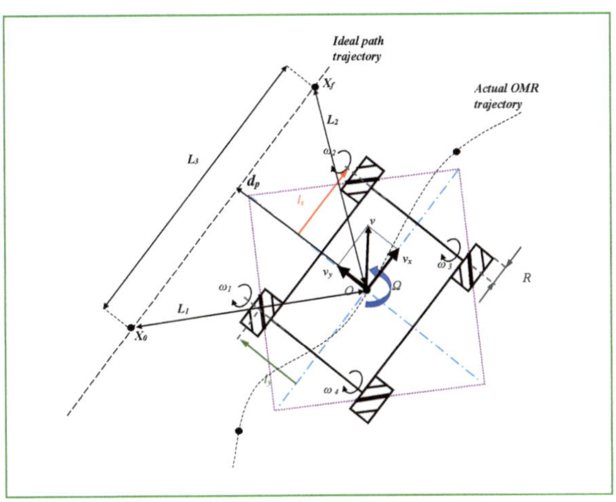

Figure 6. Coordinates system for control algorithm illustrating the used notations.

Table 5 contains the parameters of the model-predictive controller, including the penalizing factor of the cost function, the proximity threshold (*tol*) for switching cost function weights, the radius w.r.t to OMR's CG to and the prediction horizon.

Table 5. Control parameters.

Parameter	Value
Cost weight 1 of reference trajectory (w_{xy1})	0.6
Cost weight 1 of projection length to ideal path (w_{p_1})	0.01
Cost weight 2 of reference trajectory (w_{xy2})	0.05
Cost weight 2 of projection length to ideal path (w_{p_2})	2.0
Cost weight of orientation angle (w_θ)	0.3
Terminal cost weight of reference trajectory (w_{Tx})	0.8
Terminal cost weight of reference trajectory (w_{Ty})	0.8
Threshold for switching cost weights *tol*	−0.1 [m]
Maximum distance from CG to objects (d_{max})	2.5 [m]
Prediction horizon (H)	10 [samples] ($T_{horizon}$ = 0.2 s)
Map cell size	10 [cm]

4. Results

4.1. Object Detection Results

The performance of the selected object detection solutions (ssd-mobilenet-v1, ssd mobilenet-v2-lite, ssd-vgg16, and YoloV5) was evaluated on a testing subset with image resolutions varying between 720 × 404 and 2048 × 1024 pixels. The neural networks were tested on the Nvidia Jetson AGX mobile platform with the same input.

All models are optimized for Jetson Xavier AGX with the TensorRT framework from CUDA for Nvidia cards. The run time of the three selected architectures from the SSD family and the five main YoloV5 [29] is presented in Table 6. Architectures with fewer parameters performed better in terms of frames per second. Being the lightest model, the Nano YoloV5 is six times faster than the Extra Large model, the largest we considered for the Jetson platform. This highlights the importance of considering the specific hardware platform and the model's complexity for deploying object detection algorithms on mobile robots.

The two largest YoloV5 models did not bring any improvements for the overall precision and the precision per class compared to the Medium architecture; therefore, they were not considered for Table 6. A comparison between the precision of the models can be made based on the figures presented in Table 7. All architectures were trained for 150 epochs to evaluate the mean Average Precision. SSD Mobilenet v2 lite and SSD VGG16 reach a similar mAP@0.5 of 98–99%, while SSD Mobilenet v1 has a lower precision on the test subset, 86%.

Based on the results from Tables 6 and 7, we can draw the conclusion that the best model for our OMR object detection use cases is YoloV5 Medium which has a mAP comparable to SSD-VGG16, with the benefit of being twice as fast. Detection examples with the neural network models tested in the OMR environment are shown in Figure 7.

Table 6. Inference time.

Architecture	FPS on Jetson Xavier AGX	Number of Parameters
SSD Mobilenet v1	120	4.2M
SSD Mobilenet v2 lite	130	3.4M
SSD VGG16	50	35M
yoloV5 Nano	270	1.9M
yoloV5 Small	225	7.2M
yoloV5 Medium	109	21.2M
yoloV5 Large	72	46.5M
yoloV5 Extra Large	44	86.7M

Table 7. Average precision on test subset.

Model	Overall mAP@0.5	Per Class mAP@0.5				
		Conveyor Type 1	Conveyor Type 2	Rosy	Charging Station	Cone
SSD Mobilenet v1	0.860	0.909	0.891	0.816	0.717	0.969
SSD Mobilenet v2 lite	0.989	0.998	0.980	0.974	0.998	0.992
SSD VGG16	0.997	0.998	0.996	0.990	0.998	0.998
YoloV5 Nano	0.989	0.993	0.985	0.992	0.982	0.995
YoloV5 Small	0.992	0.995	0.989	0.994	0.99	0.995
YoloV5 Medium	0.994	0.995	0.992	0.995	0.991	0.995

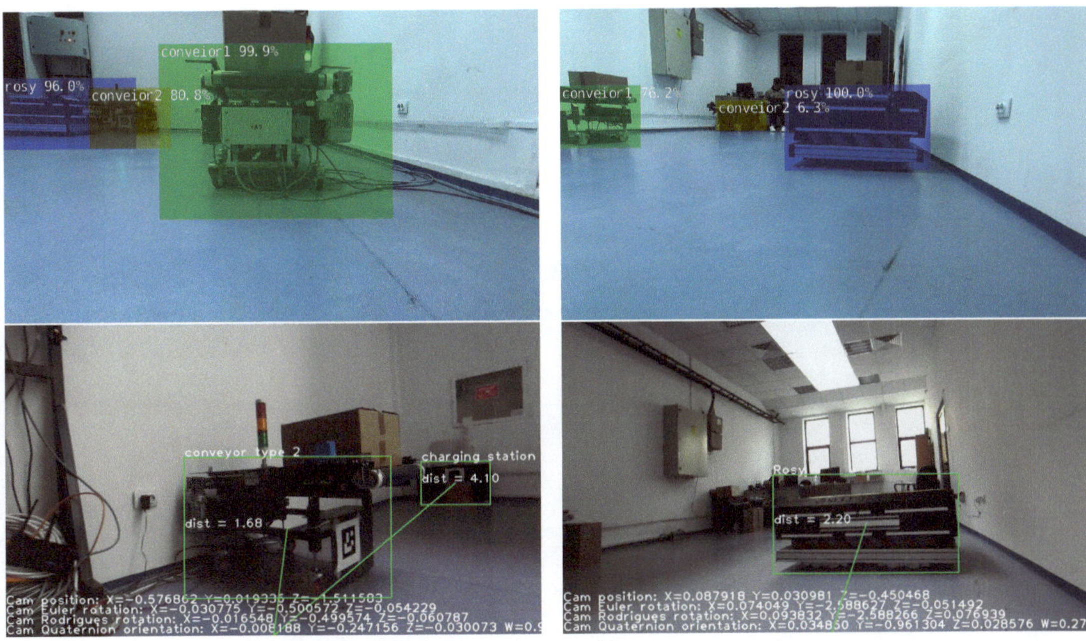

Figure 7. Detected objects with SSD architectures (**1st row**) and with YOLOv5 architecture (**2nd row**).

4.2. Simulation Results

To evaluate the control performances, we considered scenarios where the initial and final positions varied throughout the room so that obstacles blocked the OMR path. We perform numerical simulations on real data acquired from the perception module. We evaluate the steady-state error, the possible constraint violations, the cost function, and the optimization run-time.

In the first test case considered in Figure 8, the final resting position X_f is reached after avoiding the two obstacles on the circumference of virtual circles centered around the objects. The inequality geometric inequality constrained $C_o < 0$, and the actuator constraints are satisfied $C_a < 0$ with an acceptable tolerance. Generally, the tolerance is within the expected margin of 1.0×10^{-3}. The steady-state error of the controlled position (x, y) is less than 1% as measured around moment $t = 10.2$ s. The transient time is limited by the upper and lower bounds of the wheel speed, in this case, ± 10 rad/s. The orientation θ changes at each sample time as the vehicle travels towards X_f. Hence, the tracking is decent, with a peak error of 17 degrees noticeably on the roundabouts of the objects since the optimizer is more constrained. The cost function decreases as the vehicle evolves across the map. In the proximity of an object, the cost function is purposely increased to avoid local minima by amplifying the deviation from the reference trajectory and decreasing the penalizing weight for the projection to the ideal path to allow solutions on the circumference of the encircled object. The maximum number of iterations was 79 with a run-time of 0.8945 s, and the minimum number of iterations was 2 with a run-time of 0.0204 s (CPU Intel i7 7500u, dual-core, 7th generation). The mean number of iterations was 6.526, with an execution time of 0.0647 s. It must be mentioned that the run-time is less relevant since in MEX mode (Matlab executable), the run-time can be reduced considerably (in MEX mode, the average run-time was 0.0507 s, while in normal mode 0.0647 s). The execution time is platform dependent.

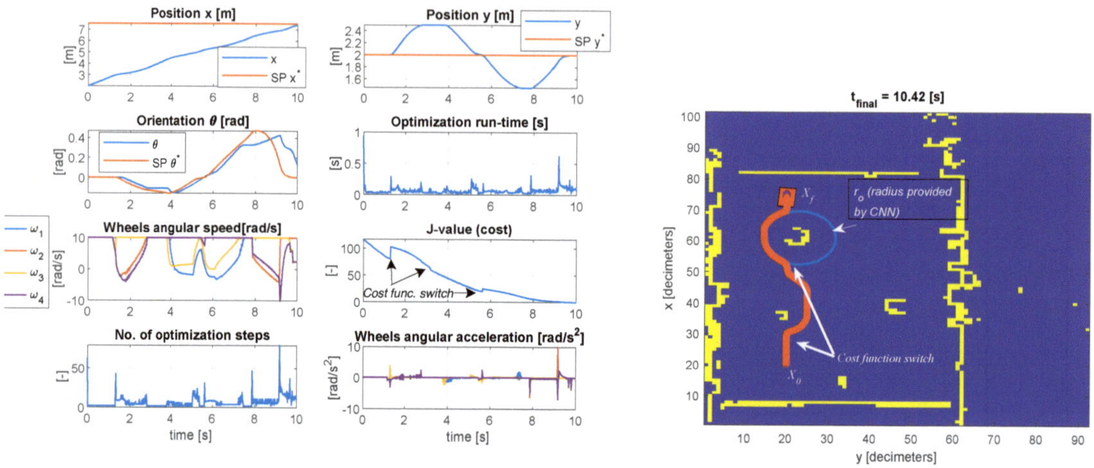

Figure 8. Simulation results of the model−predictive controller with LiDAR data and simulation of camera detection (test case I).

In the second scenario presented in Figure 9, the behavior is similar concerning the constraint tolerances. The violation of the object boundaries is within the expected limit, and the steady-state error of the controlled pose (x, y, θ) is less than 1%. In this case, the actuator constraints limit the transient time, ± 10 rad/s. The maximum number of iterations was 66 with a run-time of 0.9923 s, and the minimum number of iterations was 2 with a run-time of 0.022 s (same CPU as mentioned in test case I). The mean number of iterations was 8.5658, with a mean execution time of 0.0814 s. In MEX mode, the maximum run-time was 0.5681 s, the minimum 0.0039 s, and the average 0.0356 s. Generally, the behavior is as expected, and the run-time proves the applicability of the control structure.

Similar behavior is obtained in the third test case presented in Figure 10, but the maximum run-time is slightly higher at 1.6 s, the maximum number of iterations is 210, and the minimum is 2. The minimum run-time was 0.0191 s. However, the average run-time in MEX mode is 0.0358 s with a maximum of 0.3176 s (instead of 1.6 s as in normal mode) and a minimum of 0.0043 s.

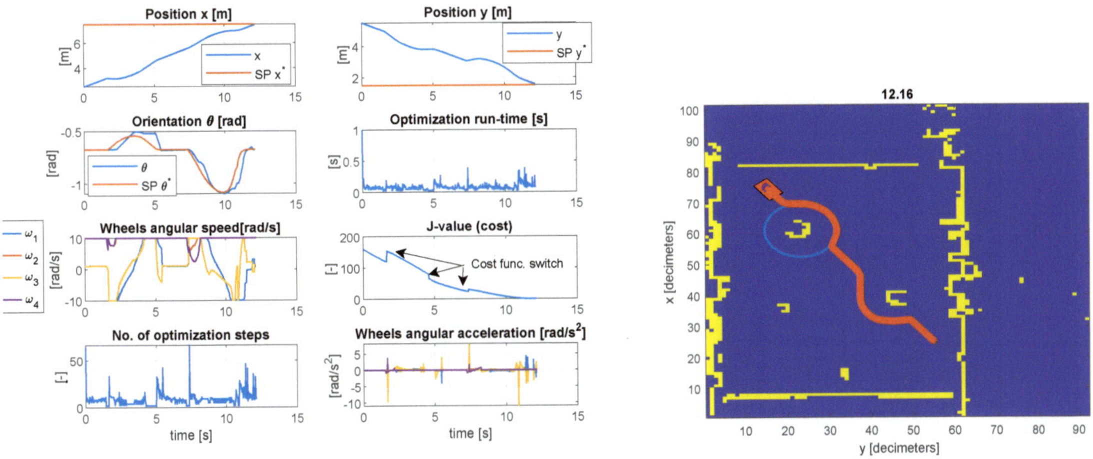

Figure 9. Simulation results of the model−predictive controller with LiDAR data and simulation of camera detection (test case II).

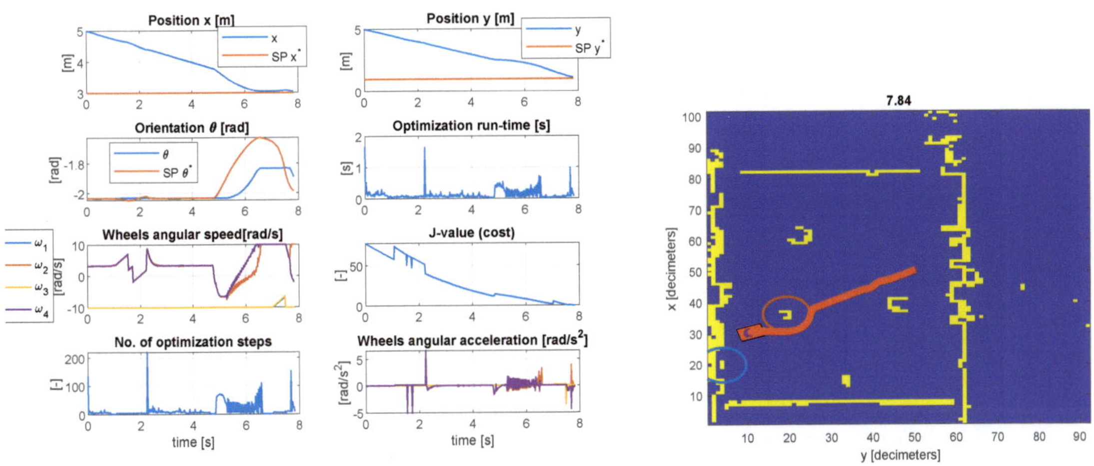

Figure 10. Simulation results of the model−predictive controller with LiDAR data and simulation of camera detection (test case III).

5. Conclusions and Future Work

The use of CNNs for object detection in mobile robot navigation provides benefits such as accuracy, robustness, and adaptability, which are desirable for the navigation of mobile robots in a logistic environment.

The paper proves the use of an object detector for a better understanding of the OMR working environment. To overcome this challenge, we also acquired a dataset for domain-specific object detection that was made public. It contains all objects of interest for the working environment, such as fixed or mobile conveyors, charging stations, other robots, and boundary cones. The results show a detection accuracy of 99% using the selected lightweight model, which was optimized to run on the available mobile platform already installed on the OMR at about 109 frames per second. The detection results offer a better understanding of the LiDAR map by assigning a name to obstacles and objects within the working environment, allowing the control model constraints to be adjusted on the fly.

This paper also demonstrates the model-predictive control of the OMR in logistic environments with actuator and geometric constraints. We avoid local minima by using variable cost function weights to navigate around obstacles while still achieving the overall objective of reducing travel distance. The execution runtime of the optimizer allows for practical implementation while the control performance is within the expected margin.

Future work is also expected to involve the deployment of the OMR controller and testing in a controlled environment and then in an automated logistic warehouse. One of the short-term goals is to collect and annotate more instances of domain-specific objects so that the intraclass variety is better covered and the detector can extrapolate on new data.

Author Contributions: Conceptualization, S.-D.A., R.M., A.-T.P. and C.-C.D.; methodology, S.-D.A., R.M., A.-T.P. and C.-C.D.; software, S.-D.A. and R.M.; validation, S.-D.A., R.M., A.-T.P. and C.-C.D.; formal analysis, S.-D.A. and R.M.; investigation, S.-D.A.; resources, S.-D.A.; data curation, S.-D.A.; writing—original draft preparation, S.-D.A. and R.M.; writing—review and editing, S.-D.A., R.M., A.-T.P. and C.-C.D.; visualization, S.-D.A. and R.M.; supervision, A.-T.P. and C.-C.D.; funding acquisition, S.-D.A. and R.M. All authors have read and agreed to the published version of the manuscript.

Funding: This paper was realized with the support of "Institutional development through increasing the innovation, development, and research performance of TUIASI–COMPETE", project funded by contract no. 27PFE/2021, financed by the Romanian Government.

Institutional Review Board Statement: Not applicable.

Informed Consent Statement: Not applicable.

Data Availability Statement: The obtained data set for object detection is publicly available.

Conflicts of Interest: The authors declare no conflict of interest.

Abbreviations

The following abbreviations are used in this manuscript:

AP	Average Precision
APF	Artificial Potential Field
AR	Average Recall
CNN	Convolutional Neural Network
CPU	Central Processing Unit
FN	False Negative
FoV	Field of View
FP	False Positive
FPS	Frames Per Second
GPU	Graphics Processing Unit
GT	Ground Truth
IMU	Inertial Measurement Unit
IoU	Intersection over Union
IR	Infrared
LiDAR	Light Detection and Ranging
mAP	mean Average Precision
mAR	mean Average Recall
MEX	Matlab Executable
MS COCO	Microsoft Common Objects in COntext
OMR	Omnidirectional Mobile Robots
OROD	Omnidirectional Robot Object Detection
SP	Set Point
SSD	Single Shot Detector
TN	True Negative
TP	True Positive
TPU	Tensor Processing Unit
YOLO	You Only Look Once the algorithm

References

1. Dosoftei, C.C.; Popovici, A.T.; Sacaleanu, P.R.; Gherghel, P.M.; Budaciu, C. Hardware in the Loop Topology for an Omnidirectional Mobile Robot Using Matlab in a Robot Operating System Environment. *Symmetry* **2021**, *13*, 969. [CrossRef]
2. Rubio, F.; Valero, F.; Llopis-Albert, C. A review of mobile robots: Concepts, methods, theoretical framework, and applications. *Int. J. Adv. Robot. Syst.* **2019**, *16*, 1729881419839596. [CrossRef]
3. Patle, B.; Babu L, G.; Pandey, A.; Parhi, D.; Jagadeesh, A. A review: On path planning strategies for navigation of mobile robot. *Def. Technol.* **2019**, *15*, 582–606. [CrossRef]
4. Ngwenya, T.; Ayomoh, M.; Yadavalli, S. Virtual Obstacles for Sensors Incapacitation in Robot Navigation: A Systematic Review of 2D Path Planning. *Sensors* **2022**, *22*, 6943. [CrossRef] [PubMed]
5. Wang, Z.; Tian, G. Hybrid offline and online task planning for service robot using object-level semantic map and probabilistic inference. *Inf. Sci.* **2022**, *593*, 78–98. [CrossRef]
6. Tătar, M.O.; Popovici, C.; Mândru, D.; Ardelean, I.; Pleşa, A. Design and development of an autonomous omni-directional mobile robot with Mecanum wheels. In Proceedings of the 2014 IEEE International Conference on Automation, Quality and Testing, Robotics, Cluj-Napoca, Romania, 22–24 May 2014; IEEE: Piscataway, NJ, USA, 2014; pp. 1–6.
7. Sheikhlar, A.; Fakharian, A. Adaptive optimal control via reinforcement learning for omni-directional wheeled robots. In Proceedings of the 2016 4th International Conference on Control, Instrumentation, and Automation (ICCIA), Qazvin, Iran, 27–28 January 2016; IEEE: Piscataway, NJ, USA, 2016; pp. 208–213.
8. Popovici, A.T.; Dosoftei, C.C.; Budaciu, C. Kinematics Calibration and Validation Approach Using Indoor Positioning System for an Omnidirectional Mobile Robot. *Sensors* **2022**, *22*, 8590. [CrossRef] [PubMed]
9. Peng, T.; Qian, J.; Zi, B.; Liu, J.; Wang, X. Mechanical design and control system of an omni-directional mobile robot for material conveying. *Procedia CIRP* **2016**, *56*, 412–415. [CrossRef]
10. Lu, X.; Zhang, X.; Zhang, G.; Jia, S. Design of adaptive sliding mode controller for four-Mecanum wheel mobile robot. In Proceedings of the 2018 37th Chinese Control Conference (CCC), Wuhan, China, 25–27 July 2018; IEEE: Piscataway, NJ, USA, 2018; pp. 3983–3987.
11. Sun, Z.; Xie, H.; Zheng, J.; Man, Z.; He, D. Path-following control of Mecanum-wheels omnidirectional mobile robots using nonsingular terminal sliding mode. *Mech. Syst. Signal Process.* **2021**, *147*, 107128. [CrossRef]
12. Lu, X.; Zhang, X.; Zhang, G.; Fan, J.; Jia, S. Neural network adaptive sliding mode control for omnidirectional vehicle with uncertainties. *ISA Trans.* **2019**, *86*, 201–214. [CrossRef]
13. Zimmermann, K.; Zeidis, I.; Abdelrahman, M. Dynamics of mechanical systems with mecanum wheels. In *Applied Non-Linear Dynamical Systems*; Springer: Berlin/Heidelberg, Germany, 2014; pp. 269–279.
14. Santos, J.; Conceição, A.G.; Santos, T.L. Trajectory tracking of omni-directional mobile robots via predictive control plus a filtered smith predictor. *IFAC-PapersOnLine* **2017**, *50*, 10250–10255. [CrossRef]
15. Conceição, A.S.; Oliveira, H.P.; e Silva, A.S.; Oliveira, D.; Moreira, A.P. A nonlinear model predictive control of an omni-directional mobile robot. In Proceedings of the 2007 IEEE International Symposium on Industrial Electronics, Vigo, Spain, 4–7 June 2007; IEEE: Piscataway, NJ, USA, 2007; pp. 2161–2166.
16. Wang, D.; Wei, W.; Yeboah, Y.; Li, Y.; Gao, Y. A robust model predictive control strategy for trajectory tracking of omni-directional mobile robots. *J. Intell. Robot. Syst.* **2020**, *98*, 439–453. [CrossRef]
17. Teatro, T.A.V.; Eklund, J.M.; Milman, R. Nonlinear model predictive control for omnidirectional robot motion planning and tracking with avoidance of moving obstacles. *Can. J. Electr. Comput. Eng.* **2014**, *37*, 151–156. [CrossRef]
18. Wang, C.; Liu, X.; Yang, X.; Hu, F.; Jiang, A.; Yang, C. Trajectory tracking of an omni-directional wheeled mobile robot using a model predictive control strategy. *Appl. Sci.* **2018**, *8*, 231. [CrossRef]
19. Rosenfelder, M.; Ebel, H.; Eberhard, P. Cooperative distributed nonlinear model predictive control of a formation of differentially-driven mobile robots. *Robot. Auton. Syst.* **2022**, *150*, 103993. [CrossRef]
20. Zhang, H.; Wang, S.; Xie, Y.; Wu, H.; Xiong, T.; Li, H. Nonlinear Model Predictive Control of an Omnidirectional Mobile Robot with Self-tuned Prediction Horizon. In Proceedings of the 2022 IEEE 17th Conference on Industrial Electronics and Applications (ICIEA), Chengdu, China, 16–19 December 2022.
21. Ren, C.; Li, C.; Hu, L.; Li, X.; Ma, S. Adaptive model predictive control for an omnidirectional mobile robot with friction compensation and incremental input constraints. *Trans. Inst. Meas. Control* **2022**, *44*, 835–847. [CrossRef]
22. Santos, J.C.; Gouttefarde, M.; Chemori, A. A nonlinear model predictive control for the position tracking of cable-driven parallel robots. *IEEE Trans. Robot.* **2022**, *38*, 2597–2616. [CrossRef]
23. Kim, H.; Kim, B.K. Minimum-energy trajectory generation for cornering with a fixed heading for three-wheeled omni-directional mobile robots. *J. Intell. Robot. Syst.* **2014**, *75*, 205–221. [CrossRef]
24. Ge, S.S.; Cui, Y.J. New potential functions for mobile robot path planning. *IEEE Trans. Robot. Autom.* **2000**, *16*, 615–620. [CrossRef]
25. Li, Z.; Zhao, S.; Duan, J.; Su, C.Y.; Yang, C.; Zhao, X. Human cooperative wheelchair with brain–machine interaction based on shared control strategy. *IEEE/ASME Trans. Mechatron.* **2016**, *22*, 185–195. [CrossRef]
26. Rösmann, C.; Hoffmann, F.; Bertram, T. Timed-elastic-bands for time-optimal point-to-point nonlinear model predictive control. In Proceedings of the 2015 European Control Conference (ECC), Linz, Austria, 15–17 July 2015; IEEE: Piscataway, NJ, USA, 2015; pp. 3352–3357.

27. Rösmann, C.; Hoffmann, F.; Bertram, T. Integrated online trajectory planning and optimization in distinctive topologies. *Robot Auton. Syst.* **2017**, *88*, 142–153. [CrossRef]
28. Smith, J.S.; Xu, R.; Vela, P. egoteb: Egocentric, perception space navigation using timed-elastic-bands. In Proceedings of the 2020 IEEE International Conference on Robotics and Automation (ICRA), Paris, France, 31 May–31 August 2020; IEEE: Piscataway, NJ USA, 2020; pp. 2703–2709.
29. Glenn, J.; Alex, S.; Jirka, B. *YOLOv5*; Zenodo: Geneva, Switzerland, 2020.
30. Sandler, M.; Howard, A.; Zhu, M.; Zhmoginov, A.; Chen, L.C. MobileNetV2: Inverted Residuals and Linear Bottlenecks. In Proceedings of the 2018 IEEE/CVF Conference on Computer Vision and Pattern Recognition, Salt Lake City, UT, USA, 18–22 June 2018; pp. 4510–4520. [CrossRef]
31. Howard, A.G.; Zhu, M.; Chen, B.; Kalenichenko, D.; Wang, W.; Weyand, T.; Andreetto, M.; Adam, H. MobileNets: Efficient Convolutional Neural Networks for Mobile Vision Applications. *arXiv* **2017**, arXiv:1704.04861.
32. Liu, S.; Deng, W. Very deep convolutional neural network based image classification using small training sample size. In Proceedings of the 2015 3rd IAPR Asian Conference on Pattern Recognition (ACPR), Kuala Lumpur, Malaysia, 3–6 November 2015; pp. 730–734. [CrossRef]
33. Bochkovskiy, A.; Wang, C.; Liao, H.M. YOLOv4: Optimal Speed and Accuracy of Object Detection. *arXiv* **2020**, arXiv:2004.10934
34. Paszke, A.; Gross, S.; Massa, F.; Lerer, A.; Bradbury, J.; Chanan, G.; Killeen, T.; Lin, Z.; Gimelshein, N.; Antiga, L.; et al. PyTorch An Imperative Style, High-Performance Deep Learning Library. In *Proceedings of the Advances in Neural Information Processing Systems*; Wallach, H., Larochelle, H., Beygelzimer, A., d'Alché-Buc, F., Fox, E., Garnett, R., Eds.; Curran Associates, Inc.: Red Hook, NY, USA, 2019; Volume 32.
35. Liu, W.; Anguelov, D.; Erhan, D.; Szegedy, C.; Reed, S.E.; Fu, C.; Berg, A.C. SSD: Single Shot MultiBox Detector. *arXiv* **2015**, arXiv:1512.02325.
36. Szegedy, C.; Liu, W.; Jia, Y.; Sermanet, P.; Reed, S.; Anguelov, D.; Erhan, D.; Vanhoucke, V.; Rabinovich, A. Going deeper with convolutions. In Proceedings of the 2015 IEEE Conference on Computer Vision and Pattern Recognition (CVPR), Boston, MA, USA, 7–12 June 2015; pp. 1–9. [CrossRef]
37. Lin, T.; Maire, M.; Belongie, S.J.; Bourdev, L.D.; Girshick, R.B.; Hays, J.; Perona, P.; Ramanan, D.; Dollár, P.; Zitnick, C.L. Microsoft COCO: Common Objects in Context. *arXiv* **2014**, arXiv:1405.0312.
38. Cordts, M.; Omran, M.; Ramos, S.; Rehfeld, T.; Enzweiler, M.; Benenson, R.; Franke, U.; Roth, S.; Schiele, B. The Cityscapes Dataset for Semantic Urban Scene Understanding. *arXiv* **2016**, arXiv:1604.01685.
39. Cordts, M.; Omran, M.; Ramos, S.; Scharwächter, T.; Enzweiler, M.; Benenson, R.; Franke, U.; Roth, S.; Schiele, B. The Cityscapes Dataset. In Proceedings of the CVPR Workshop on the Future of Datasets in Vision, Boston, MA, USA, 11 June 2015.
40. Seo, J.; Chae, S.; Shim, J.; Kim, D.; Cheong, C.; Han, T.D. Fast contour-tracing algorithm based on a pixel-following method for image sensors. *Sensors* **2016**, *16*, 353. [CrossRef]

sensors

MDPI

Article

High-Dynamic-Range Tone Mapping in Intelligent Automotive Systems

Ivana Shopovska *,†, Ana Stojkovic †, Jan Aelterman, David Van Hamme and Wilfried Philips

IMEC-IPI-Ghent University, 9000 Ghent, Belgium; ana.stojkovic@ugent.be (A.S.); jan.aelterman@ugent.be (J.A.); david.vanhamme@ugent.be (D.V.H.); wilfried.philips@ugent.be (W.P.)

* Correspondence: ivana.shopovska@ugent.be

† These authors contributed equally to this work.

Abstract: Intelligent driver assistance systems are becoming increasingly popular in modern passenger vehicles. A crucial component of intelligent vehicles is the ability to detect vulnerable road users (VRUs) for an early and safe response. However, standard imaging sensors perform poorly in conditions of strong illumination contrast, such as approaching a tunnel or at night, due to their dynamic range limitations. In this paper, we focus on the use of high-dynamic-range (HDR) imaging sensors in vehicle perception systems and the subsequent need for tone mapping of the acquired data into a standard 8-bit representation. To our knowledge, no previous studies have evaluated the impact of tone mapping on object detection performance. We investigate the potential for optimizing HDR tone mapping to achieve a natural image appearance while facilitating object detection of state-of-the-art detectors designed for standard dynamic range (SDR) images. Our proposed approach relies on a lightweight convolutional neural network (CNN) that tone maps HDR video frames into a standard 8-bit representation. We introduce a novel training approach called detection-informed tone mapping (DI-TM) and evaluate its performance with respect to its effectiveness and robustness in various scene conditions, as well as its performance relative to an existing state-of-the-art tone mapping method. The results show that the proposed DI-TM method achieves the best results in terms of detection performance metrics in challenging dynamic range conditions, while both methods perform well in typical, non-challenging conditions. In challenging conditions, our method improves the detection F2 score by 13%. Compared to SDR images, the increase in F_2 score is 49%.

Keywords: high dynamic range; tone mapping; deep learning; object detection; autonomous driving

Citation: Shopovska, I.; Stojkovic, A.; Aelterman, J.; Van Hamme, D.; Philips, W. High-Dynamic-Range Tone Mapping in Intelligent Automotive Systems. *Sensors* **2023**, *23*, 5767. https://doi.org/10.3390/s23125767

Academic Editors: Constantin Caruntu and Ciprian Romeo Comşa

Received: 18 April 2023
Revised: 12 June 2023
Accepted: 19 June 2023
Published: 20 June 2023

1. Introduction

According to studies by the Governors Highway Safety Association (GHSA) [1] and the National Highway Traffic Safety Administration (NHTSA) [2], between 50% and 75% of fatal traffic accidents involving pedestrians occur at night, despite the decreased traffic flow. One of the main contributing factors to nighttime accidents is impaired visibility due to insufficient light. Although the human eye is adaptable to a wide range of light conditions, it has trouble adjusting to low-light conditions. Blinding lights are also a significant challenge that impairs a driver's vision.

Increasing the vehicle's autonomy is expected to improve traffic safety by reducing the impact of human error [3,4]. The intermediate steps towards achieving full autonomy include the incorporation of increasingly intelligent sensing and control technologies in vehicles to assist human drivers [5,6], referred to as "advanced driver assistance systems" (ADAS).

In the context of challenging light conditions, the ratio between the brightest and the darkest perceivable light level is defined as the dynamic range (DR). Digital cameras used in commercial automotive vision systems are more limited in dynamic range than the eye, typically at around 70 dB. This is insufficient for a number of driving conditions. For example, oncoming car headlights at night or the sun low at the horizon are blinding to a typical sensor. Entering and exiting a tunnel introduces challenges of the stark contrast between dark and bright, where the sensitivity of a standard sensor can adapt well only to the bright or dark areas of the scene, but not both simultaneously. Such conditions impair visibility and pose a danger to drivers and other road users.

For ADAS and autonomous vehicles, using a high-dynamic-range (HDR) sensor with a low sensitivity threshold and a high saturation capacity is beneficial to increasing the signal-to-noise ratio (SNR) of the captured images. To encode HDR data, a representation of more than the standard 8 bits is required. For example, the automotive HDR sensor of the Sony IMX490 (Sony, Tokyo, Japan) uses 24 bits. However, processing high-bit-depth data requires more powerful computational resources than what is available on onboard computers. Another challenge is the limited availability of perception algorithms that work with high bit-depth data. The majority of state-of-the-art computer vision methods are optimized for 8-bit standard dynamic range (SDR) images, which makes them incompatible with high-bit-depth HDR data.

One solution is to adapt the existing algorithms to work directly with HDR input. The limitations of such a strategy are the need to redesign the state-of-the-art methods to a non-standard data representation, as well as the need to acquire sufficient new data to retrain the algorithms. We approach this problem from the perspective of adapting the input data instead, reasoning that it has a greater potential for facilitating the integration of HDR cameras in intelligent perception systems. Our approach is to non-linearly compress (tone map) the HDR data into a standard 8-bit representation, for application in object detection tasks for environment perception in traffic.

In this study, we demonstrated improved detection capabilities over a conventional tone-mapping-based detection system by specifically optimizing the tone mapping process for the detection of traffic-related objects. We compressed the intensity range of HDR images through object-aware tone mapping into an 8-bit/color channel representation while preserving the important image information. In comparison with other tone mapping operators developed for perceptual picture quality, our method helps to improve object detection while maintaining a natural image appearance.

Section 2 presents an overview of relevant literature related to high-dynamic-range imaging and tone mapping that optimizes the visual quality, as well as optimization of algorithms designed for more specific tasks that use HDR inputs. In Section 3, we describe the proposed algorithm for tone mapping using a deep neural net, optimized with respect to object detection performance for automotive applications. The experiments and results are presented in Section 4, and Section 5 concludes the paper and discusses future prospects.

2. Related Work

In the research literature, the vast majority of state-of-the-art tone mapping methods are optimized for enhancement of the image appearance for visualization purposes, based on models of the human visual system (HVS) or for subjective quality and artistic photography purposes [7–17]. The majority of these methods consider either local or global image features, but the most successful approaches typically analyze the input on multiple scales [9]. Global processing analyzes the overall illumination range, allowing preservation of the relative intensity differences without introducing artifacts such as halos and contrast reversal. A local analysis, on the other hand, helps preserve sharpness and enhance details. Nevertheless, ranking tone-mapping methods is not straightforward and depends on the evaluation criteria, as found by [11,12]. A recent method, iCAM06-m [10], builds upon the perceptual appearance model iCAM06 [8], and focuses on improving the color representation of tone-mapped HDR images. Subjective and objective evaluation experiments were carried out to evaluate the perceptual image quality of the results, achieving slightly better results for the modified model. However, these metrics do not adequately capture the effectiveness of the system for object detection purposes.

Another limitation of the aforementioned methods is that the performance is highly dependent on adequate hyperparameter tuning. To improve the robustness to changing conditions, more recent efforts rely on training a neural network for adapting to a broader range of scene contexts [14–17]. The limitations of these methods are that they are relatively computationally complex for implementation in real-time processing, as well as that they aim for general solutions with perceptual image quality goals. Furthermore, these goals are often subjective, without a precise and consistent performance metric.

To the knowledge of the authors, the quality of tone mapping operators has not been thoroughly evaluated with respect to object detection accuracy, and only a few studies exist that focus on object detection performance in HDR content [18–20]. Furthermore, the effectiveness of HDR imaging in various illumination conditions has not been extensively investigated in the automotive vision context. In this work, we aimed to optimize the process of tone mapping of HDR images to facilitate the detection of traffic-related objects in challenging light conditions.

In Ref. [18], the object detection performance was evaluated when seven different tone mapping operators (TMOs) were applied to HDR images. Moreover, the authors retrained three different neural-network-based object detector heads on a dataset of tone-mapped HDR images. The training dataset comprises a carefully selected subset of publicly available HDR images, manually annotated with labels for six object categories. The conclusions of this paper show that using tone-mapped HDR data results in better detection performance compared to using SDR data and that, in the majority of situations, most of the TM operators result in similar detection performance. However, the analysis does not differentiate between non-challenging and challenging light conditions, whereas the impact of different tone mapping methods is mostly observed in relatively few but safety-important challenging scenarios.

In continuation of this work, in [19], the authors explored the feasibility of training object detectors directly using HDR images. Due to the limited availability of HDR data for training, they created a pseudo-HDR dataset by applying a dynamic range expansion operator [21] on a set of SDR images. For evaluation, a small set of true HDR images was used. On average, the performance of detectors trained on HDR images was similar to that of detectors that were trained for SDR images and applied to tone-mapped HDR images. However, an isolated, small-scale study using 126 images of scenes with extreme light differences indicates a significant performance gain when HDR data are used, due to the better preservation of meaningful features.

Despite being evaluated on object detection performance, the aforementioned methods do not receive feedback from a task-specific module such as an object detector. A novel idea that combines multi-task learning is presented in [20]. The paper proposes a method for exposure selection as an alternative to using HDR sensors. A model that predicts optimal exposure values for image acquisition was trained in a joint end-to-end pipeline with image processing and object detection modules. The entire pipeline is supervised only with the object detector loss at the end. The reported results indicate that the proposed method outperforms the standard luminance-based and gradient-based auto-exposure control methods in supporting the object detection task for several automotive object categories.

In Ref. [22], the authors propose a novel approach, called TMO-Det, that jointly optimizes a generative adversarial network (GAN)-based tone mapping model and an object detector. Next to the discriminator, the GAN architecture is extended with a detection branch, which also provides feedback to the generator. The training loss function is augmented to enforce not only visual similarity to a classical tone mapping method but to also maximize object detection performance on the generated images. The main conclusions of this paper are that training the tone mapping GAN model jointly with a detector can achieve higher detection accuracy compared to classical TMOs as well as compared to GAN-based tone mapping architecture without jointly training with a detector. However, according to the authors, this increase was not very significant. Nonetheless, considering both detection performance measured by average precision (AP) and image quality measured by the Tone-Mapping Quality Index (TMQI) [23], the proposed method achieves the best results among the compared TMOs.

Since the main goal of this work is the application of tone mapping in an autonomous driving context, we have evaluated the effectiveness of several successful classical state-of-the-art tone mapping methods from the literature [9,24–27]. They were evaluated in terms of object detection performance applied to the tone-mapped outputs of each method. The evaluation confirmed our assumption that quality metrics such as SNR, contrast, or color saturation are not indicative of the performance of each tone mapping method in detection applications, and methods can vary in their robustness to scene conditions. Based on a comparison between many different state-of-the-art methods, we have selected the method of Farbman et al. [9] as a state-of-the-art representative in further evaluation. This is an edge-preserving filtering technique based on weighted least-squares optimization. In HDR tone mapping, it is used for compressing the luminance while preserving the details.

In the proposed method, all algorithm design choices consider the effect on the detection of traffic road users. For evaluation purposes, we relied on HDR images simulated from images in a public SDR traffic dataset [28], as well as our multi-modal traffic dataset including true HDR data with "person" annotations [29].

3. Proposed Method

In this paper, we present a tone mapping method that generalizes to the large variety of light conditions that is encountered in driving scenarios. The optimization of the algorithm is set in a learning-based framework using a convolutional neural network (CNN) trained to perform tone mapping of HDR images while maximizing image quality and object detection objectives.

The proposed method introduces three main novelties:

- The method emphasizes accuracy in image regions containing objects of interest by increasing the sampling density of these image regions during the neural network's stochastic gradient descent training.
- During training, data augmentation techniques are used to improve the model's robustness to challenging lighting conditions, particularly in traffic situations. The augmentations are informed by an analysis of factors identified as having a significant impact on detection performance in a study we conducted for this research.
- Unlike most HDR methods in the literature, the method is evaluated using real driving data, with detection accuracy measured in the context of traffic safety. Additionally, the method's tone mapping performance is evaluated on challenging data, which is a small but important subset of traffic situations.

3.1. Network Architecture

The design of the network architecture was inspired by ExpandNet [21], which is a framework originally designed for HDR expansion (inverse tone mapping). In this work, we adapted a simplified architecture based on ExpandNet for tone mapping. The architecture utilizes multi-scale processing and draws from successful classical approaches while maintaining low complexity.

Figure 1 illustrates the structure of the proposed tone mapping method. The full-resolution HDR image is passed through convolution layers of size $3 \times 3 \times 64$ and convolutions of size $3 \times 3 \times 128$, followed by non-linear activations (ReLUs) in a local processing branch. The local branch encodes local image features related to object edges and structures, with the purpose of ensuring correct contrast compression without contrast inversion and halo artifacts. A global branch applies strided $3 \times 3 \times 64$ convolutions followed by ReLUs and a $4 \times 4 \times 64$ convolution to summarize the global illumination into a single, 64-element vector. The effect of this branch can be interpreted as analogous to calculating a global image histogram in classical methods. A final fusion branch concatenates the local features with the global representation and applies two full-resolution convolution layers: a convolution of size $3 \times 3 \times 64$ followed by a final layer of size $1 \times 1 \times 3$. The output is a tone-mapped RGB image suitable for 8-bit quantization without a significant loss of details in the relevant luminance ranges and local regions.

Computational efficiency is an important constraint for computer vision systems in autonomous driving and advanced driver assistance systems (ADAS) applications. Considering efficiency, we identified the middle "dilation" branch from the original ExpandNet architecture as the least contributing and omitted it in the simplified design shown in Figure 1 while still preserving the local and global branches. The number of trainable parameters was reduced by approximately 25%, and our experiments confirmed that the reduction did not cause a noticeable loss in learning capability.

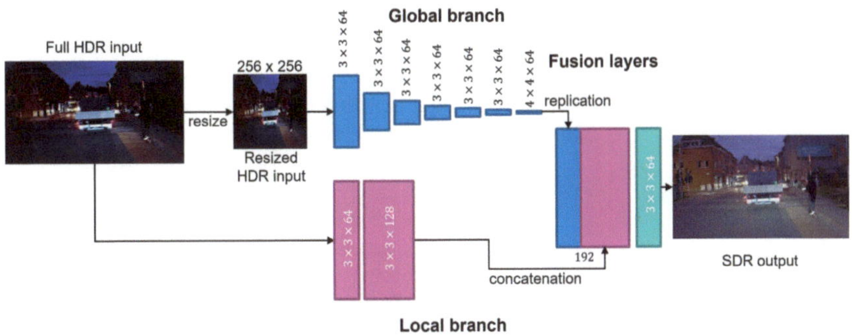

Figure 1. The proposed tone mapping architecture was inspired by ExpandNet [21], and it was simplified by discarding a branch of layers called the "dilation branch", which operates on the full resolution with a wide perceptive field and is therefore computationally complex. The network is comprised of convolutional layers followed by ReLU activations. The global branch spatially down-samples the feature maps in each subsequent layer through skip convolutions, and the local branch operates at the original resolution. The fusion layers combine the local and global features into an output tone-mapped image.

Consequently, the proposed tone mapping architecture has 340,227 trainable parameters, which is only 0.5% of the size of YOLO v3. In terms of calculations, our architecture performs 290 million floating-point operations (GFLOPs) for a 3-channel image of size 1.6 Mpixels, which is approximately 46% of the complexity of YOLO v3 for the same image size.

3.2. Training Cost Function

Our tone mapping CNN is trained to map the intensity range of the input HDR images by enforcing similarity between the output and a reference SDR image. The original similarity loss proposed in [21] is suitable in our framework as well since it ensures correct color and detail reconstruction. Following the definition in [21], the loss $\mathcal{L}(\mathbf{Z},\mathbf{Y})$ between the output image $\mathbf{Z}$ and the reference image $\mathbf{Y}$ is defined as a linear combination of the mean L_1-based loss and the mean cosine similarity-based loss L_{CS}, with a constant linear coefficient $\lambda = 5$:

$$\mathcal{L}(\mathbf{Z},\mathbf{Y}) = L_1(\mathbf{Z},\mathbf{Y}) + \lambda L_{CS}(\mathbf{Z},\mathbf{Y}), \tag{1}$$

where $L_1(\mathbf{Z},\mathbf{Y})$ is the mean absolute difference between the output image $\mathbf{Z}$ and the ground truth image $\mathbf{Y}$ averaged over all three RGB color channels and all N pixel locations. This loss was chosen due to its robustness to outliers and was found to result in sharper images compared to L_2-based loss.

The cosine similarity (CS) reflects a relative similarity between two colors and is invariant to the absolute intensity (the vector magnitude). It is calculated as the inner product of the vectors of RGB intensities at matching pixel locations in the two images, normalized by the product of their magnitudes (Equation (2)). The cosine similarity-based loss L_{CS} used in this work is inversely proportional to the cosine similarity metric:

$$L_{CS}(\mathbf{Z},\mathbf{Y}) = 1 - \frac{1}{N}\sum_{i=1}^{N} \frac{[Z_{i,R}, Z_{i,G}, Z_{i,B}] \cdot [Y_{i,R}, Y_{i,G}, Y_{i,B}]^T}{\|[[Z_{i,R}, Z_{i,G}, Z_{i,B}]\| \|[Y_{i,R}, Y_{i,G}, Y_{i,B}]\|}. \quad (2)$$

In the HDR-to-SDR mapping context, this loss is suitable for ensuring correct color reconstruction, which is less sensitive to luminance compression in the tone mapping process.

3.3. Data Pre-Processing and Augmentation

Currently, to the best of our knowledge, there is no extensive, publicly available annotated traffic dataset of high-dynamic-range images. Therefore, we relied on a public automotive SDR dataset [28], based on which we created synthetic HDR training images. The modifications that simulate different challenging situations applied as an augmentation to the training set were also applied for controlled, simulated experiments for validation. Furthermore, for evaluation in real conditions, we collected and annotated our own dataset with true HDR images [29].

After investigating the main contributing factors that negatively affect detection performance, we focused on two aspects: the robustness to noise and the image brightness in dark scenes. Modifying the intensity values in dark areas can increase the prominence of noise and negatively impact the quality and thus detection performance. Our proposed contribution to mitigating this problem is to focus on improving the contrast-to-noise ratio of the output images, as described below.

To create a training dataset, we applied different pre-processing and data augmentation steps to the original SDR images and obtained pairs of input HDR and ground truth SDR images. This process is illustrated in Figure 2 and summarized in the following steps:

1. A collection of SDR images $\mathbf{X}_1, \mathbf{X}_2, \ldots$ is used as a source for simulating input HDR images, as well as for a reference during training.
2. If a dark SDR training image is identified, its exposure is increased by simulating a higher sensor gain. In the case of a bright (daylight) image, the exposure remains unchanged. This becomes the target SDR ground truth image $\mathbf{Y}$.
3. The original SDR image is converted into an HDR image $\mathbf{H}_0$ using a dynamic range expansion operator [21]. Poisson noise is added with a random noise variance, through mosaicking/demosaicking ($\mathbf{H}_1$).
4. A contrast remapping is performed to simulate (augment) alternative and more challenging light conditions in terms of light intensity contrast ($\mathbf{H}_2$).
5. To make the model robust to various light sources and daylight conditions, a slight color temperature shift is applied to each training image randomly. This results in the final simulated HDR image ($\mathbf{H}$). The full HDR image is used as an input into the global branch of the network responsible for global illumination compression.
6. Lastly, to give higher significance to the reconstruction of objects of interest, a region is cropped from the HDR image which is sampled around the center of a known object location. This region ($\tilde{\mathbf{H}}$) is the input to the local branch of the network responsible for the local contrast mapping.

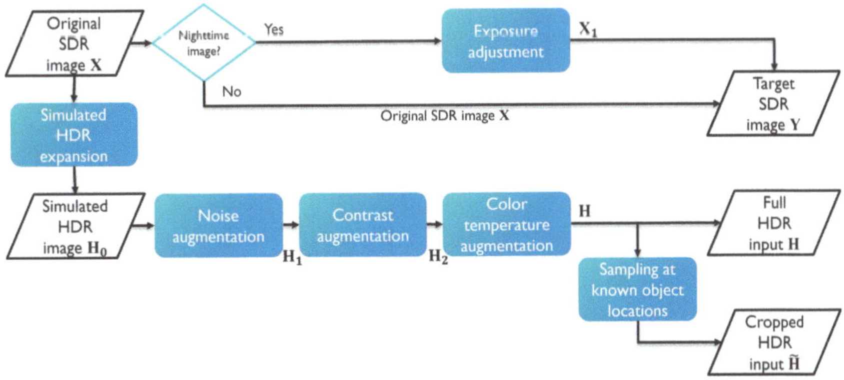

Figure 2. Flowchart of the process of creating synthetic HDR training data from SDR images. The blue blocks represent the data pre-processing steps that comprise the proposed detection-informed training procedure. It focuses on creating realistic and challenging training conditions. The training inputs that are created are an HDR input image and a crop from the same image centered at a known object location, and they are coupled with a training ground-truth (target) SDR image.

3.3.1. Datasets

The Berkeley Deep Drive (BDD) dataset [28] consists of diverse images of daylight, night-time, and dawn/dusk traffic scenes in urban scenarios. A collection of corresponding object annotations are available in the form of 2D bounding box coordinates and labels for multiple road user categories. The images in the dataset are captured using the automatic exposure settings of an SDR camera, and stored in an 8-bit representation.

For training we used 3657 images from the BDD dataset. The validation set consists of a different set of 550 images used to evaluate the model during training, and another 99 images were selected as a test set for objective evaluation of the algorithm.

Our collection of HDR test data [29] is part of a multi-modal dataset recorded in traffic conditions and annotated in a semi-automatic fashion with labels of the class "person", including pedestrians and cyclists. In this work, we used an actual high-dynamic-range camera (using the Sony IMX490 sensor) with a dynamic range of 120 dB. The image data were saved in a 24-bit format. We used approximately 370 diverse test images captured in daylight, twilight, and at night.

3.3.2. Enhancement of Dark Ground Truth Training Images

Our small-scale investigation revealed that in the majority of dynamic range conditions, object detectors are robust to a decrease in contrast. However, low contrast negatively impacts the detection confidence, and from a picture quality perspective in a driver assistance context, low contrast is also undesirable. Therefore, the proposed approach increases the brightness of nighttime ground truth images to train the network to enhance the visibility of objects. Specifically, a nighttime target SDR image $\mathbf{Y}$ at pixel location i is created by increasing the exposure of the original nighttime SDR image $\mathbf{X_1}$ by $s \in [0.5, 1.5]$ stops as:

$$Y_i = \left(X_{1,i}^{\beta} \cdot 2^s\right)^{1/\beta}, \quad (3)$$

where the exponential mapping with $\beta = 2.2$ serves as a replacement of a camera response curve, and the parameter s is determined by random sampling from a uniform distribution to increase the variability in the data and the robustness of the model.

3.3.3. Noise Augmentation

Since noise is inextricably linked to low-light conditions, we propose to incorporate a noise-augmented training procedure. Noise robustness is achieved by augmenting the training dataset with noisy images by adding different levels of signal-dependent Poisson noise in Bayer mosaicked (raw) image data. The parameters of the Poisson noise distribution are: $k = 1$, and λ is randomly sampled from a range between 0.2% of the pixel intensity $H_{0,i}$ (corresponding to low amount of noise) and 20% of the pixel intensity $H_{0,i}$ (high amount of noise). The images are then demosaicked using bilinear interpolation into full-resolution color images before providing them as input to the neural network. The mosaicking/demosaicking aspect is important because it simulates a real sensor and because this process affects the spatial correlation of the noise in neighboring pixels and therefore the local contrast. An example of the results of this procedure applied to an SDR image for visualization purposes is shown in Figure 3.

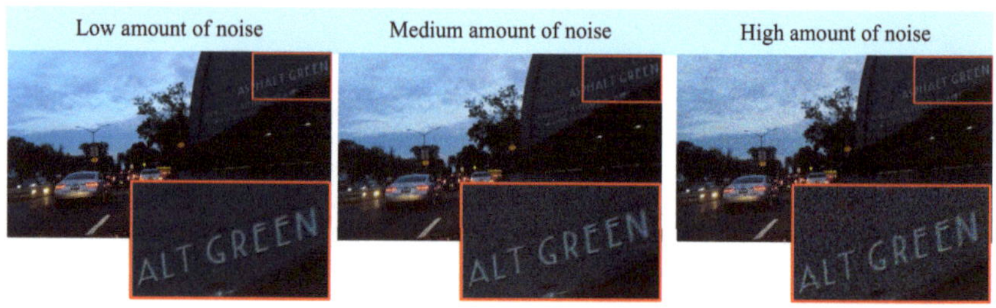

Figure 3. Example of simulating different amounts of Poisson noise to augment the training set and create a model robust to noise.

3.3.4. Contrast Augmentation Procedures

The dynamic range of real-world scenes can vary depending on light sources and atmospheric conditions. For example, a sunlit daytime urban scene is typically about five orders of magnitude brighter than an artificially illuminated street at night [20]. Similarly, the contrast between dark areas and bright lights in nighttime scenes can have a ratio of $10^4 : 1$.

One of the contributions of this paper is the data augmentation approach focusing on contrast robustness. By simulating realistic conditions, a pool of contrast mapping techniques is created, including gamma encoding, sigmoidal contrast stretching, and selective region-based non-linear mapping described further. At each iteration, for each training image, one contrast mapping function is randomly selected and applied to the simulated HDR output form ExpandNet [21].

Gamma encoding performs non-linear mapping of the image intensity values $H_{1,i}$ normalized in the range [0, 1]: $H_{2,i} = H_{1,i}^{\gamma}$. In our augmentation procedure, we applied gamma expansion with $\gamma \in [1, 3]$, selected by random sampling for each training image, to further expand the dynamic range and effectively increase the contrast by making the shadows darker. This procedure simulates variability in ambient light in the environment.

The sigmoidal contrast stretching maps the normalized luminance range into a sigmoidal shape as $H_{2,i} = \frac{H_{1,i}}{1+e^{-kH_{1,i}}}$, thus stretching the difference between the dark and the bright areas within the same bit-depth [30]. By varying $k \in [4, 12]$, which determines the steepness of the sigmoid, various dynamic range versions of the scene are created at each training iteration, simulating a sensor with more limited dynamic range capabilities. With steeper curves, the dark areas become darker and close to the lower sensitivity threshold, while the bright areas become closer to the saturation limit.

For simulating even more challenging situations in the scene, we applied a selective degradation procedure that selectively suppresses the luminance in the shadows, without significantly disturbing the bright areas. A selective mask $\mathbf{S}$ was used to identify the shadow regions in the image. Given an input HDR image $\mathbf{H_1}$ with intensity normalized in the range $[0, 1]$, the selective mask at pixel i is defined as $S_i = \frac{1}{1+e^{-20(H_{1,i}-t)}}$. The threshold t to select between shadows and bright areas is determined by uniform random sampling between 60% and 70% of the maximum possible luminance intensity $t \in [0.6, 0.7]$.

Finally, the simulation of challenging conditions is achieved as a linear combination as $H_{2,i} = (1 - S_i)cH_{1,i} + S_iH_{1,i}$, where $cH_{1,i}$ is a luminance-compressed (dark) representation of the image $\mathbf{H_1}$ at pixel i, using an experimentally chosen parameter $c = 0.004$. Such modifications correspond to challenging situations such as entering a tunnel or a camera blinded by strong direct light. Easy conditions are simulated by selecting parameters that do not increase the contrast significantly, and medium (mixed) conditions are obtained by blending an easy and a challenging version of a frame, using a random blending weight.

3.3.5. Color Temperature Change

The motivation to augment the dataset with a variable color temperature results from the observation that the ambient light throughout the day has a variable color temperature and that camera compensation for it is often inadequate or absent, especially in those challenging conditions. Taking the color temperature into consideration assists the process of learning to generalize to images of different color balances. In the augmentation process, we simulate color temperature variations from 2000 K to 10,000 K following [31]. Since state-of-the-art object detectors are highly robust to the color balance, this augmentation predominantly contributes to the visual quality of the image and only affects detection performance in corner cases which significantly distort the natural appearance of the colors.

3.4. General Training Details

As illustrated in Figure 4, the global branch summarizes the entire image into a single 64-element vector through strided convolutions. Due to the large dimensionality reduction in this branch, preservation of fine image details is not essential. Therefore, for efficiency, the image is resized to a fixed size of 256 × 256 pixels before being processed by the global branch.

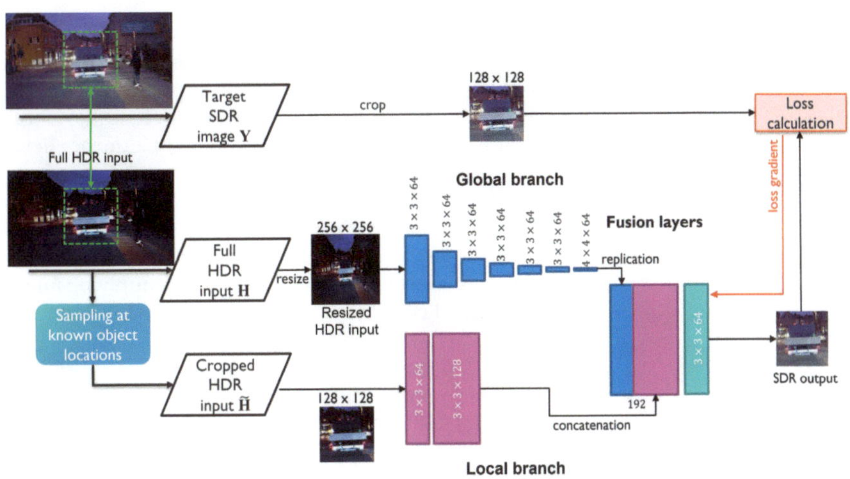

Figure 4. An illustration of the proposed training approach using crops at object-centered locations to focus on reconstruction of details. The size of the convolution kernels is indicated by the numbers at the corresponding feature maps of each layer.

The local branch looks at local image patches at each pixel location and requires full-resolution image information, without any loss of detail. To achieve a memory- and computationally-efficient method, this is implemented by cropping and processing rectangular regions taken from the full-resolution image. To that end, another contribution of this paper is the selection of regions for training that are fed into the local branch. Since the end goal is the application of the proposed method in object detection in traffic scenes, the quality of the output has the highest importance in regions with important road users.

With the aforementioned considerations, for training, we used crops of size 128 × 128 pixels extracted through weighted random sampling, focusing on areas where the presence of at least one object is known from the ground-truth labels.

The network was trained using the Adam [32] optimizer, with an initial learning rate of 10^{-3}, reduced to 10^{-4} during the final epochs, for approximately 3500 epochs, each with a random variation of the original training set. The batch size was gradually increased to 16. Training a lightweight neural network using the Adam optimizer combined with extensive data augmentation helps to mitigate a common training-time issue of vanishing gradients [33].

The proposed method was developed in the Python programming language using the PyTorch framework on a Linux operating system. The CNN training and inference processing was carried out and tested using an Nvidia GeForce RTX 2080 Ti graphics card with 11 GB of memory.

4. Experiments and Results

To evaluate the contributions of the proposed training procedure, we carried out an ablation study, where initially a baseline model was trained without the proposed novelties: the ground truth images were the original (non-enhanced) SDR images, no noise was added, and the region crops were sampled at uniformly distributed random positions as inputs to the local branch. The baseline model is used as a reference for comparison with our proposed training strategy that is informed by an empirical analysis of factors related to detection performance.

The quality of the tone mapping output was evaluated based on the object detection performance of the YOLO v3 object detector [34]. We used a YOLO model pre-trained on the COCO dataset [35], and considered only the traffic-related objects from its multi-class output.

For a comprehensive analysis, we evaluated the results in several aspects: robustness to challenging light conditions, robustness to noise, and overall detection performance by comparison with the state-of-the-art in tone mapping. After a rigorous comparison evaluation study, the algorithm by Farbman et al. [9] was selected as a representative of the state-of-the-art (SOTA), due to its strong performance in various conditions.

4.1. Evaluation with Simulated HDR Data

In the first set of experiments, the goal is to evaluate the contribution of the proposed novelties towards improving the detection quality. The comparison with the baseline model is to assess relative improvement by applying informed training and comparing it to the selected classical state-of-the-art method illustrates the absolute performance of the proposed method.

The first set of experiments focuses on evaluating the robustness of the proposed method to dynamic range (DR) conditions in the scene. To this end, three different versions of the test set were created: non-challenging, medium, and challenging set. More specifically, from the available SDR images in the Berkeley dataset [28], HDR images were simulated by dynamic range expansion [21] and applying the selective degradation procedure described in Section 3.3.4 with increasing degradation levels.

The results in Table 1 demonstrate that the proposed tone mapping method DI-TM significantly outperforms that of Farbman et al. [9] in the challenging cases. We reason that the strength of the proposed network lies in the training on a versatile set of realistic conditions specific to traffic situations, such as blinding headlights or street lights at night and direct sunlight also reflecting from the road. The two methods perform similarly well on the medium and the non-challenging sets, showing that our model is robust to a range of different illumination levels while mitigating the need for initial parameter adaptation. In this experiment, the comparison with our baseline model verifies that the proposed contrast augmentation steps are effective for training a robust model.

Table 1. Object detection performance (average F2 score for "person", "car", and "traffic light") of YOLO v3 [34] applied to the outputs of the classical-state-of-the-art method of Farbman et al. [9] on our baseline model and the outputs from the proposed detection-informed method DI-TM in scenes of different dynamic range. The bold numbers indicate the best performance among the listed methods in each category.

	Average F2 Score per Scene Type		
Method	**Non-Challenging**	**Medium**	**Challenging**
Farbman et. al [9]	0.48	0.48	0.32
Baseline model	0.49	0.46	0.37
Proposed DI-TM	**0.50**	**0.49**	**0.44**

As an illustration, Figure 5 provides an example of the different levels of simulated dynamic range conditions and shows the output of the comparison with the method of Farbman et al. [9]. The proposed DI-TM method produces more consistent tone mapping results across a range of challenging conditions thanks to the robust training strategy. Furthermore, in the output of the proposed method, more objects are correctly detected in challenging conditions due to better detail preservation and contrast adaptation.

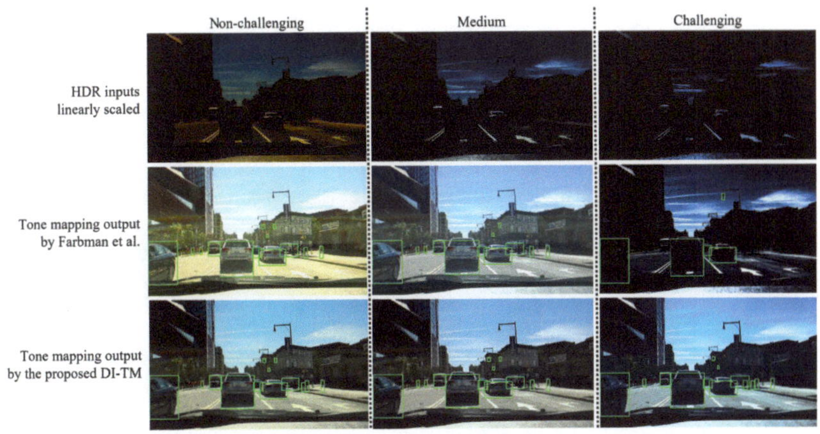

Figure 5. Example of the performance of the reference SOTA tone mapping method Farbman et al. [9] and the proposed DI-TM in variable scene dynamic range conditions. The green bounding boxes represent correct detection outputs (true positives) for "person", "car", and "traffic light" combined. The proposed method DI-TM is more robust in variable and extreme contrast conditions.

From a qualitative perspective, we evaluated the proposed method in terms of its robustness to noise. In this experiment, three variants of the test set were obtained by increasing the Poisson noise strength (see Figure 3) that is added to the HDR test set. The results in Table 2 indicate that, due to the denoising properties of the proposed method, object detection is less affected by noise in all three categories compared to the method of Farbman et al. [9]. Furthermore, compared to the baseline model, the noise augmentation approach proves to be beneficial for detection since it helps the network learn to generate less noisy images.

Table 2. Object detection performance of YOLO v3 [34] on the outputs of the classical-state-of-the-art method of Farbman et al. [9], our baseline model and the proposed detection-informed method DI-TM in variable degrees of Poisson noise. The bold numbers indicate the best performance among the listed methods in each category.

	Average F2 Score per Noise Level		
Method	**Low**	**Medium**	**High**
Farbman et. al [9]	0.43	0.35	0.22
Baseline model	0.43	0.36	0.26
Proposed DI-TM	**0.45**	**0.40**	**0.34**

4.2. Evaluation with True HDR Data

To assess the performance in real-world conditions, we relied on a diverse subset of 372 annotated frames of our own captured dataset using a pair of SDR and HDR cameras, aligned and synchronized to match their viewpoints. In total, this subset contains 1398 "person" objects. Comparing to SDR helps to evaluate the benefits of using an HDR sensor in the automotive context. For objective evaluation of the proposed method, we relied on the comparison with the method of Farbman et al. [9].

The first experiment focused on the difficulty of the light conditions in terms of dynamic range, similar to the experiment in Section 4.1, where high-contrast scenes are considered more challenging for visibility than low-contrast scenes. To keep the original pixel intensity distribution unchanged, the distinction between easy and challenging subsets was achieved by manually identifying images of low and high contrast. Due to its higher representation in the dataset as well as the object detection reliability, this experiment evaluated only the category "person".

In Table 3, the detection performance of the object detector for the class "person" is presented for easy and challenging dynamic range conditions. The results confirm that using HDR data is beneficial, especially in challenging conditions. Furthermore, the results show that adapting the tone mapping method specifically for object detection in the automotive context brings an additional contribution to the detection quality. We believe that this is a result of the extensive data augmentation process combining variable contrast, illumination, and noise simulation. Such conditions are more scarce in typical driving conditions; however, they are critical for the safety and reliability of intelligent vehicles.

Table 3. Object detection performance of YOLO v3 [34] on matching SDR frames, on the outputs of Farbman et al. [9], and on the proposed detection-informed method DI-TM in selected subsets of easy (lower dynamic range) and hard (high dynamic range) examples from our own collected dataset for the category "person". The bold numbers indicate the best performance among the listed methods in each category.

	F2 Score per Scene Type	
Method	**Easy**	**Challenging**
SDR data	0.68	0.41
Farbman et al. [9]	0.73	0.54
Proposed DI-TM	**0.75**	**0.61**

An example of a challenging night-time scene is presented in Figure 6. The cropped regions were further manually enhanced only for visualization purposes. It can be noticed that due to the high contrast, the pedestrians in the darkness were missed in the SDR image, while they were detected in the tone-mapped output using the proposed DI-TM. Due to the highly variable augmentation, our method is robust to changing scene conditions without any manual interventions, unlike the method of Farbman et al. [9]

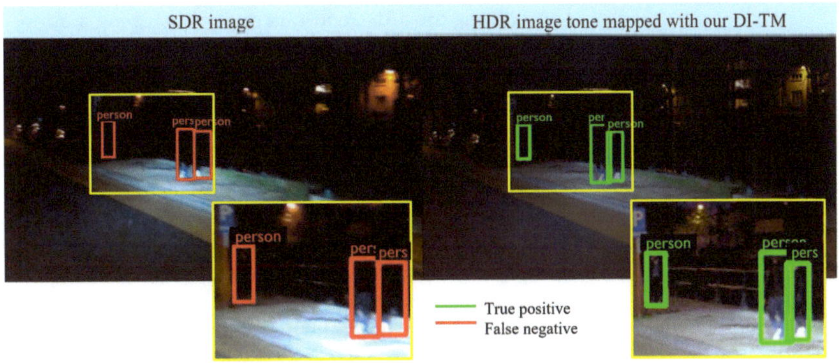

Figure 6. An example of a challenging scene in our dataset of SDR and true HDR images. In an SDR representation, much of the contrast at the object edges is lost, and the objects in the darkness are invisible to the detector. HDR images preserve fine intensity differences, and our tone mapping method enhances the details such that they become visible for the detector, as well as for visualization to a human driver.

Finally, as a qualitative evaluation, we present an experiment focusing on extreme dynamic range conditions. In this evaluation, the model was applied to samples from our recorded true HDR dataset. Examples of this evaluation are presented in Figure 7. The proposed method can suppress the noise, and it is more robust to variable and highly challenging conditions compared to the representative of the classical state-of-the-art.

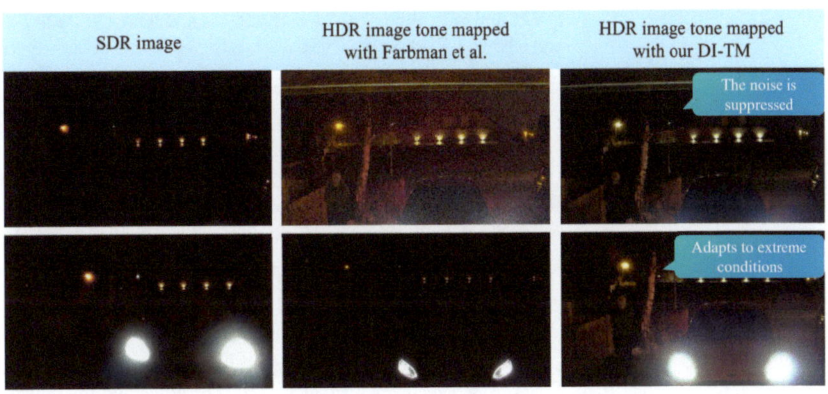

Figure 7. Qualitative evaluation of the robustness of Farbman et al. [9] vs. proposed model DI-TM1 in extremely challenging high-contrast night-time scenes

Both the qualitative and the quantitative evaluations confirm the effectiveness of the image pre-processing strategies involved in data augmentation. The ideas were motivated by realistic driving conditions and observations following an investigation into the performance of object detectors in various high-contrast conditions. On the other hand, the method of Farbman et al. [9] is among the state-of-the-art in classical tone mapping algorithms; however, it is optimized for aesthetic aspects of the image quality in photography and cinematic applications.

5. Conclusions

In this study, our focus lies in the design of an HDR tone mapping model aimed at enhancing image quality while considering the objectives of object detection. We believe that this approach holds significant potential for facilitating the integration of HDR cameras into ADAS and autonomous vehicle perception systems. We investigated different factors contributing towards good-quality tone mapping, informed by the performance of objec detectors on the tone mapping outputs.

We proposed a detection-informed strategy for tone mapping, referred to as DI-TM and evaluated its performance in multiple aspects, including image quality and detection accuracy. Our findings demonstrate that the proposed approach outperforms classical state-of-the-art tone mappers in easy and challenging lighting conditions while maintaining robustness across different noise levels. Compared to using an SDR sensor, the detection performance improvement in challenging light conditions was 49% as measured by the F_2 score, which highlights the advantage of integrating HDR cameras in intelligent vehicles. Compared to a state-of-the-art tone mapping method optimized for visual quality, with our detection-optimized approach, we achieved an improvement of 13% in detection accuracy.

These results suggest the promising potential for informed, task-based algorithm design improvements in the already very researched field of HDR tone mapping, with future applications in enriching the vision systems of autonomous driving platforms.

Future research will focus on exploring the robustness of the proposed training approach in multiple-vehicle perception tasks, examining its effectiveness across various scenarios. Additionally, there is a need to explore the tradeoff between computational efficiency and performance, especially in the context of adapting existing computer vision tools or developing novel techniques tailored to handle high-dynamic-range (HDR) data. This investigation will contribute to the optimization and advancement of more accurate environment perception systems by application of HDR data in the automotive domain.

Author Contributions: Conceptualization, J.A., D.V.H. and W.P.; Data curation, I.S., A.S. and D.V.H.; Formal analysis, I.S., A.S. and J.A.; Funding acquisition, W.P.; Investigation, I.S. and A.S.; Methodology, I.S., A.S. and J.A.; Project administration, J.A., D.V.H. and W.P.; Resources, W.P.; Software, I.S. and A.S.; Supervision, J.A. D.V.H. and W.P.; Validation, I.S. and A.S.; Visualization, I.S.; Writing—original draft, I.S.; Writing—review and editing, J.A. All authors have read and agreed to the published version of the manuscript.

Funding: This research received no external funding.

Institutional Review Board Statement: Not applicable.

Informed Consent Statement: Not applicable.

Data Availability Statement: The data presented in this study are available on request from the corresponding author. The data are not publicly available due to privacy restrictions.

Acknowledgments: This work was carried out in collaboration with HiSilicon.

Conflicts of Interest: The authors declare no conflict of interest.

References

1. Macek, K. *Pedestrian Traffic Fatalities by State: 2021 Preliminary Data*; Technical Report; Governors Highway Safety Association (GHSA): Washington, DC, USA, 2022.
2. NHTSA's National Center for Statistics and Analysis. *Pedestrians: 2017 Data. Traffic Safety Facts Report No. DOT HS 812 681*; U.S. Department of Transportation: Washington, DC, USA, 2019.
3. Teoh, E.R.; Kidd, D.G. Rage against the machine? Google's self-driving cars versus human drivers. *J. Saf. Res.* **2017**, *63*, 57–60. [CrossRef] [PubMed]
4. Kalra, N.; Paddock, S.M. Driving to safety: How many miles of driving would it take to demonstrate autonomous vehicle reliability? *Transp. Res. Part A Policy Pract.* **2016**, *94*, 182–193. [CrossRef]
5. Di, X.; Shi, R. A survey on autonomous vehicle control in the era of mixed-autonomy: From physics-based to AI-guided driving policy learning. *Transp. Res. Part C Emerg. Technol.* **2021**, *125*, 103008. [CrossRef]
6. Commission, E. New Rules to Improve Road Safety and Enable Fully Driverless Vehicles in the EU. 2022. Available online: https://ec.europa.eu/commission/presscorner/detail/en/IP_22_4312 (accessed on 17 April 2023).
7. Reinhard, E.; Stark, M.; Shirley, P.; Ferwerda, J. Photographic tone reproduction for digital images. In Proceedings of the 29th Annual Conference on Computer Graphics and Interactive Techniques, San Antonio, TX, USA, 23–26 July 2002; pp. 267–276.
8. Kuang, J.; Johnson, G.M.; Fairchild, M.D. iCAM06: A refined image appearance model for HDR image rendering. *J. Vis. Commun. Image Represent.* **2007**, *18*, 406–414. [CrossRef]
9. Farbman, Z.; Fattal, R.; Lischinski, D.; Szeliski, R. Edge-preserving decompositions for multi-scale tone and detail manipulation. *ACM Trans. Graph. (TOG)* **2008**, *27*, 1–10. [CrossRef]

10. Li, Y.; Liao, N.; Wu, W.; Deng, C.; Li, Y.; Fan, Q.; Liu, C. Tone Mapping Operator for High Dynamic Range Images Based on Modified iCAM06. *Sensors* **2023**, *23*, 2516. [CrossRef] [PubMed]
11. Goswami, A.; Ak, A.; Hauser, W.; Le Callet, P.; Dufaux, F. Reliability of Crowdsourcing for Subjective Quality Evaluation of Tone Mapping Operators. In Proceedings of the 2021 IEEE 23rd International Workshop on Multimedia Signal Processing (MMSP), Tampere, Finland, 6–8 October 2021; pp. 1–6.
12. Cerdá-Company, X.; Párraga, C.A.; Otazu, X. Which tone-mapping operator is the best? A comparative study of perceptual quality. *J. Opt. Soc. Am. A* **2018**, *35*, 626–638. [CrossRef] [PubMed]
13. Su, C.C.; Wang, R.; Lin, H.J.; Liu, Y.L.; Chen, C.P.; Chang, Y.L.; Pei, S.C. Explorable tone mapping operators. In Proceedings of the 2020 25th International Conference on Pattern Recognition (ICPR), Milan, Italy, 10–15 January 2021; pp. 10320–10326.
14. Rana, A.; Singh, P.; Valenzise, G.; Dufaux, F.; Komodakis, N.; Smolic, A. Deep tone mapping operator for high dynamic range images. *IEEE Trans. Image Process.* **2019**, *29*, 1285–1298. [CrossRef] [PubMed]
15. Panetta, K.; Kezebou, L.; Oludare, V.; Agaian, S.; Xia, Z. Tmo-net: A parameter-free tone mapping operator using generative adversarial network, and performance benchmarking on large scale hdr dataset. *IEEE Access* **2021**, *9*, 39500–39517. [CrossRef]
16. Wang, C.; Chen, B.; Seidel, H.P.; Myszkowski, K.; Serrano, A. Learning a self-supervised tone mapping operator via feature contrast masking loss. In *Proceedings of the Computer Graphics Forum*; Wiley Online Library: Hoboken, NJ, USA, 2022; Volume 41, pp. 71–84.
17. Zhang, J.; Wang, Y.; Tohidypour, H.; Pourazad, M.T.; Nasiopoulos, P. A Generative Adversarial Network Based Tone Mapping Operator for 4K HDR Images. In Proceedings of the 2023 International Conference on Computing, Networking and Communications (ICNC), Honolulu, HI, USA, 20–22 February 2023; pp. 473–477.
18. Mukherjee, R.; Melo, M.; Filipe, V.; Chalmers, A.; Bessa, M. Backward compatible object detection using hdr image content. *IEEE Access* **2020**, *8*, 142736–142746. [CrossRef]
19. Mukherjee, R.; Bessa, M.; Melo-Pinto, P.; Chalmers, A. Object detection under challenging lighting conditions using high dynamic range imagery. *IEEE Access* **2021**, *9*, 77771–77783. [CrossRef]
20. Onzon, E.; Mannan, F.; Heide, F. Neural auto-exposure for high-dynamic range object detection. In Proceedings of the IEEE/CVF Conference on Computer Vision and Pattern Recognition, Nashville, TN, USA, 20–25 June 2021; pp. 7710–7720.
21. Marnerides, D.; Bashford-Rogers, T.; Hatchett, J.; Debattista, K. Expandnet: A deep convolutional neural network for high dynamic range expansion from low dynamic range content. In *Proceedings of the Computer Graphics Forum*; Wiley Online Library: Hoboken, NJ, USA, 2018; Volume 37, pp. 37–49.
22. Koçdemir, İ.H.; Koz, A.; Akyuz, A.O.; Chalmers, A.; Alatan, A.; Kalkan, S. Tmo-Det: Deep Tone-Mapping Optimized with and for Object Detection. 2022. 4132028. Available online: https://papers.ssrn.com/sol3/papers.cfm?abstract_id=4132028 (accessed on 18 June 2023).
23. Yeganeh, H.; Wang, Z. Objective quality assessment of tone-mapped images. *IEEE Trans. Image Process.* **2012**, *22*, 657–667. [CrossRef] [PubMed]
24. Reinhard, E.; Devlin, K. Dynamic range reduction inspired by photoreceptor physiology. *IEEE Trans. Vis. Comput. Graph.* **2005**, *11*, 13–24. [CrossRef] [PubMed]
25. Mantiuk, R.; Myszkowski, K.; Seidel, H.P. A perceptual framework for contrast processing of high dynamic range images. *ACM Trans. Appl. Percept. (TAP)* **2006**, *3*, 286–308. [CrossRef]
26. Drago, F.; Myszkowski, K.; Annen, T.; Chiba, N. Adaptive logarithmic mapping for displaying high contrast scenes. In *Proceedings of the Computer Graphics Forum*; Wiley Online Library: Hoboken, NJ, USA, 2003; Volume 22, pp. 419–426.
27. Durand, F.; Dorsey, J. Fast bilateral filtering for the display of high-dynamic-range images. In Proceedings of the 29th Annual Conference on Computer Graphics and Interactive Techniques, San Antonio, TX, USA, 23–26 July 2002; pp. 257–266.
28. Yu, F.; Chen, H.; Wang, X.; Xian, W.; Chen, Y.; Liu, F.; Madhavan, V.; Darrell, T. Bdd100k: A diverse driving dataset for heterogeneous multitask learning. In Proceedings of the IEEE/CVF Conference on Computer Vision and Pattern Recognition, Seattle, WA, USA, 13–19 June 2020; pp. 2636–2645.
29. Dimitrievski, M.; Shopovska, I.; Van Hamme, D.; Veelaert, P.; Philips, W. Automatic labeling of vulnerable road users in multi-sensor data. In Proceedings of the 2021 IEEE International Intelligent Transportation Systems Conference (ITSC), Indianapolis, IN, USA, 19–22 September 2021; pp. 2623–2630.
30. Saruchi, S. Adaptive sigmoid function to enhance low contrast images. *Int. J. Comput. Appl.* **2012**, *55*, 45–49. [CrossRef]
31. Helland, T. How to Convert Temperature (K) to RGB: Algorithm and Sample Code. 2012. Available online: https://tannerhelland.com/2012/09/18/convert-temperature-rgb-algorithm-code.html (accessed on 18 June 2023).
32. Kingma, D.P.; Ba, J. Adam: A method for stochastic optimization. *arXiv* **2014**, arXiv:1412.6980.
33. Abuqaddom, I.; Mahafzah, B.A.; Faris, H. Oriented stochastic loss descent algorithm to train very deep multi-layer neural networks without vanishing gradients. *Knowl.-Based Syst.* **2021**, *230*, 107391. [CrossRef]
34. Redmon, J.; Farhadi, A. Yolov3: An incremental improvement. *arXiv* **2018**, arXiv:1804.02767.
35. Lin, T.Y.; Maire, M.; Belongie, S.; Hays, J.; Perona, P.; Ramanan, D.; Dollár, P.; Zitnick, C.L. Microsoft coco: Common objects in context. In Proceedings of the 13th European Conference on Computer Vision, Zurich, Switzerland, 6–12 September 2014; pp. 740–755.

sensors

MDPI

Article

Control Architecture for Connected Vehicle Platoons: From Sensor Data to Controller Design Using Vehicle-to-Everything Communication

Razvan-Gabriel Lazar, Ovidiu Pauca, Anca Maxim and Constantin-Florin Caruntu *

Department of Automatic Control and Applied Informatics, "Gheorghe Asachi" Technical University of Iasi, 700050 Iasi, Romania; lazar.razvan@ac.tuiasi.ro (R.-G.L.); pauca.ovidiu@ac.tuiasi.ro (O.P.); anca.maxim@ac.tuiasi.ro (A.M.)
* Correspondence: caruntuc@ac.tuiasi.ro

Abstract: A suitable control architecture for connected vehicle platoons may be seen as a promising solution for today's traffic problems, by improving road safety and traffic flow, reducing emissions and fuel consumption, and increasing driver comfort. This paper provides a comprehensive overview concerning the defining levels of a general control architecture for connected vehicle platoons, intending to illustrate the options available in terms of sensor technologies, in-vehicle networks, vehicular communication, and control solutions. Moreover, starting from the proposed control architecture, a solution that implements a Cooperative Adaptive Cruise Control (CACC) functionality for a vehicle platoon is designed. Also, two control algorithms based on the distributed model-based predictive control (DMPC) strategy and the feedback gain matrix method for the control level of the CACC functionality are proposed. The designed architecture was tested in a simulation scenario, and the obtained results show the control performances achieved using the proposed solutions suitable for the longitudinal dynamics of vehicle platoons.

Keywords: control architecture; connected vehicle platoons; V2X communication; CACC; DMPC

Citation: Lazar, R.-G.; Pauca, O.; Maxim, A.; Caruntu, C.-F. Control Architecture for Connected Vehicle Platoons: From Sensor Data to Controller Design Using Vehicle-to-Everything Communication. *Sensors* **2023**, *23*, 7576. https://doi.org/10.3390/s23177576

Academic Editor: Enrico Meli

Received: 13 July 2023
Revised: 24 August 2023
Accepted: 29 August 2023
Published: 31 August 2023

1. Introduction

Nowadays, with the ever-increasing number of vehicles on the highways, there is a stringent need to improve the driving experience quality through autonomous driving, by making use of the available vehicle connectivity. There is much interest in the research community to explore this topic, a testimony given by the following highly cited review works. In [1], a survey on the control of connected and automated vehicles (CAVs), with emphasis on control solutions for improving the energy efficiency of different powertrain architectures is given. In [2], a comprehensive survey on urban traffic signal control for CAVs, with both deterministic and stochastic approaches, is provided.

A typical control framework for connected vehicles (CVs) is given by the connected cruise control (CCC) architecture, which is suitable for a vehicle group consisting of both autonomous and human-driven vehicles, connected with a vehicle-to-vehicle (V2V) communication network [3]. In [4], a fuzzy support vector machine (SVM) method for CCC, using radar and V2V communication to detect the lane change of a side vehicle, is proposed. In [5], a deep reinforcing learning (DRL) solution to solve a CCC problem with communication delays and dynamic traffic changes is given. In [6], a nonlinear range policy for a CCC application with merging capabilities is provided.

Vehicle platoons are control applications for groups of CVs, which are designed using the advantages of vehicle-to-infrastructure (V2I) and V2V connectivity. To perform a desired common task for the entire platoon, each vehicle needs to exchange the relevant local measured data with the other participants. Usually, a vehicle platoon task is to travel with a velocity imposed by the leader vehicle of the platoon, while maintaining a desired safe distance between the follower vehicles [7]. One of the most promising functionalities for CAVs is cooperative adaptive cruise control (CACC). The most defining things for CACC are the use of sensors and communication technologies. As an extension of adaptive cruise control (ACC), in CACC systems, CAVs use V2V communications to exchange information with other CAVs in an autonomous manner, as well as V2I to provide information on traffic conditions and traffic management [8].

There are several design frameworks that have recently been researched for the platoon formation of multiple CVs, such as non-cooperative differential games [9], estimation of the communication delays via an adaptive switched predictor [10], model-based predictive control (MPC) with switching communication topology [11], robust feedback control [12], or a cooperative adaptive sliding mode [13], among others. In [14], an LMI-based optimisation problem for a cooperative optimal control method for a CAV platoon is given. In [15], a consensus-based impulse control method, which simplifies the communication exchange in a platoon of CVs, is proposed. In [16], security for the distributed platooning control of CAVs, subject to denial-of-service (DoS) attacks, is discussed. In [17], an ACC strategy for CAVs in a platoon formation subject to cyber-attacks and communication delays is given.

However, in the state-of-the-art literature, there are few studies that detail the design of a complete control architecture for a connected vehicle group. Thus, this work presents a detailed survey regarding the components required by a control architecture for a vehicle group, by describing the multitude of options available in terms of sensors, control, and communication requirements. Moreover, the proposed study of the control architecture can be used as a tutorial in designing a control solution for an automated vehicle group.

The main contributions of this paper are the following:

- A detailed survey was carried out on the solutions available in the literature of general control architectures for the CAV platoon, from the point of view of the constructive levels, the elements that define each level, and the links between them;
  Moreover, in relation to other works on this topic [1,18,19], this paper presents a more detailed description of the sensors and V2V communication standards essential for the CAV concept;
- Starting from the proposed architecture, a suitable control architecture is defined for a platoon of connected vehicles based on the CACC strategy, by presenting the necessary sensors, the suitable types of communication means, and the control solutions;
- Finally, two control methods are proposed for the longitudinal dynamics suitable for a CAV platoon framework. Thus, a state-space distributed model-based predictive control (DMPC) method suitable for vehicle platooning is described and tested. Moreover, a second control method is proposed, designed using off-line optimisation to compute a feedback gain control matrix. Both methods are compared using a CAV platoon application.

The remainder of this paper is structured as follows. Section 2 deals with a general presentation of architecture design for connected vehicle platoons, by describing the main subsystems, their purpose, and component elements. In Section 3, different aspects of the vehicle communication systems are presented, synthesised under intra-vehicle communication and inter-vehicular communications. Section 4 presents the proposed control architecture for a specific case involving a CACC strategy and model used to describe the longitudinal dynamics of a vehicle platoon. Section 5 illustrates the simulation results obtained using the proposed control solutions, and Section 6 includes a thorough analysis based on their resulting performances. In the last section, one can find the conclusions and future research directions.

2. Architecture Design for Connected Vehicle Platoons

The control architecture design defines the necessary stages in the transformation of an ordinary car into a connected one by adding additional components, including different sensors that allow the vehicle to detect the environment and communicate with other traffic participants and with the intelligent infrastructure, as well as adequate control strategies for controlling the vehicle's mobility.

A general control architecture is presented in Figure 1, where the communication process between the most important subsystems of the architecture is illustrated. At a high level, the architecture has five subsystems, responsible for defining the autonomy process of a connected vehicle, such as sensing interface, perception, planning, decision, and control. The proposed architecture is used by each vehicle regardless of the position within the platoon. Depending on the type of the vehicle (leader or follower), the general control architecture presented can be used differently, from the point of view of the component elements and the functions performed by each of its subsystems.

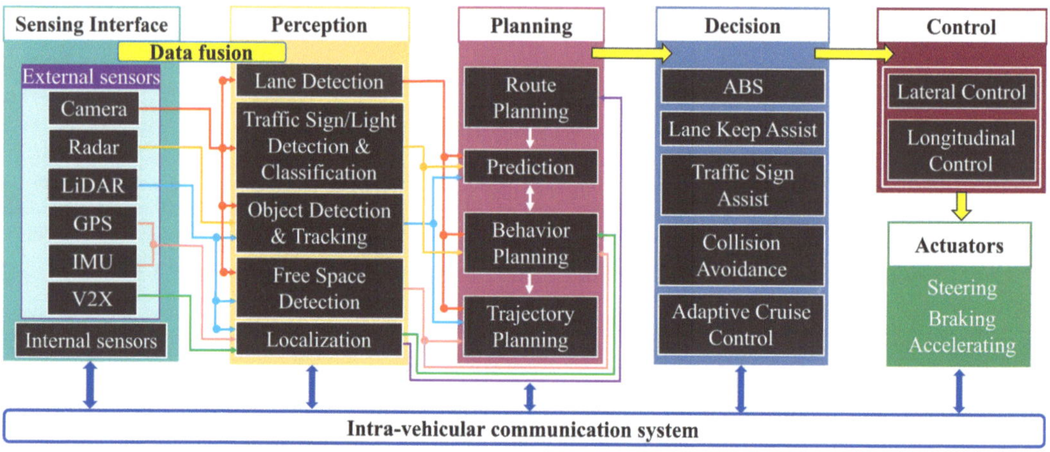

Figure 1. Example of a general control architecture.

According to the work in [20], these subsystems, which are described in detail in the following subsections, have the purpose of (1) retrieving information from the real environment around the vehicle (i.e., the sensing interface), (2) fusing the data with the purpose of detection and localisation (i.e., perception), (3) choosing the route (i.e., planning), (4) predicting the behaviour of other traffic participants and planning the optimal trajectory (i.e., decision), and based on the taken decision, (5) controlling the vehicle by operating the various responsible actuators (i.e., control). In order to exchange the necessary information between them, these subsystems are connected with a car-level communication system, more precisely, the intra-vehicular communication network. For the exchange of information between several connected vehicles, each with its own control architecture, an inter-vehicular communication network is used, synthesised under vehicle-to-everything (V2X) [21].

2.1. Sensing Interface

This subsystem shows how the information is captured from the vehicle's environment, such as the detection of its position in relation to the surroundings, but also information about the other traffic participants. The sensing interface consists of different sensors for data collection. These sensors can be classified into two categories: (1) internal sensors that provide information only about the state of the vehicle and (2) external sensors used to capture data from the outside of the vehicle. All of them are better exemplified in Figure 2, where one can see the sensors' position, the coverage area, and the performed functions.

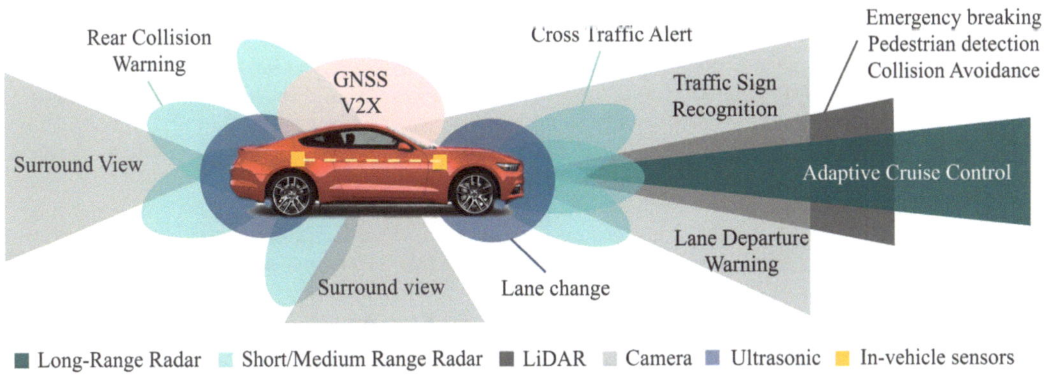

Figure 2. Typical types of sensors and their functionalities.

2.1.1. Literature Review

In the following, the most important external sensors are presented in detail:

(1) The camera is one of the key sensors of a vehicle, used for perceiving the visual environment, lane and traffic sign recognition, object tracking, and much more [22]. Cameras may be mono, stereo, or full surround, and placed in areas such as dashboards and windshields. These types of cameras used for autonomous vehicles are described in detail in [7]. Depending on the quality of the lens, the maximum working distance of the camera is around 250 m. The main advantages of a camera are related to the accuracy of the colour distribution, the contour of the surroundings, and the texture [23]. As a disadvantage, cameras are sensitive to low-intensity lights and can be affected by weather conditions [24].

(2) Radar is the most common sensor used in vehicles to identify and locate objects in the presence of various interferences, such as noise, clutter, and jamming [25]. To measure the distance, the time Of flight (TOF) method [26] is used, whereas to measure the relative velocity, the Doppler shift [23] is used. Thus, radars lend themselves very well for obstacle detection [27] and pedestrian and vehicle recognition [28,29]. Also, some functionalities of the radars are blind spot detection, rear collision warning, emergency braking, and cross-traffic alert. Radar sensors operate in the millimetre-wave (mm-Wave) spectrum, using different frequency bands, such as 24, 60, 77, and 79 GHz, being able to measure a range from 5 to 200 m [30]. Depending on the type of radar and the application for which it is used, radar sensors are divided into short, medium, or long-range ones. The most important characteristics of the types of radars used in the automotive field are presented in [25]. Radar sensors also offer the benefits of high availability and low cost [31]. Moreover, compared to cameras, they are less affected by the weather and the low lighting environment [26]. Disadvantages include lack of precision, receded field of view (FOV), and the production of false positives by rejecting emitted signals [32].

(3) A light detection and ranging (LiDAR) sensor is a technology used to determine precise information about the distance and size of objects [33]. It uses a remote sensing technique, producing pulses of infrared or laser light and measuring the time it takes for the pulses to be reflected [34], a principle known as TOF, and it is similar to how the radar sensor operates. The range of LiDAR is about 200 m on average [35], using 905 nm and 1550 nm spectra [36]. There are different types of LiDAR sensors, these being 2D, 3D, and solid-state [37]. The general specifications for each type of LiDAR sensor are presented in [34]. Compared to the camera, the LiDAR sensor has better detection capabilities in terms of range, with bad weather and low lighting affecting this sensor less than the camera [26]. Compared to radar, it has a higher accuracy and precision, but also a superior 3D perception competence.

As a disadvantage, the LiDAR sensor is affected by severe weather conditions, such as snow, fog, and rain [38]. Moreover, in terms of cost and low availability, the LiDAR is less competitive than the other two types of sensors.

(4) An ultrasonic sensor is the most diligent and cheap sensor, used for short-range obstacle detection, proximity sensing for lane change, and parking functions [39] These sensors use ultrasonic waves to measure the distance to objects by calculating the TOF of the emitted wave. These sensors operate in the 20–40 KHz range, with a detection range of generally less than 11 m [40], and are used at low speeds. Also, these sensors are easier to implement, and work satisfactorily in bad weather conditions and in dusty environments [41]. The main disadvantages are the disturbances in the sound waves and the tendency to produce false positives in the measurements, as well as the need to use multiple sensors to obtain a complete view; thus, mutual interference is produced between them [42].

(5) Global Positioning System (GPS) and inertial measurement unit (IMU) technologies are used for navigation and localisation purposes, by determining the exact position of the vehicle and helping it to navigate. The GPS is a system used to obtain information about geolocation, speed, and time, each vehicle containing a GPS receiver that connects to GPS satellites [43]. The position of the vehicle is given by the GPS coordinates, but the accuracy with which these are extracted depends on several factors. Therefore, position errors can be obtained with an average value of 3 m, and with a deviation of 1 m [44], and can reach up to 20 m depending on the environment. In urban environments, the GPS performances are lower [45]. It presents advantages in terms of cost and the way of managing the accumulation of errors over time. The disadvantages would be related to precision, which is reduced to one metre for current vehicles, but also the inability to operate in environments where the view of the sky is obstructed, such as tunnels [26]. An IMU is an electronic device that measures and reports the body's specific force, angular rate, and sometimes the magnetic field surrounding the body, using a combination of accelerometers and gyroscopes, sometimes also magnetometers [46]. Therefore, with the help of these data, the linear velocity and angular positions for the vehicle can be calculated. The IMU sensor can be combined with the GPS, as a complementary sensor, because the IMU can not give the position error by itself, but also for the performance qualities of the IMU sensor in tunnels [47]. Moreover, to improve the estimation of the vehicle's position, different techniques are used in which GPS and IMU data are fused [48].

2.1.2. Summary

Table 1 illustrates the comparison between external sensors from the point of view of the most relevant metrics. Thus, it can be concluded which of them lends itself best depending on the functionality chosen for CAVs from the platoon.

Due to the fact that the leader vehicle is in front of the platoon, it imposes the travel velocity and direction for all members, and has also the role of detecting the lanes and the various obstacles on the road. To fulfil the leader's tasks, it can be equipped with the following sensors: (i) camera (for lane, traffic signs/lights, and obstacle size detections), and (ii) radar and LiDAR (for obstacle detection at large distances). In other words, because the follower vehicles require high accuracy for distance measurement, each follower can be equipped with radar, and with a camera for lane detection. The GPS and IMU sensors should be used by each vehicle from the platoon to measure their velocity and obtain their position.

Table 1. Comparison of different sensing technologies.

Metrics	Camera	Radar	LiDAR	Ultrasonic
Technology	Lights	Radio waves	Laser beams	Ultrasound waves
Range	≅250 m	5–200 m	≅200 m	Up to 10 m
Data per second	20–40 MB	10–100 KB	10–70 MB	10–100 KB
Bad weather functionality	Poor	Good	Fair	Good
Low lighting functionality	Fair	Good	Good	Good
Speed detection	Poor	Very good	Good	Poor
Distance detection	Poor	Very good	Good	Good
Resolution	Very good	Average	Good	Poor

2.2. *Perception*

The perception subsystem is composed of software elements that receive the information from the sensors and combine and structure it in a simpler form, in order to classify it. Sensor fusion is a necessary process for this stage, and involves the combination of information from all the available sensors in the vehicle [49]. Thus, a complete assessment of the environment can be carried out and more precise information can be obtained.

Literature Review

In practice, different algorithms are used for the fusion process, such as Kalman and Bayesian filters [20]. In [23], several data fusion methods are presented, based on the following strategies: discernible units, complementary features, target attributes, and the decision making of different sensors.

The perception subsystem is in charge of two key tasks: the localisation of the ego vehicle and the detection of other traffic participants and other elements of interest from the surrounding environment. Within the localisation task, the location of the vehicle relative to a map is computed. More precisely, the vehicle's position is determined using data received from different sensors, such as GPS, IMU, LiDAR, and V2X communication, combined with the use of maps. The work in [50] presents a localisation system for urban and indoor scenarios, where LiDAR, IMU, and GPS sensors are integrated. A multitude of combinations of different sensors for data fusion in localisation and mapping are presented in [51], with an emphasis on limitations without fusion and fusion advantages. The detection process uses camera, radar, LiDAR, and ultrasonic sensors to collect the necessary data for the identification and classification of various elements of the vehicle's external environment. In [52,53], a fusion between camera and LiDAR used for pedestrian detection is given, whereas in [54,55], the same is employed for road detection. Moreover, for vehicle and lane detection, information from camera and radar are used in [56]. The perception subsystem is used by both leaders and followers to fuse sensor measurements.

2.3. *Planning and Decision*

The planning subsystem uses information from the perception subsystem to find the most suitable route for the vehicle, from the origin to the destination, both for short-term and long-term planning. The GPS navigation system has the role of a global planner, being used to plan routes, but considering the current requirements, it does not ensure the safety of the user [57]. In this context, according to Figure 1, a traditional planner structure for a self-driving car consists of a route planner, behaviour planner, and trajectory planner.

The decision subsystem, which is the next step after planning, makes a decision and sends all the information to the control subsystem by assuming a compact data form received from the previous subsystems. A number of the decisions available for connected vehicles are illustrated in Figure 1, such as anti-lock braking systems (ABS), lane keep assist (LKA), traffic sign assist (TSA), collision avoidance (CA), adaptive cruise control (ACC), and more. The decision-making process is based on the information from the previous subsystems available at the current moment, but also uses past information. Furthermore, real-time data from maps, traffic models, and additional information from the driver are also used as information. Thus, depending on the type of decision chosen, based on the collected information, it is forwarded to the control subsystem, which in turn will choose the appropriate control type and optimal method.

Literature Review

Route planning, at a general level, involves the use of a route planner, whose purpose is to identify the path that a vehicle must travel between two cardinal points. Moreover, route planning includes several dynamic parameters, such as congestion level, spontaneous indicators, meteorological conditions, and others [58]. An optimal route consists of a continuous adjustment of the planning process, in which the vehicles decide their routes in an adaptive manner. Considering the unexpected changes that may appear along the route (e.g., traffic barricades or lane obstructions), an efficient planning subsystem employs dynamic optimisation techniques at each discrete moment of time. For this type of planning, updated maps and data provided by the localisation stage from the previous subsystem are basically used. In the case of connected vehicles, a distributed route design strategy is used, where each vehicle collects its own traffic data and calculates its own route, thus improving the overall calculation time for the route planner [59].

Behaviour planning involves the use of a behaviour planner, which is closely related to a predictor. The prediction component evaluates the behaviour of other traffic participants such as vehicles and pedestrians, but also other intervening obstacles, in order to obtain risk and road traffic management [60]. Moreover, another source of information for this planner is sensor fusion data from the perception subsystem. As such, the behaviour planner uses the information from lane detectors, traffic lights, and traffic signs, as well as detected objects, but also information from the localisation part. All this information is used to plan their own safe handling behaviour. Therefore, having all these data as inputs, certain decisions are issued for the vehicle, such as maintaining or changing lanes, maintaining the current distance from the vehicle in front, and maintaining the speed by braking or accelerating.

Trajectory planning uses a trajectory planner in order to generate a series of trajectories based on the behaviour planner, taking into account several aspects, such as driver comfort, various road limitations, and vehicle dynamics [61]. The most used methods to design the trajectory planner are based on polynomial equations [62,63], Bézier curves [64,65], and MPC algorithms [66,67]. Thus, taking into account the previously mentioned aspects, the desired trajectory is determined and sent to the decision subsystem [68].

The planning subsystem is mainly intended for the leader vehicle; it computes the global route using localisation functionality based on GPS and V2X communication. Based on the information from the perception subsystem, the leader determines the general behaviour of the platoon by choosing the most suitable action. After that, it must inform the followers about the chosen decision (maintaining or changing lanes, maintaining the speed by braking or accelerating, and more). Finally, based on the previously mentioned information, the leader uses trajectory planning to compute a path so that the platoon can follow the global route and avoid collisions that may occur.

The follower vehicles receive the decision taken by the leader, and depending on this, the following cases result: (i) maintaining or reducing the speed to ensure an imposed distance to the vehicle in front (longitudinal dynamics); and (ii) maintaining or changing lanes to minimise the lateral position error between followers and the vehicle in front.

2.4. Control

The control subsystem receives its task and all related information from the decision subsystem. Thus, starting from the desired trajectory determined by the planning subsystem, and the imposed driving action, implements the best decision for the vehicle. Following the description from Figure 1, once the decision is taken, an appropriate control strategy is selected.

Literature Review

Let us assume that a collision avoidance action is requested. This can happen if an unexpected obstacle is detected on the travel lane in front of the vehicle. Prior to this decision, at the stage of the planning subsection, the trajectory planner computes the optimal trajectory to avoid the obstacle, starting from the current position measured by the sensors to the final position for the vehicle, which usually is on the neighbouring available lane. In the control subsystem, using this information as input data, the lateral control strategy is implemented. In this case, the end result of the control action is a steering movement, i.e., the vehicle's wheels are moved with the desired steering angle while following the planned trajectory [68]. The lateral control supposes that vehicles are equipped with GPS and IMU sensors to measure their position and orientation; also, the vehicles have to be equipped with LiDAR and camera sensors to measure their position and orientation with respect to the neighbour vehicles or an obstacle. In the case of an ego vehicle, the lateral control function uses the measurements of these sensors and inputs received from the planning and decision subsystems to steer the vehicle so that it follows the imposed trajectory and avoids collision with obstacles or other vehicles. In the case of a vehicle group (e.g., platoon), there are three preferred approaches: (i) follower vehicles do not use information about the vehicle in front and they only follow the road; (ii) a follower vehicle receives from the vehicles in front information about their lateral references and uses it to determine its own future trajectory; and (iii) the case in which the follower vehicle is following the lateral trajectory of the vehicle in front. In the last two cases, the use of vehicle communication can improve performances due to the fact that the follower is not using only measurements from its own sensors, but also information received from its neighbour vehicles. The most used approaches for lateral control are based on the MPC strategy [69–71], LQR algorithm [72,73], adaptive control [74], and optimal control [75]. The lateral control of the leader involves steering the vehicle according to the trajectory from the planning subsystem. Also, the follower vehicles use lateral control to track the trajectory of the vehicle in front and to maintain the lanes (if no other functionality is chosen, e.g., collision avoidance).

Moreover, when ACC-based travel is decided, the most suitable control strategy is longitudinal control. This means that the vehicle must travel with an imposed longitudinal velocity (i.e., cruise control) while maintaining a safe distance with respect to the vehicle in front (i.e., headway control). Here, the control action is either braking, if the current velocity is greater than the desired velocity, or accelerating, if the measured velocity is lower than the imposed target [76]. The cruise control (CC) functionality is specific for an ego vehicle or the leader vehicle from a platoon. The vehicles use sensors like GPS and IMU to measure their velocity, which is used afterwards to compute the error to the imposed velocity. The longitudinal controller uses these measurements and errors to calculate the control inputs that command the brake or acceleration. In the cases of ACC functionality, follower vehicles have to use sensors like radar or LiDAR to determine the distance between vehicles. Moreover, if the vehicles can exchange information through communication networks about their velocity, acceleration, or position, then the CACC functionality can be used to ensure improved performances obtained by the ACC. The most used approaches for the longitudinal control are also based on the MPC algorithm [62,77,78], LQR [79], and PID controllers [80]. The advantage of using the MPC strategy and vehicle communication compared to the other methods consists of the possibility to use the future actions' predictions of a neighbouring vehicle. The leader vehicle uses longitudinal control to travel with the imposed velocity, and the follower vehicles use it to maintain the imposed distance from the vehicle in front.

3. Vehicle Communication

This section presents the different aspects of the vehicle communication systems by introducing several communication standards used for the exchange of information between the vehicle components (i.e., intra-vehicle communication), as well as for the inter-vehicular communications, between the vehicle and other traffic participants or intelligent infrastructure, known as V2X communication.

3.1. Intra-Vehicle Communication

Intra-vehicle communication is an absolutely necessary requirement to take into account in the development of new model cars. Thus, for proper vehicle operation, strict information must be exchanged in real time between different nodes/modules. Depending on the communication architecture, the amount of data to be transmitted, bandwidth, reliability, and security, several networks can be distinguished.

3.1.1. Literature Review

The most important intra-vehicle protocols are presented in the following:

(1) The controller area network (CAN) protocol is an automotive-specific bus standard, usually used for powertrain and body control applications. Thus, from the data rate point of view, two networks can be distinguished: (i) high-speed CAN (500 Kb/s) for real-time control for chassis and power-train electronic control units (ECUs), and (ii) low-speed CAN (125 Kb/s) for body and comfort electronics. The CAN is a multi-master serial bus that uses multiplexed communication between several ECUs in the vehicle [81]. Related to CAN arbitration, bus access conflicts are resolved by bit-level arbitration using the carrier sense multiple access with bitwise arbitration (CSMA/BA) technique. The CAN also contains five mechanisms for error detection, three for the message level: cyclic redundancy check (CRC), frame check, and acknowledgement (ACK) bit; and two for the bit level: bus monitoring and bit stuffing [82]. The advantages of CAN are reliability, robustness, low cost, high flexibility, and low network complexity. As disadvantages, it would be that it is not deterministic, it is not suitable for safety-critical applications, and it has a low bandwidth [83].

(2) The local interconnect network (LIN) is a low-cost network used for simple, less time-critical applications, especially used for connecting sensors and actuators. The LIN protocol uses the master–slave architecture; the master sends a frame header and the slave node must respond with a frame response. For low-cost requirements, a single wire is used at the physical level, thus resulting in a limited data rate of 19.2 Kb/s [84]. To detect incorrect messages in the network, LIN uses parity bits and checksum. The advantages of the LIN network are related to the ease of use, low implementation costs, and its deterministic characteristic when compared to other networks. As disadvantages, it is not as reliable as CAN, has a lower bandwidth, and less effective bus access, and cannot be used for time-critical applications [85].

(3) The FlexRay protocol was developed by the FlexRay consortium, and it is used for time-critical applications in the advanced chassis control area [86]. During data transmission, each node uses two parallel communication channels, the exchange of information being performed based on a communication schedule. Regarding the bus access principle, two methods are used: time division multiple access (TDMA) and flexible TDMA (FTDMA) [87]. A FlexRay frame consists of three parts: the header, the payload segment, and the trailer CRC. For error protection, checksums and redundancy mechanisms are used [88]. The advantages of the FlexRay protocol are its flexibility, higher data rate, and deterministic behaviour. Moreover, it offers constant latency and scalable fault tolerance, which makes it suitable for "drive-by-

wire" applications. The disadvantages of this protocol can be summed up in the very high implementation costs and the high complexity compared to CAN [89].

(4) The Media Oriented Systems Transport (MOST) protocol is a multimedia network developed for infotainment applications. The bandwidth is up to 150 Mb/s, supporting both synchronous and asynchronous transmission. The MOST network can manage up to 64 devices using a ring topology, which can be easily connected and removed using the plug-and-play functionality. Moreover, different end-user applications can be connected to this network, such as radios, GPS, and entertainment systems [90]. Although this protocol satisfies the requirements for infotainment applications, there is a limitation in terms of bandwidth for certain requirements [91].

(5) Automotive Ethernet is a communications bus that successfully serves high-bandwidth applications in the field of autonomous driving and connected cars. Ethernet technology has several uses besides in-vehicle communication, such as measurement and calibration, but also diagnostics over IP (DoIP) [92]. The Ethernet standards used for automotive requirements are 100Base-T1 and 1000Base-T1. Automotive Ethernet is implemented with a single twisted cable pair, obtaining data rates from 10 Mb/s to 10 Gb/s. Ethernet lends itself very well to the requirements of applications in the advanced driver-assistance system (ADAS) field, which requires the use of large bandwidth for the sensors used. Moreover, related to diagnostics, Ethernet has started to replace CAN, offering a much shorter time for flashing procedures [89]. Supporting a switched network technology, another advantage is the reduced cost of cabling. The main disadvantages are related to the high costs, resulting in a more expensive physical-level interface [90]. Besides this, Ethernet does not offer deterministic and real-time communication, and this is the main reason why automotive Ethernet does not completely replace the CAN protocol.

3.1.2. Summary

A comparison of in-vehicle network protocols from the point of view of technical characteristics is illustrated in Table 2, where can be observed the essential aspects that can classify each protocol according to its advantages, as it can be concluded which is more suitable depending on the chosen architecture.

Table 2. Classification of intra-vehicle network protocols.

Intra-Vehicle Network	Bit Rate	Data Length	Access Control	Messaging	Network Topology	Error Detection
LIN	19.2 Kb/s	8 bytes	Polling	Master–Slave	Bus	8-bit Checksum
CAN	125 Kb/s–1 Mb/s	0–8 bytes	CSMA/CA	Multi-Master	Bus Star	15-bit CRC
FlexRay	Up to 10 Mb/s	0–254 bytes	TDMA FTDMA	Multi-Master	Bus Star Multi-star	24-bit CRC
MOST	Up to 150 Mb/s	Up to 364 Bytes	TDMA Support for (a)synchronous	Master–Slave Streams	Ring	16-bit CRC
Automotive Ethernet	Up to 10 Gb/s	Up to 1500 bytes	CSMA/CR	Based on IP	Bus Star	32-bit CRC

3.2. *V2X Communication*

V2X communication technology has major importance in the implementation of an intelligent transport system (ITS), offering a level of automated driving and intelligent mobility. Furthermore, this technology involves the exchange of information between a vehicle and other entities of the traffic system.

3.2.1. Literature Review

The V2X includes four modes of communication: vehicle-to-vehicle (V2V), vehicle-to-infrastructure (V2I), vehicle-to-pedestrian (V2P), and vehicle-to-network (V2N) [93]. Each of these communication modes is illustrated in Figure 3 and exemplified in the following.

(1) V2V communication allows for the exchange of information between vehicles in proximity, exchanging useful information about vehicle location, traffic accidents, speed, and traffic dynamics [94]. Each vehicle is equipped with an on-board unit (OBU). Communication between vehicles is achieved by forming a mesh network and connecting them as nodes to the network [95]. Therefore, for the exchange of information between nodes, messages are used with the aim of creating a more efficient decision-making system. Thus, if used properly, V2V communication has the benefits of increased driver safety and road capacity, improving fuel efficiency and preventing possible accidents [96].

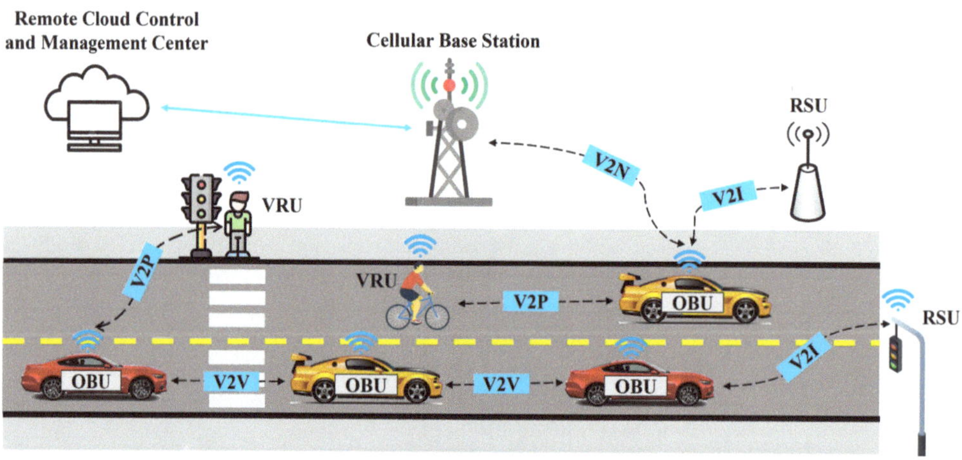

Figure 3. V2X communication modes.

(2) V2I communication refers to the exchange of information between the vehicle and various equipment installed on the road infrastructure [97]. V2I communication can be ad hoc, wireless, or bidirectional [98]. The vehicles collect information from a road side unit (RSU), which is a stationary unit installed along the roads. This information is used for traffic management [99]. Thus, useful information can be obtained about traffic congestion, available parking, the most efficient routes, and road conditions [100]. All these are used to obtain reduced fuel consumption, increase mobility, and reduce polluting emissions [22].

(3) V2P involves real-time, wireless communication between vehicles and vulnerable road users (VRUs), such as pedestrians, bicyclists, and more [101]. Each VRU has user equipment (UE), usually a mobile phone, which makes it possible to exchange information with vehicles. Thus, messages and alerts are sent about the location, speed, and direction of VRUs [102]. Using V2P, communication between vehicles and VRUs can be achieved even in unfavourable weather conditions [93]. Therefore, the benefits of this type of communication refer to the improvement of pedestrian safety and the reduction in traffic accidents in which VRUs are involved.

(4) V2N communication allows the vehicle to access the network, through a server, for various cloud-based services. This type of communication can be made directly between the vehicle and the network or indirectly through a node installed in the road infrastructure, depending on the distance between the vehicle and the network infrastructure [103]. Vehicles can receive broadcast alerts regarding various aspects

of traffic, such as accidents ahead, traffic congestion, or support for planning the best route. All this leads to increased vehicle safety, better route planning, and better traffic efficiency [104].

V2V communication is used for all platoon members to exchange information related to position, velocity, acceleration, and decision. Moreover, V2P communication is also necessary for leaders and followers according to safety considerations for vulnerable road users. The V2I and V2N communications are used only by the leader vehicle to obtain information about traffic management, traffic congestion, road conditions, accidents ahead, and support for planning the best route.

From the point of view of the communication standards used for V2X communication, two important categories of communication technologies are distinguished: dedicated short-range communication (DSRC) and Cellular-V2X (C-V2X) technology. Each of these is presented in detail in the following:

(1) DSRC is a specific communication standard for V2X technology, which allows for wireless communication between connected vehicles, but also with road infrastructure. The DSRC involves short-range bidirectional wireless communication, and it is used for V2V and V2I communications [8]. The DSRC system is based on a series of IEEE and SAE standards. For the physical (PHY) and medium access control (MAC) layers, DSRC uses the IEEE 802.11p standard, for requirements related to authentication, data transmission, and high mobility challenges. The network and security services are defined in the IEEE 1609.x family of standards [105,106]. In the U.S., the Federal Communications Commission (FCC) has allocated for DSRC a 75 MHz spectrum, divided into 10 MHz channels, in the 5.9 GHz frequency band [8]. DSRC-based V2X is successfully used in applications such as traffic safety, traffic management, and commercial vehicle applications [107]. Thus, this standard comes with the following benefits: low latency, high reliability, data rates from 3 Mbps up to 27 Mbps, and ad hoc communications. Cooperative awareness messages (CAMs) and event-triggered warnings, i.e., decentralised environmental notification message (DENM)-type messages, were established in the IEEE 802.11p standard by The European Telecommunications Standards Institute (ETSI). On the other side, in the U.S., the basic safety message (BSM) set messages have been defined by the Society of Automotive Engineers (SAE) [108]. The CAM and BSM are periodical messages sent between vehicles and between vehicles and the infrastructure. These contain information about the status information on heading, speed, position, and acceleration. Moreover, for V2X applications, the transmission frequency of CAM messages is standardised between 1 to 10 Hz, and the broadcast rate of BSM messages is 10 Hz. The DENM messages are warnings transmitted in emergency situations. These are decentralised and information is transmitted directly between vehicles, without the involvement of a centralised infrastructure [109].

(2) C-V2X technology is based on cellular systems, merging the traditional V2X network with the cellular network [110]. According to the 3rd Generation Partnership Project (3GPP) unified global standards, this communication technology uses long-term evolution (LTE)–V2X for assisted driving and 5G New Radio (NR)–V2X for autonomous driving [111]. The working frequency for C-V2X is the same as in the case of the DSRC, operating in the 5.9 Hz frequency band, a band allocated for communications in the intelligent transportation system (ITS) area. LTE-V2X uses single-carrier frequency division multiple access (SC-FDMA) and supports 10 MHz and 20 MHz channels. The communication channel for LTE-V2X uses resource blocks (RBs) of 180 kHz; this implies 12 subcarriers of 15 kHz each. Moreover, from the point of view of time, the channel is divided into sub-frames of 1 ms [112]. LTE-V2X uses communication modes 3 and 4 for resource allocation. The initial advantages of LTE-V2X were improving road safety and reducing traffic congestion. This is possible by periodically broadcasting a CAM message between connected vehicles and LTE-V2X supporting in-coverage, out-of-coverage, and partial-coverage scenarios.

The 5G NR-V2X technology started to be developed from Release 16, coming as a complement to LTE-V2X. For 5G NR, two frequency ranges are defined in which it can operate: frequency range 1 (410 MHz–7.126 GHz) and frequency range 2 (24.25–52.6 GHz). This results in bandwidth for the channel in both bands of 10, 20, 30, and 40 MHz. Besides these, 5G NR-V2X supports various frequency division multiplexing (OFDM) methodologies [113]. For 5G NR-V2X, two communication modes are defined, as in the case of LTE-V2X, mode 1 and mode 2. The advantages of 5G NR-V2X technology include increased capacity and speed, as well as reliability but also a considerable decrease in latency [114]. The C-V2X uses two complementary transmission modes: the Uu and PC5 interfaces. Modes 1 and 3 correspond to the Uu interface, this being a traditional radio interface that accesses terminals through a base station, using uplink (UL) and downlink (DL) transmissions. Within Uu, C-V2X applications operate in traditional mobile broadband licensed spectrum. This interface is used in V2N communication, for long-range applications. The PC5 interface corresponds to modes 2 and 4 and involves direct communication between traffic entities, unassisted by the base station. It is used for V2V, V2I, and V2P communications, the exchange of information being carried out with the help of sidelink (SL) transmission. Within PC5, C-V2X applications operate in the 5.9 GHz ITS band for short-range applications, on a distance of less than 1 km [115].

3.2.2. Summary

The technical characteristics of DSRC and C-V2X standards are illustrated comparatively in Table 3, where can be observed the properties of each communication technology: IEEE 802.11p, LTE, and 5G NR. Therefore, depending on the V2X communication requirements for each use case, the appropriate communication standard can be chosen.

Table 3. Comparison between communication standards used for V2X communication.

Features	DSRC	LTE-V2X	NR-V2X
Communication technology	IEEE 802.11p	LTE	5G NR
Frequency bands	5.9 GHz	5.9 GHz	5.9–52.6 GHz including mmWave
Data rates	3–27 Mb/s	150 Mb/s	1–10 Gb/s
Latency	Up to 150 ms	10–100 ms	<5 ms
Communication modes	Broadcast	Broadcast	Broadcast, unicast, and multicast
Mobility support	252 km/h	350 km/h	500 km/h
Transmission time	0.4 ms	1 ms	1 ms
Retransmission	None	Blind	HARQ-based
Sub-carrier spacing	156.25 KHz	15 KHz	Sub–6 GHz: 15, 30, 60 KHz; mmWave: 60, 120 KHz

4. Cooperative Adaptive Cruise Control for Vehicle Platoon

Interconnected vehicle systems are built around the ability of multiple vehicles to establish a local network and communicate with one another their mobility characteristics so that cooperative maneuvers can be performed such as maintaining lanes, maintaining a constant speed, changing lanes, and many others. In recent years, cooperative adaptive cruise control (CACC) has emerged as a promising technology in vehicular safety applications. CACC-based platoons involve a group of vehicles that are connected through wireless communication and are capable of performing coordinated driving manoeuvres, such as acceleration, deceleration, and lane changing. The use of CACC in platooning applications has been shown to improve traffic efficiency, reduce fuel consumption, and enhance safety by mitigating the effects of human error [8].

This section presents a proposal for a control architecture targeting a specific case involving a CACC strategy for a vehicle platoon, as illustrated in Figure 4. The topology used is predecessor–follower; thus, the vehicles periodically transmit their current state information, such as location, speed, and acceleration. The proposed control architecture that implements the CACC functionality has the following components:

- Sensors:
  - The leader is equipped with a long-range radar for obstacle detection;
  - The follower vehicles are equipped with a short-range radar to measure the distance to the preceding vehicle;
  - Each vehicle is equipped with a camera for lane and obstacle detection, and GPS and IMU to determine the position and velocity of the vehicle.
- Vehicle communication:
  - The communication channel consists of a V2V link between the vehicles. Thus, vehicles are equipped with DSRC technology for short-range communication using CAM messages. These are sent with a frequency of 10 Hz, with each vehicle sending 10 messages per second, which is the minimum required by the CACC functionality [116]. The bitrate for CAM messages is set to 6 Mbit/s, which means an optimal value for vehicular scenarios [117];
  - For each vehicle, the data from the sensors are processed and then transferred to the vehicle's central control unit for fusion using a CAN bus. Also, the automotive Ethernet bus is preferred for the camera sensor according to its required high bandwidth.
- Control solutions:
  - Specifically for lateral control, vehicles can use an LQR algorithm to implement the lane keep assist functionality;
  - For the longitudinal control, two methods based on the DMPC algorithm and feedback gain matrix are proposed.

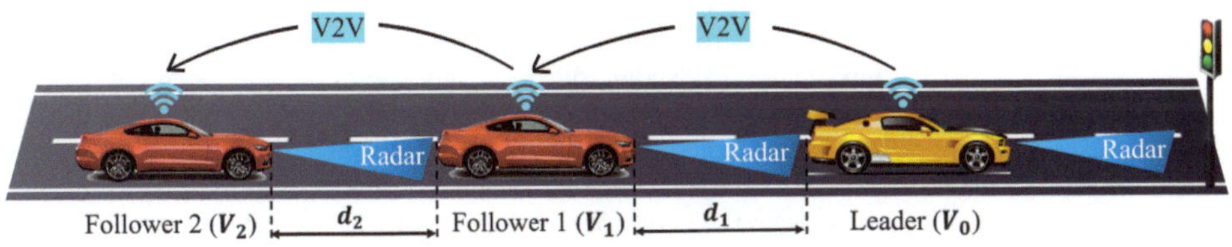

Figure 4. Cooperative adaptive cruise control for a vehicle platoon.

In what follows, the modelling method and the two control strategies proposed for the longitudinal dynamics are detailed.

4.1. Vehicle Platoon Modelling

The leader vehicle (denoted with V_0) is in front of the platoon and leads the platoon with a desired travel velocity. The follower vehicles are tracking the vehicle in front while keeping an imposed distance from it. Each vehicle uses GPS, IMU, camera, and LiDAR to measure the velocity, acceleration, position, and velocity errors, and receives information about the states of its in-front neighbour via V2V communication. Moreover, the control solution for CACC functionality assumes that in front of the leader is a "virtual leader" vehicle that is moving with the desired acceleration. In this way, the leader vehicle can be modelled as a follower that has to follow the virtual leader. The model that describes the longitudinal dynamics [118] is given by (1). This model describes the relationship between states of the vehicle, position and velocity errors and inputs. Moreover, the model illustrates the coupling between two consecutive vehicles from the platoon :

$$\begin{bmatrix} \dot{e}_{pi}(t) \\ \dot{e}_{vi}(t) \\ \dot{a}_i(t) \end{bmatrix} = \begin{bmatrix} 0 & 1 & -\delta \\ 0 & 0 & -1 \\ 0 & 0 & -1/\tau \end{bmatrix} \begin{bmatrix} e_{pi}(t) \\ e_{vi}(t) \\ a_i(t) \end{bmatrix} + \begin{bmatrix} 0 \\ 0 \\ 1/\tau \end{bmatrix} u_i(t) + \begin{bmatrix} 0 \\ 1 \\ 0 \end{bmatrix} a_{i-1}(t), \quad (1)$$

where e_{pi} represents the longitudinal position error of vehicle i to the vehicle in front $i-1$; e_{vi} represents the velocity error; a_i represents the acceleration; u_i represents the input, i.e., acceleration request; $\delta = 0.7$ s represents time headway; and $\tau = 0.1$ s represents the time constant. Notice that for the leader vehicle $i = 0$, the acceleration a_{-1} represents the virtual leader's imposed acceleration a_r.

To control the vehicle platoon, this study proposes two control solutions based on the distributed model-based predictive control (DMPC) strategy and the feedback gain matrix method. In the latter, the control law of each vehicle is a linear combination of its states and states of the vehicle in front:

$$u_i = K_{i,i}\zeta_i + K_{i,i-1}\zeta_{i-1}, \quad (2)$$

where $\zeta_i = \begin{bmatrix} e_{pi}(t) \\ e_{vi}(t) \\ a_i(t) \end{bmatrix}$ represents the vector of system states and $K_{i,j}$ represents a real array. The model presented in this section will be used in the design phase of the two control methods (i.e., DMPC strategy and feedback gain matrix method), but also to simulate the longitudinal dynamics of the platoon.

4.2. Communication Topologies

The use of vehicular communication in designing the CACC solution improves driving performance, safety, and stability. Vehicles can obtain information about velocities, accelerations, and positions from vehicles in front and use them to decide the new actions so that the imposed constraints and targets are respected. The main advantages of V2V communication are represented by [19,119]: (i) improving safety, (ii) optimising the use of roads by reducing the space between vehicles, (iii) reducing fuel consumption and pollution by minimising the accelerations, (iv) improving control performances, and (v) ensuring string stability. The most studied communication typologies are represented by (i) predecessor–follower communication, where each follower i receives information from the vehicle in front $i-1$; (ii) leader–follower, where each follower i receives information from the leader vehicle $i = 0$; (iii) leader–predecessor–follower, where each follower i receives information from the vehicle in front $i-1$ and from the leader $i = 0$; (iv) bidirectional communication, where each vehicle i receives information from its neighbour vehicles $i+1$ and $i-1$. However, the performance of the vehicle platoons that use these communication topologies depends on the model and the chosen control solution. In the case of the MPC algorithm, some studies show that the control performances are quite similar for these topologies [68,120]. Also, solutions that ensure string stability and performances for a vehicle platoon's lateral and longitudinal dynamics are proposed in [121–123]. These solutions use predecessor–follower communication in cases where the dynamics of a follower is described taking into account the position of the vehicle in front, and leader–follower communication when the dynamics of a follower is described taking into account the position of the leader or both of them.

Due to the fact that the model (1) describes the position of a vehicle to the vehicle in front and a follower i is coupled with follower $i-1$ through the acceleration, this work uses the predecessor–follower communication topology.

4.3. Distributed Model-Based Predictive Control Method

This section presents the DMPC strategy used to control the velocity and distance between vehicles. The algorithm is used by each vehicle to compute its control inputs (i.e., acceleration requests). The method uses a model of the vehicle to predict its behaviour and to determine a sequence of optimal control inputs so that a cost function is minimised. Also, due to the coupling between two consecutive vehicles, the method supposes communication between vehicles regarding their prediction of acceleration. With this, the prediction of vehicle future states is improved, which implies better performance.

Consider a system in chain architecture described by:

$$\zeta_i(k+1) = A_i\zeta_i(k) + B_iu_i(k) + A_{i,i-1}\zeta_{i-1}(k), \tag{3}$$

where $A_i \in \mathbb{R}^{n\times n}$, $B_i \in \mathbb{R}^{m\times n}$, $A_{i,i-1} \in \mathbb{R}^{n\times n_{i,i-1}}$, n represents the number of states for subsystem i, m represents the number of inputs for subsystem i, $n_{i,i-1}$ represents the number of states through which the subsystems i and $i-1$ are coupled, and $i = 0,\ldots,M$, $M+1$ represents the number of subsystems.

The optimal sequence of control inputs is determined by solving at each sample time the DMPC problem 1. This problem assumes the minimisation of a cost function that has three types of terms: (i) terms that minimise the prediction of vehicle position and velocity errors, (ii) terms that minimise the control efforts, and (iii) terms that minimise the error between the prediction of the states of vehicle $\zeta_i(\cdot)$ and the assumed prediction of the states of the vehicle in front $\tilde{\zeta}_{i-1}(\cdot)$. Moreover, the DMPC problem 1 takes into account the imposed constraints on vehicle inputs and states.

Problem 1. *At each discrete step k, starting from an initial state $\zeta_i(k) = \zeta_0$ and using the system model* (3) *to predict the states of the vehicles, compute a finite horizon optimal input sequence minimising the cost function:*

$$\begin{aligned} J_i(k,\zeta_i(k),U_i(k)) =&\zeta_i(N|k)^T Q_i\zeta_i(N|k) + (\zeta_i(N|k) - \tilde{\zeta}_{i-1}(N|k))^T W_i(\zeta_i(N|k) - \tilde{\zeta}_{i-1}(N|k)) + \\ &+ \sum_{j=0}^{N-1}[\zeta_i(j|k)^T Q_i\zeta_i(j|k) + u_i^T(j|k)R_iu_i(j|k) + \\ &+ (\zeta_i(j|k) - \tilde{\zeta}_{i-1}(j|k))^T W_i(\zeta_i(j|k) - \tilde{\zeta}_{i-1}(j|k))] \end{aligned} \tag{4}$$

over $U_i(k)$, subject to the following constraints:

$$\begin{aligned} U_i^{min} &\le U_i(j|k) \le U_i^{max}, \\ \zeta_i^{min} &\le \zeta_i(j|k) \le \zeta_i^{max}, \end{aligned} \tag{5}$$

where Q_i, R_i, and W_i represent the weighting matrices, $W_0 = 0$ (for the leader), N represents the prediction horizon, $U_i(k) = [u_i(0|k),\ldots,u_i(N-1|k)]^T$ is the sequence of control inputs, and $\tilde{\zeta}_{i-1} = [\zeta_{i-1}(2|k-1),\ldots,\zeta_{i-1}(N|k-1),\zeta_{i-1}(N|k-1)]^T$ represents the prediction of the states for vehicle $i-1$, computed at step $k-1$ and sent to vehicle i.

Remark 1. *Note that for the leader vehicle, with index $i = 0$, the vehicle in front is considered to be the "virtual leader" that is driving with the imposed acceleration, so $\tilde{\zeta}_{-1}$ represents the imposed acceleration for the leader for the next N steps.*

4.4. Feedback Gain Matrix Method

The second method uses a feedback gain matrix to compute the control inputs instead of a complex algorithm, such as DMPC. This solution has the advantage of requiring a low computational effort and being easier to be implemented on hardware with limited storage and computational capabilities. In what follows, the method used to determine this control matrix is detailed.

The model (3) can be rewritten as

$$\zeta_{\aleph}(k+1) = A_{\aleph}\zeta_{\aleph}(k) + B_{\aleph}u_{\aleph}(k) + A_{\aleph,0}\zeta_r(k), \tag{6}$$

where $\zeta_{\aleph} = [\zeta_0^T, \ldots, \zeta_M^T]^T$ and $u_{\aleph} = [u_0^T, \ldots, u_m^T]^T$ aggregate all states and inputs of each subsystem $i = 0, \ldots, M$, $\zeta_r \in \mathbb{R}^{n_r}$ represents the imposed reference for subsystem 0 (i.e., the leader), $A_{\aleph,0} = \begin{bmatrix} A_{n,n_r} \\ O_{n,n_r} \\ \cdots \\ O_{n,n_r} \end{bmatrix}$, O_{n,n_r} represents the null matrix of size $(n \times n_r)$.

Notice that the matrix A_{n,n_r} corresponds to the matrix $A_{0,-1}$, and ζ_r to x_{-1} from (3). Then, the control law of the whole system can be defined as:

$$u_{\aleph}(k) = K\zeta_{\aleph}(k), \tag{7}$$

where $K \in \mathbb{R}^{(M+1)\times n(M+1)}$ represents the feedback gain matrix.

Moreover, each vehicle receives via V2V communication the states of the vehicle in front. Based on this, the control law of each follower is considered as in (2). The leader vehicle does not have a vehicle in front (except the virtual leader), so its control law is defined as $u_0 = K_{0,0}\zeta_0$. The control matrix K has non-zero elements only on the sub-block $(1,1)$ corresponding to the leader and on the sub-block $\{(i+1,i),(i+1,i+1)\}$ corresponding to the follower vehicles. Notice that a sub-block (i,j) refers to the elements from matrix K represented by line i and columns $\{j, j+1, \ldots, j+n-1\}$. Also, the notation $K_{i,j}$ is referring to the sub-block $(i+1, j+1)$ in matrix K.

The control matrix K is obtained by solving Problem 2. The method assumes the use of a set of reference states, and the control matrix is computed so that the error between the reference states and model states (6) is minimised:

Problem 2. *Starting from a set of reference states, compute the K matrix so that the following cost function is minimised:*

$$V(K) = \sum_{j=1}^{L} V_j(k), \tag{8}$$

$$V_j(k) = \sum_{k=0}^{t_{kf}} ||\zeta_{\aleph}^r(j,k) - \zeta_{\aleph}(k)||_2^2, \tag{9}$$

over K, subject to the following constraints:

$$\begin{aligned} \zeta_{\aleph}(k+1) &= A_{\aleph}\zeta_{\aleph}(k) + B_{\aleph}u_{\aleph}(k) + A_{\aleph,0}\zeta_r(j,k) \\ u_{\aleph}(k) &= K\zeta_{\aleph}(k), \\ \zeta_{\aleph}(0) &= \zeta_{\aleph}^r(j,0), \end{aligned} \tag{10}$$

where $\zeta_{\aleph}^r$ represents the set of the reference states, t_{kf} represents the length of a reference, L represents the number of states from the set.

Note that the first solution, i.e., CACC based on the DMPC strategy, has the advantage that at each sample time, the method computes the command minimising a cost function that takes into account the prediction of the vehicle states, i.e., position and velocity errors and acceleration, as well as imposed constraints on the states and inputs, and also takes into account the information about the prediction of acceleration received from the vehicle in front. However, it has the disadvantage of requiring a high computing power to solve optimisation Problem 1. The second method requires a significant computational effort, but only in the phase of computing the feedback gain matrix K. After that, the command is computed at each sample time using Equation (2). The disadvantages of this method are represented by the possibility of violating imposed constraints, and also by the fact that it does not use a prediction of the vehicle's state. Moreover, the method does not receive the prediction of the acceleration of the vehicle in front as it is using the method based on the DMPC approach. But the last disadvantage can be minimised in the phase of computing the matrix K by "training" the feedback gain matrix, so that the error between the two methods is minimised by choosing the set of references as the solution of the DMPC algorithm [124] (see Problem 2).

5. Illustrative Results

This section presents the simulation results obtained using the proposed control solutions. The platoon is formed by a leader, followed by three follower vehicles. As it was previously mentioned, the leader and follower vehicles use their sensors to obtain information about their velocity, acceleration, position, and velocity errors. These measurements are used by the control methods to obtain the prediction of vehicle states and to compute the control input. Also, the vehicles receive from their vehicle in-front information that contains the prediction of their accelerations (DMPC method) and their states (feedback gain matrix method). Note that the leader vehicle does not have a real vehicle in front, which means that model (1) can be used to compute the position and velocity error states.

The parameters used by the DMPC controller are represented by prediction horizon $N = 50$ time samples, $Q_0 = Q_1 = \ldots = Q_M = diag\{1, 10, 0.1\}$, $R_0 = R_1 = \ldots = R_M = 0.1$, $W_0 = 0$, $W_i = diag\{3, 3, 3\}$, $i = 1, \ldots, M$, $M = 3$. The limits imposed on the input u and longitudinal error e_p are represented by $u^{min} = -2$ m/s^2, $u^{max} = 2$ m/s^2, $e_{pi}^{min} = -0.7$ m, $e_{pi}^{max} = 0.7$ m. The used sample time is $T_s = 0.1$ s. For the second method based on the feedback matrix, the set of the reference state is formed by $L = 100$ references computed using the DMPC strategy. These reference states were obtained using a set of 100 reference accelerations illustrated in Figure 5. The length of each reference is $t_{kf} = 200/T_s$.

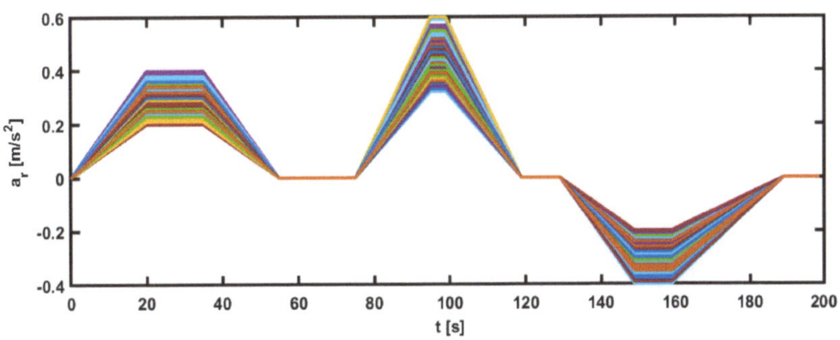

Figure 5. Set of 100 accelerations used by the second control method to design the control feedback matrix for the platoon.

To test the proposed control methods, a simulation scenario was designed. The reference of the leader consists of a series of changes in imposed acceleration to test the methods in various situations. The reference and acceleration of vehicles are illustrated by Figure 6. Based on these figures, it can be noticed that all vehicles follow the acceleration and deceleration behaviour of the vehicle in front, but the first method (based on the DMPC algorithm) has a smoother acceleration compared to the second method (based on the feedback gain matrix). But this difference between methods is not noticed in the graph of velocity; see Figure 7, where all follower vehicles travel with the same velocity in both cases. All vehicles succeed in following the vehicle in front with imposed distance and with small errors, as can be observed in Figure 8. For all methods, the errors decrease in the upstream direction. Moreover, for the second method, the maximum of the absolute values of eigenvalues is $\rho = max(|A_{\aleph} + B_{\aleph}K|) = 0.9774 < 1$, which means the platoon is globally stable. The control inputs are illustrated in Figure 9. The difference between the two methods is the following: the first method obtained a higher value for the requested acceleration compared to the second method, which implies higher fuel consumption. Also, all commands and position errors respect the imposed constraints.

Remark 2. *Note that for each reference acceleration from those* 100*, the evolution of the states of the vehicle platoon was computed by imposing the reference accelerations for the platoon and having the DMPC strategy as a control solution. As a result, a set of* 100 *reference states, $\zeta^r_{\aleph}$, was obtained. The control feedback matrix K was computed by solving Problem 2. By finding a control law* (7) *so that the cost function* (8) *is minimised, the states of the platoon are led to follow the dynamics of the states controlled by the DMPC strategy. This means that the behaviour of the platoon (controlled with* (7)*) is close to the behaviour of the platoon controlled by the DMPC algorithm. Moreover, using a random large set of references, the platoon is tested in multiple operating points, ensuring that no bias from a particular case influences the calculus of matrix K.*

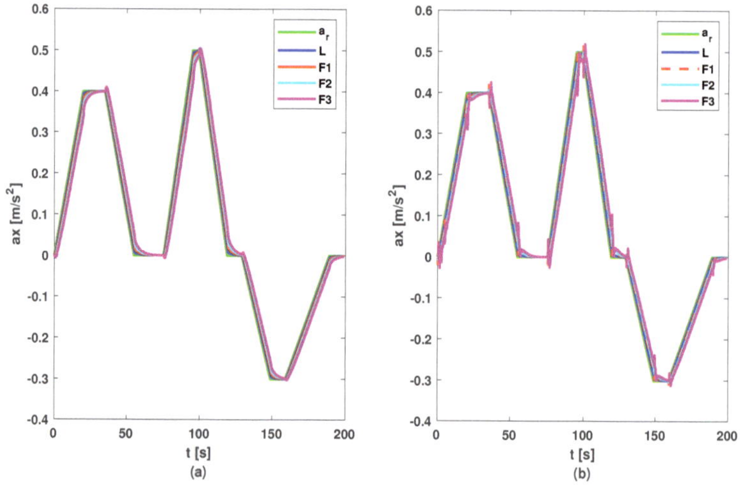

Figure 6. Accelerations of vehicles: (**a**) DMPC method; (**b**) feedback gain matrix method.

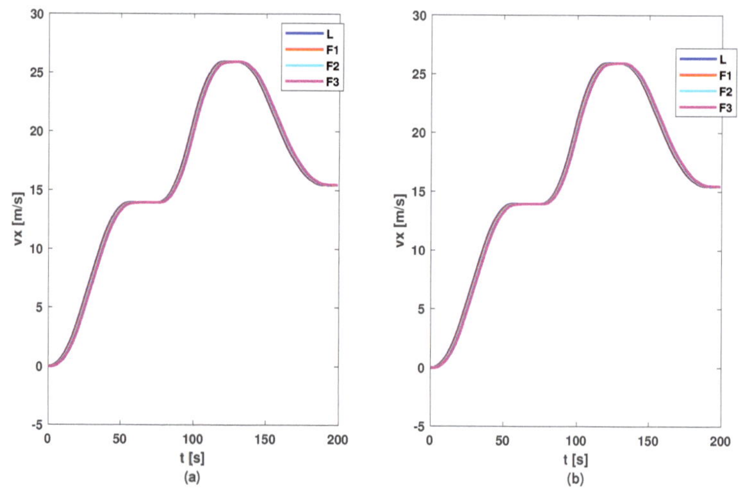

Figure 7. Velocities of vehicles: (**a**) DMPC method; (**b**) feedback gain matrix method.

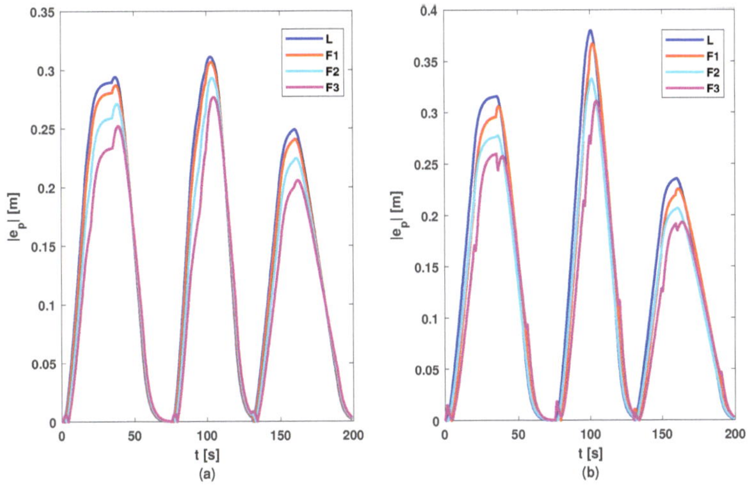

Figure 8. Position errors of vehicles: (**a**) DMPC method; (**b**) feedback gain matrix method.

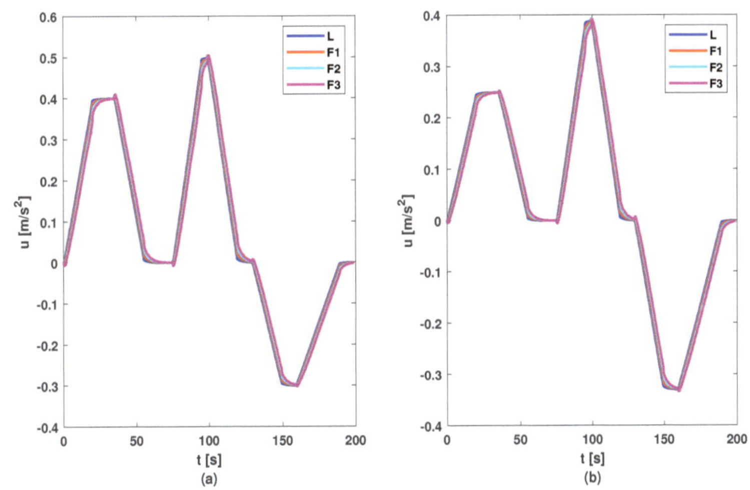

Figure 9. Control inputs of vehicles: (**a**) DMPC method; (**b**) feedback gain matrix method.

6. Discussion

In order to ensure a fair comparison between the two methods, each proposed control solution was tested for 100 random references for acceleration (other than those used in the phase of computing matrix K, see Figure 5). Figures 10–13 contain the position error of the leader and follower vehicles obtained from 100 simulated cases. Moreover, for each vehicle, the average error is computed and illustrated with a black continuous line. From these result, it can be noticed that the maximum average position error for each vehicle are the following: (i) method based on DMPC algorithm—$max(e_{p0}) = 0.297$ m, $max(e_{p1}) = 0.291$ m, $max(e_{p2}) = 0.276$ m, $max(e_{p3}) = 0.258$ m; (ii) method based on feedback gain matrix—$max(e_{p0}) = 0.338$ m, $max(e_{p1}) = 0.326$ m, $max(e_{p2}) = 0.296$ m, $max(e_{p3}) = 0.276$ m.

Furthermore, using the results obtained from these 100 cases, a cumulative cost was used to evaluate better the performances of the two methods. This cost takes into account the position errors and input efforts of the vehicles:

$$J = \frac{1}{t_{kf}} \sum_{s=1}^{100} \sum_{i=0}^{3} \sum_{j=1}^{t_{kf}} (e_{pi}^2(j) + u_i^2(j)). \tag{11}$$

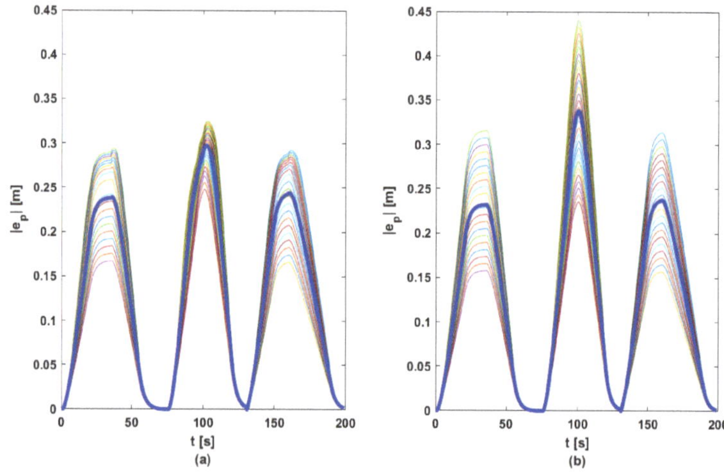

Figure 10. Position errors sets for L: (**a**) DMPC method; (**b**) feedback gain matrix method. Thin line—100 case; bold line—mean of $|e_p|$.

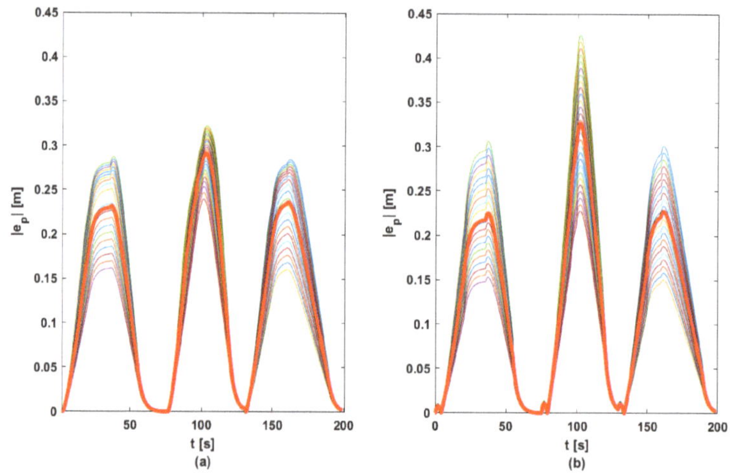

Figure 11. Position errors sets for F1: (**a**) DMPC method; (**b**) feedback gain matrix method. Thin line—100 case; bold line—mean of $|e_p|$.

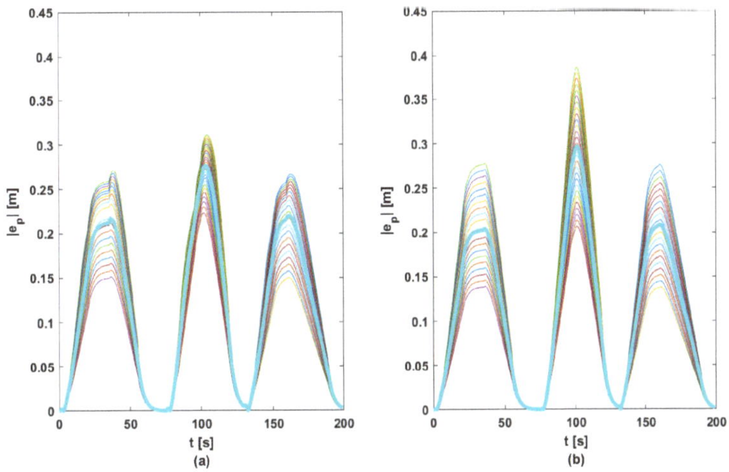

Figure 12. Position errors sets for F2: (**a**) DMPC method; (**b**) feedback gain matrix method. Thin line—100 case; bold line—mean of $|e_v|$.

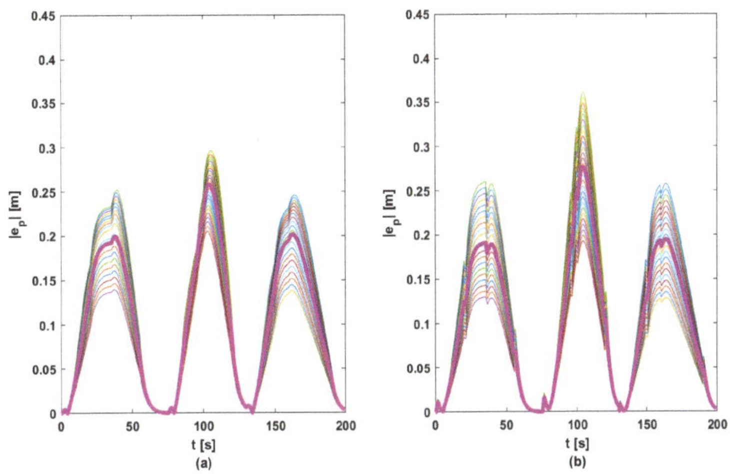

Figure 13. Position errors sets for F3: (**a**) DMPC method; (**b**) feedback gain matrix method. Thin line—100 case; bold line—mean of $|e_p|$.

The method based on DMPC obtains a cost equal to $J_{DMPC} = 6.18$, and the second method, based on the feedback gain matrix, obtains a cost equal to $J_K = 5.828$. Based on these results, the second method obtains control performances quite similar to the one based on the DMPC approach. As can be seen from these results, by designing the feedback matrix K using the proposed solution, it can be obtained a control solution that has a behaviour close to the DMPC algorithm but did not require specialised software and online optimisation to be implemented. These results made the second method proper for real-time implementation. The first method, based on the DMPC strategy, has the advantage that it can take into account online constraints imposed for the states and commands compared with the second method. Also, the first method, using a model of the system, can predict the state's evolution. Although the used model usually has modelling errors, DMPC can obtain better performances. But the second control method is more suitable for real-time implementation because it does not require high computational power or complex optimisation algorithms to be implemented online.

The simulations were performed using MATLAB R2022b on Windows 10, 64-bit Operating System with a laptop Intel Core i7-8750H CPU @ 2.200 GHz and 8 GB RAM.

7. Conclusions and Future Work

This paper presents in detail the related levels of a highly advanced vehicle control architecture (sensing interface, perception, planning, decision, and control) and their element components. Moreover, the proposed control architecture was used to design a control solution for the longitudinal dynamics of a vehicle platoon. Also, for the control level, two methods were proposed: (i) the first one uses a complex control algorithm represented by DMPC; and (ii) the second method uses feedback gain matrices computed using results from the first method. The simulation results prove that the second method obtains similar performance compared to the one based on the DMPC algorithm, and this fact made it suitable for real-time implementation due to the simplicity of the control law.

Future work will focus on using the proposed control architecture in a laboratory scale real-time vehicle platooning system.

Author Contributions: Conceptualisation, R.-G.L., O.P., A.M. and C.-F.C.; methodology, R.-G.L., O.P. and A.M.; software, O.P.; writing—original draft preparation, R.-G.L., O.P. and A.M.; writing—review and editing, R.-G.L., O.P., A.M. and C.-F.C.; supervision, C.-F.C.; project administration, C.-F.C. All authors have read and agreed to the published version of the manuscript.

Funding: The work of R.G.L. was supported by the project "Collaborative environment for developing OpenStack-based cloud architectures with applications in RTI" SMIS 124998 from The European Regional Development Fund through the Competitiveness Operational Program 2014–2020, priority axis 1: Research, technological development and innovation (RTI)—the POC/398/1/1 program. The work of O.P. and A.M. was supported by "Institutional development through increasing the innovation, development and research performance of TUIASI—COMPETE 2.0", project funded by contract no. 27PFE /2021, financed by the Romanian government. The work of A. M. was also supported by "Gheorghe Asachi" Technical University of Iasi (TUIASI), through the Project "Performance and excellence in postdoctoral research 2022".

Institutional Review Board Statement: Not applicable.

Informed Consent Statement: Not applicable.

Data Availability Statement: Not applicable.

Conflicts of Interest: The authors declare no conflict of interest. The funders had no role in the design of the study; in the collection, analyses, or interpretation of data; in the writing of the manuscript; or in the decision to publish the results.

References

1. Guanetti, J.; Kim, Y.; Borrelli, F. Control of connected and automated vehicles: State of the art and future challenges. *Annu. Rev. Control* **2018**, *45*, 18–40. [CrossRef]
2. Guo, Q.; Li, L.; Ban, X. Urban traffic signal control with connected and automated vehicles: A survey. *Transp. Res. Part C* **2019**, *110*, 313–334. [CrossRef]
3. Orosz, G. Connected cruise control: Modelling, delay effects, and nonlinear behaviour. *Veh. Syst. Dyn.* **2016**, *54*, 1147–1176. [CrossRef]
4. Ma, Z.; Huo, Q.; Yang, X.; Zhao, X. Safety Cruise Control of Connected Vehicles Using Radar and Vehicle-to-Vehicle Communication. *IEEE Syst. J.* **2020**, *14*, 4602–4613. [CrossRef]
5. Wang, Z.; Jin, S.; Liu, L.; Fang, C.; Li, M.; Guo, S. Design of Intelligent Connected Cruise Control With Vehicle-to-Vehicle Communication Delays. *IEEE Trans. Veh. Technol.* **2022**, *74*, 9011–9025. [CrossRef]
6. Zhang, T.; Zou, Y.; Zhang, X.; Guo, N.; Wang, W. A Cruise Control Method for Connected Vehicle Systems Considering Side Vehicles Merging Behavior. *IEEE Access* **2019**, *7*, 6922–6936. [CrossRef]
7. Li, Y.; Moreau, J.; Ibanez-Guzman, J. Unconventional Visual Sensors for Autonomous Vehicles. *arXiv* **2022**, arXiv:2205.09383.
8. Lazar, R.G.; Varga, V.; Caruntu, C.F. TrueTime-based Analysis of a Distributed Generalized Predictive Control Architecture for CACC Systems. In Proceedings of the 25th International Conference on System Theory, Control and Computing (ICSTCC), Iasi, Romania, 20–23 October 2021; pp. 612–617.
9. Jond, H.; Yildiz, A. Connected and automated vehicle platoon formation control via differential games. *IET Intell. Transp. Syst.* **2023**, *17*, 312–326. [CrossRef]

10. Zhang, H.; Liu, J.; Wang, C.; Huang, C.; Yan, H. Adaptive Switched Control for Connected Vehicle Platoon with Unknown Input Delays. *IEEE Trans. Cybern.* **2023**, *53*, 1511–1521. [CrossRef] [PubMed]
11. Wang, P.; Deng, H.; Zhang, J.; Wang, L.; Zhang, M.; Li, Y. Model Predictive Control for Connected Vehicle Platoon Under Switching Communication Topology. *IEEE Trans. Intell. Transp. Syst.* **2022**, *23*, 7817–7830. [CrossRef]
12. Xu, Z.; Jiao, X. Robust Control of Connected Cruise Vehicle Platoon with Uncertain Human Driving Reaction Time. *IEEE Trans. Intell. Veh.* **2022**, *7*, 368–376. [CrossRef]
13. Liu, G.; Zhao, R.; Su, M. Cooperative Adaptive Sliding Mode Platoon Control of Intelligent Connected Vehicles under Communication Interruption. *Int. J. Automot. Technol.* **2023**, *24*, 513–525. [CrossRef]
14. Chen, J.; Li, J.; Xu, Z.; Wu, X. Cooperative optimal control for connected and automated vehicles platoon. *Phys. A Stat. Mech. Its Appl.* **2022**, *607*, 128197. [CrossRef]
15. Wu, Z.; Sun, J.; Xu, R. Consensus-based connected vehicles platoon control via impulsive control method. *Phys. A Stat. Mech. Its Appl.* **2021**, *580*, 126190. [CrossRef]
16. Zhang, D.; Shen, Y.P.; Zhou, S.Q.; Dong, X.W.; Yu, L. Distributed Secure Platoon Control of Connected Vehicles Subject to DoS Attack: Theory and Application. *IEEE Trans. Syst. Man Cybern. Syst.* **2021**, *51*, 7269–7278. [CrossRef]
17. Petrillo, A.; Pescapé, A.; Santini, S. A Secure Adaptive Control for Cooperative Driving of Autonomous Connected Vehicles in the Presence of Heterogeneous Communication Delays and Cyberattacks. *IEEE Trans. Cybern.* **2021**, *51*, 1134–1149. [CrossRef]
18. Naus, G.J.L.; Vugts, R.P.A.; Ploeg, J.; van de Molengraft, M.J.G.; Steinbuch, M. String-Stable CACC Design and Experimental Validation: A Frequency-Domain Approach. *IEEE Trans. Veh. Technol.* **2010**, *59*, 4268–4279. [CrossRef]
19. Wang, Z.; Wu, G.; Barth, M.J. A Review on Cooperative Adaptive Cruise Control (CACC) Systems: Architectures, Controls, and Applications. In Proceedings of the 21st International Conference on Intelligent Transportation Systems (ITSC), Maui, HI, USA, 4–7 November 2018; pp. 2884–2891.
20. Ahangar, M.N.; Ahmed, Q.Z.; Khan, F.A.; Hafeez, M. A Survey of Autonomous Vehicles: Enabling Communication Technologies and Challenges. *Sensors* **2021**, *21*, 706. [CrossRef]
21. Dey, K.; Yan, L.; Wang, X.; Wang, Y.; Shen, H.; Chowdhury, M.; Yu, L.; Qiu, C.; Soundararaj, V. A Review of Communication, Driver Characteristics, and Controls Aspects of Cooperative Adaptive Cruise Control (CACC). *IEEE Trans. Intell. Transp. Syst.* **2015**, *17*, 491–509. [CrossRef]
22. Skrickij, V.; Šabanovič, E.; Žuraulis, V. Autonomous Road Vehicles: Recent Issues and Expectations. *IET Intell. Transp. Syst.* **2020**, *14*, 471–479. [CrossRef]
23. Wang, Z.; Wu, Y.; Niu, Q. Multi-Sensor Fusion in Automated Driving: A Survey. *IEEE Access* **2020**, *8*, 2847–2868. [CrossRef]
24. Campbell, S.; O'Mahony, N.; Krpalcova, L.; Riordan, D.; Walsh, J.; Murphy, A.; Ryan, C. Sensor Technology in Autonomous Vehicles: A review. In Proceedings of the 29th Irish Signals and Systems Conference (ISSC), Belfast, UK, 21–22 June 2018; pp. 1–4.
25. Steinbaeck, J.; Steger, C.; Holweg, G.; Druml, N. Next generation radar sensors in automotive sensor fusion systems. In Proceedings of the Sensor Data Fusion: Trends, Solutions, Applications (SDF), Bonn, Germany, 10–12 October 2017; pp. 1–6.
26. Liu, L.; Lu, S.; Zhong, R.; Wu, B.; Yao, Y.; Zhang, Q.; Shi, W. Computing Systems for Autonomous Driving: State of the Art and Challenges. *IEEE Internet Things J.* **2020**, *8*, 6469–6486. [CrossRef]
27. Alencar, F.A.; Rosero, L.A.; Filho, C.M.; Osório, F.S.; Wolf, D.F. Fast Metric Tracking by Detection System: Radar Blob and Camera Fusion. In Proceedings of the 12th Latin American Robotics Symposium and 3rd Brazilian Symposium on Robotics (LARS-SBR), Uberlandia, Brazil, 29–31 October 2015; pp. 120–125.
28. Etinger, A.; Balal, N.; Litvak, B.; Einat, M.; Kapilevich, B.; Pinhasi, Y. Non-Imaging MM-Wave FMCW Sensor for Pedestrian Detection. *IEEE Sens J.* **2014**, *14*, 1232–1237. [CrossRef]
29. Lee, S.; Yoon, Y.J.; Lee, J.E.; Kim, S.C. Human–vehicle classification using feature-based SVM in 77-GHz automotive FMCW radar. *IET Radar Sonar Navig.* **2017**, *11*, 1589–1596. [CrossRef]
30. Farooq, A.; Ahmed, Q.; Alade, T. Indoor two way ranging using mm-Wave for future wireless networks. In Proceedings of the Emerging Tech (EMiT) Conference, Huddersfield, UK, 9–11 April 2019.
31. Rosique, F.; Navarro, P.J.; Fernández, C.; Padilla, A. A Systematic Review of Perception System and Simulators for Autonomous Vehicles Research. *Sensors* **2019**, *19*, 648. [CrossRef]
32. Alluhaibi, O.; Ahmed, Q.Z.; Pan, C.; Zhu, H. Hybrid Digital-to-Analog Beamforming Approaches to Maximise the Capacity of mm-Wave Systems. In Proceedings of the IEEE 85th Vehicular Technology Conference (VTC Spring), Sydney, NSW, Australia, 4–7 June 2017; pp. 1–5.
33. Alaba, S.Y.; Ball, J.E. A Survey on Deep-Learning-Based LiDAR 3D Object Detection for Autonomous Driving. *Sensors* **2022**, *22*, 9577. [CrossRef] [PubMed]
34. Ignatious, H.A.; Hesham-El-Sayed; Khan, M. An overview of sensors in Autonomous Vehicles. *Procedia Comput. Sci.* **2022**, *198*, 736–741. [CrossRef]
35. Lopac, N.; Jurdana, I.; Brnelić, A.; Krljan, T. Application of Laser Systems for Detection and Ranging in the Modern Road Transportation and Maritime Sector. *Sensors* **2022**, *22*, 5946. [CrossRef]
36. Wojtanowski, J.; Zygmunt, M.; Kaszczuk, M.; Mierczyk, Z.; Muzal, M. Comparison of 905 nm and 1550 nm semiconductor laser rangefinders' performance deterioration due to adverse environmental conditions. *Opto-Electron. Rev.* **2014**, *22*, 183–190. [CrossRef]

37. Li, Y.; Ibanez-Guzman, J. Lidar for Autonomous Driving: The Principles, Challenges, and Trends for Automotive Lidar and Perception Systems. *IEEE Signal Process. Mag.* **2020**, *37*, 50–61. [CrossRef]
38. Rasshofer, R.; Spies, M.; Spies, H. Influences of weather phenomena on automotive laser radar systems. *Adv. Radio Sci.* **2011**, *9*, 49–60. [CrossRef]
39. Gruyer, D.; Magnier, V.; Hamdi, K.; Claussmann, L.; Orfila, O.; Rakotonirainy, A. Perception, information processing and modeling: Critical stages for autonomous driving applications. *Annu. Rev. Control.* **2017**, *44*, 323–341. [CrossRef]
40. Zhang, Y.; Carballo, A.; Yang, H.; Takeda, K. Perception and sensing for autonomous vehicles under adverse weather conditions: A survey. *ISPRS J. Photogramm. Remote. Sens.* **2023**, *196*, 146–177. [CrossRef]
41. Stiawan, R.; Kusumadjati, A.; Aminah, N.S.; Djamal, M.; Viridi, S. An Ultrasonic Sensor System for Vehicle Detection Application. *J. Phys. Conf. Ser.* **2019**, *1204*, 012017. [CrossRef]
42. Lim, B.S.; Keoh, S.L.; Thing, V.L.L. Autonomous vehicle ultrasonic sensor vulnerability and impact assessment. In Proceedings of the IEEE 4th World Forum on Internet of Things (WF-IoT), Singapore, 5–8 February 2018; pp. 231–236.
43. Rahiman, W.; Zainal, Z. An overview of development GPS navigation for autonomous car. In Proceedings of the IEEE 8th Conference on Industrial Electronics and Applications (ICIEA), Melbourne, VIC, Australia, 19–21 June 2013; pp. 1112–1118.
44. Hughes, W.J.; Team, E. *Global Positioning System (GPS) Standard Positioning Service (SPS) Performance Analysis Report*; Tech Report NSTB/WAAS T and E Team; U.S. Department of Transportation: Washington, DC, USA, 2014.
45. Ahmad, K.A.B.; Sahmoudi, M.; Macabiau, C. Characterization of GNSS receiver position errors for user integrity monitoring in urban environments. In Proceedings of the ENC-GNSS 2014, European Navigation Conference, Rotterdam, The Netherlands, 15–17 April 2014.
46. Ahmad, N.; Ghazilla, R.A.B.R.; Khairi, N.M.; Kasi, V. Reviews on Various Inertial Measurement Unit (IMU) Sensor Applications. In Proceedings of the IEEE Workshop on Signal Processing Systems, Taipei City, Taiwan, 16–18 October 2013.
47. Liu, S.; Li, L.; Tang, J.; Wu, S.; Gaudiot, J.L. *Creating Autonomous Vehicle Systems*; Synthesis Lectures on Computer Science; Springer: Berlin/Heidelberg, Germany, 2017; Volume 6; pp. 1–186.
48. Zhao, S.; Chen, Y.; Farrell, J.A. High-Precision Vehicle Navigation in Urban Environments Using an MEM's IMU and Single-Frequency GPS Receiver. *IEEE Trans. Intell. Transp. Syst.* **2016**, *17*, 2854–2867. [CrossRef]
49. Katsaggelos, A.K.; Bahaadini, S.; Molina, R. Audiovisual Fusion: Challenges and New Approaches. *Proc. IEEE* **2015**, *103*, 1635–1653. [CrossRef]
50. Gao, Y.; Liu, S.; Atia, M.M.; Noureldin, A. INS/GPS/LiDAR Integrated Navigation System for Urban and Indoor Environments Using Hybrid Scan Matching Algorithm. *Sensors* **2015**, *15*, 23286–23302. [CrossRef] [PubMed]
51. Fayyad, J.; Jaradat, M.A.; Gruyer, D.; Najjaran, H. Deep Learning Sensor Fusion for Autonomous Vehicle Perception and Localization: A Review. *Sensors* **2020**, *20*, 4220. [CrossRef]
52. Melotti, G.; Premebida, C.; Goncalves, N.M.M.d.S.; Nunes, U.J.C.; Faria, D.R. Multimodal CNN Pedestrian Classification: A Study on Combining LIDAR and Camera Data. In Proceedings of the 21st International Conference on Intelligent Transportation Systems (ITSC), Maui, HI, USA, 4–7 November 2018.
53. Schlosser, J.; Chow, C.K.; Kira, Z. Fusing LIDAR and images for pedestrian detection using convolutional neural networks. In Proceedings of the IEEE International Conference on Robotics and Automation (ICRA), Stockholm, Sweden, 16–21 May 2016; pp. 2198–2205.
54. Gu, S.; Lu, T.; Zhang, Y.; Alvarez, J.M.; Yang, J.; Kong, H. 3-D LiDAR + Monocular Camera: An Inverse-Depth-Induced Fusion Framework for Urban Road Detection. *IEEE Trans. Intell. Veh.* **2018**, *3*, 351–360. [CrossRef]
55. Caltagirone, L.; Bellone, M.; Svensson, L.; Wahde, M. LIDAR-camera fusion for road detection using fully convolutional neural networks. *Robot. Auton. Syst.* **2018**, *111*, 125–131. [CrossRef]
56. Bertozzi, M.; Bombini, L.; Cerri, P.; Medici, P.; Antonello, P.C.; Miglietta, M. Obstacle detection and classification fusing radar and vision. In Proceedings of the IEEE Intelligent Vehicles Symposium, Eindhoven, Netherlands, 4–6 June 2008; pp. 608–613.
57. Bast, H.; Delling, D.; Goldberg, A.V.; Müller-Hannemann, M.; Pajor, T.; Sanders, P.; Wagner, D.; Werneck, R.F. Route Planning in Transportation Networks. In *Algorithm Engineering*; Springer: Berlin/Heidelberg, Germany, 2015.
58. Zhang, M.Y.; Yang, S.C.; Feng, X.J.; Chen, Y.Y.; Lu, J.Y.; Cao, Y.G. Route Planning for Autonomous Driving Based on Traffic Information via Multi-Objective Optimization. *Appl. Sci.* **2022**, *12*, 11817. [CrossRef]
59. Liu, B.; Long, J.; Deng, M.; Yang, X.; Shi, Y. An Adaptive Route Planning Method of Connected Vehicles for Improving the Transport Efficiency. *ISPRS Int. J.-Geo-Inf.* **2022**, *11*, 39. [CrossRef]
60. Wei, J.; Snider, J.M.; Gu, T.; Dolan, J.M.; Litkouhi, B. A behavioral planning framework for autonomous driving. In Proceedings of the IEEE Intelligent Vehicles Symposium Proceedings, Dearborn, MI, USA, 8–11 June 2014; pp. 458–464.
61. Mir, I.; Gul, F.; Mir, S.; Khan, M.A.; Saeed, N.; Abualigah, L.; Abuhaija, B.; Gandomi, A.H. A Survey of Trajectory Planning Techniques for Autonomous Systems. *Electronics* **2022**, *11*, 2801. [CrossRef]
62. Jond, H.; Nabiyev, V.; Benveniste, R. Trajectory Planning Using High Order Polynomials under Acceleration Constraint. *J. Optim. Ind. Eng.* **2016**, *10*, 1–6.
63. Pauca, O.; Maxim, A.; Caruntu, C.F. Trajectory Planner based on Third-order Polynomials applied for Platoon Merging and Splitting. In Proceedings of the 29th Mediterranean Conference on Control and Automation (MED), Puglia, Italy, 22–25 June 2021; pp. 83–88.

64. Zheng, L.; Zeng, P.; Yang, W.; Li, Y.; Zhan, Z. Bézier curve-based trajectory planning for autonomous vehicles with collision avoidance. *IET Intell. Transp. Syst.* **2020**, *14*, 1882–1891. [CrossRef]
65. Song, B.; Wang, Z.; Zou, L. An improved PSO algorithm for smooth path planning of mobile robots using continuous high-degree Bezier curve. *Appl. Soft Comput.* **2021**, *100*, 106960. [CrossRef]
66. Dixit, S.; Montanaro, U.; Dianati, M.; Oxtoby, D.; Mizutani, T.; Mouzakitis, A.; Fallah, S. Trajectory Planning for Autonomous High-Speed Overtaking in Structured Environments Using Robust MPC. *IEEE Trans. Intell. Transp. Syst.* **2020**, *21*, 2310–2323. [CrossRef]
67. Zuo, Z.; Yang, X.; Li, Z.; Wang, Y.; Han, Q.; Wang, L.; Luo, X. MPC-Based Cooperative Control Strategy of Path Planning and Trajectory Tracking for Intelligent Vehicles. *IEEE Trans. Intell. Veh.* **2021**, *6*, 513–522. [CrossRef]
68. Pauca, O.; Maxim, A.; Caruntu, C. Control Architecture for Cooperative Autonomous Vehicles Driving in Platoons at Highway Speeds. *IEEE Access* **2021**, *9*, 153472–153490. [CrossRef]
69. Zhang, S.; Liu, X.; Deng, G.; Ou, J.; Yang, E.; Yang, S.; Li, T. Longitudinal and Lateral Control Strategies for Automatic Lane Change to Avoid Collision in Vehicle High-Speed Driving. *Sensors* **2023**, *23*, 5301. [CrossRef]
70. Gao, Y. Model Predictive Control for Autonomous and Semiautonomous Vehicles. Ph.D. Thesis, UC Berkeley, Berkeley, CA, USA, 2014.
71. Keviczky, T.; Falcone, P.; Borrelli, F.; Asgari, J.; Hrovat, D. Predictive control approach to autonomous vehicle steering. In Proceedings of the 2006 American Control Conference, Minneapolis, MN, USA, 14–16 June 2006; Volume 15, p. 6.
72. Alfatti, F.; Montani, M.; Favilli, T.; Annicchiarico, C.; Berzi, L.; Pierini, M.; Pugi, L.; Capitani, R. Implementation and Performances Evaluation of Advanced Automotive Lateral Stability Controls on a Real-Time Hardware in the Loop Driving Simulator. *Appl. Sci.* **2023**, *13*, 6592. [CrossRef]
73. Yang, T.; Bai, Z.; Li, Z.; Feng, N.; Chen, L. Intelligent Vehicle Lateral Control Method Based on Feedforward + Predictive LQR Algorithm. *Actuators* **2021**, *10*, 228. [CrossRef]
74. Moreno-Gonzalez, M.; Artuñedo, A.; Villagra, J.; Join, C.; Fliess, M. Speed-Adaptive Model-Free Path-Tracking Control for Autonomous Vehicles: Analysis and Design. *Vehicles* **2023**, *5*, 698–717. [CrossRef]
75. Xu, Q.; Lu, X.; Xu, J. Optimized Active Collision Avoidance Algorithm of Intelligent Vehicles. *Electronics* **2023**, *12*, 2451. [CrossRef]
76. Maxim, A.; Lazar, C.; Caruntu, C.F. Distributed Model Predictive Control Algorithm with Communication Delays for a Cooperative Adaptive Cruise Control Vehicle Platoon. In Proceedings of the 28th Mediterranean Conference on Control and Automation, Saint-Raphael, France, 15–18 September 2020; pp. 909–914.
77. Guo, J.; Wang, Y.; Chu, L.; Bai, C.; Hou, Z.; Zhao, D. Adaptive Cruise System Based on Fuzzy MPC and Machine Learning State Observer. *Sensors* **2023**, *23*, 5722. [CrossRef] [PubMed]
78. Braband, M.; Scherer, M.; Voos, H. Global Sensitivity Analysis of Economic Model Predictive Longitudinal Motion Control of a Battery Electric Vehicle. *Electronics* **2022**, *11*, 1574. [CrossRef]
79. Wang, Y.; Jin, P.J. Model predictive control policy design, solutions, and stability analysis for longitudinal vehicle control considering shockwave damping. *Transp. Res. Part C Emerg. Technol.* **2023**, *148*, 104038. [CrossRef]
80. Kebbati, Y.; Ait-Oufroukh, N.; Vigneron, V.; Ichalal, D.; Gruyer, D. Optimized self-adaptive PID speed control for autonomous vehicles. In Proceedings of the 26th International Conference on Automation and Computing (ICAC), Portsmouth, UK, 2–4 September 2021; pp. 1–6.
81. Lazar, R.G.; Caruntu, C.F. Simulator for the Automotive Diagnosis System on CAN using Vector CANoe Environment. In Proceedings of the 24th International Conference on System Theory, Control and Computing (ICSTCC), Sinaia, Romania, 8–10 October 2020; pp. 705–710.
82. Rohit, M.; Ritesh, S.; Gunjan, M. An Analytical Study of Communication Protocols Used in Automotive Industry. *Int. J. Eng. Res. Technol.* **2014**, *2*, 287–292.
83. Bozdal, M.; Samie, M.; Jennions, I. A Survey on CAN Bus Protocol: Attacks, Challenges, and Potential Solutions. In Proceedings of the International Conference on Computing, Electronics & Communications Engineering (iCCECE), Southend, UK, 16–17 August 2018; pp. 201–205.
84. Páez, F.; Kaschel, H. Design and Testing of a Computer Security Layer for the LIN Bus. *Sensors* **2022**, *22*, 6901. [CrossRef]
85. Brandl, M.; Kellner, K. Performance Evaluation of Power-Line Communication Systems for LIN-Bus Based Data Transmission. *Electronics* **2021**, *10*, 85. [CrossRef]
86. Rathore, R.S.; Hewage, C.; Kaiwartya, O.; Lloret, J. In-Vehicle Communication Cyber Security: Challenges and Solutions. *Sensors* **2022**, *22*, 6679. [CrossRef]
87. Dvořák, J.; Hanzálek, Z. Using Two Independent Channels With Gateway for FlexRay Static Segment Scheduling. *IEEE Trans. Ind. Inform.* **2016**, *12*, 1887–1895. [CrossRef]
88. Shaw, R.; Jackman, B. An introduction to FlexRay as an industrial network. In Proceedings of the IEEE International Symposium on Industrial Electronics, Cambridge, UK, 30 June–2 July 2008; pp. 1849–1854.
89. Tuohy, S.; Glavin, M.; Hughes, C.; Jones, E.; Trivedi, M.; Kilmartin, L. Intra-Vehicle Networks: A Review. *IEEE Trans. Intell. Transp. Syst.* **2015**, *16*, 534–545. [CrossRef]
90. Lo Bello, L.; Patti, G.; Leonardi, L. A Perspective on Ethernet in Automotive Communications - Current Status and Future Trends. *Appl. Sci.* **2023**, *13*, 1278. [CrossRef]

91. Sumorek, A.; Buczaj, M. New elements in vehicle communication "media oriented systems transport" protocol. *Teka Kom Motoryz. Energ. Rol. PAN* **2012**, *12*, 275–279.
92. Hank, P.; Müller, S.; Vermesan, O.; Van Den Keybus, J. Automotive Ethernet: In-vehicle networking and smart mobility. In Proceedings of the Design, Automation & Test in Europe Conference & Exhibition (DATE), Grenoble, France, 18–22 March 2013 pp. 1735–1739.
93. Kawser, M.; Sajjad, S.; Fahad, S.; Ahmed, S.; Rafi, H. The Perspective of Vehicle-to-Everything (V2X) Communication towards 5G *Int. J. Comput. Sci. Netw. Secur.* **2019**. *19*, 146–155.
94. Arena, F.; Pau, G. An Overview of Vehicular Communications. *Future Internet* **2019**, *11*, 27. [CrossRef]
95. Anaya, J.J.; Talavera, E.; Jiménez, F.; Zato, J.G.; Gómez, N.; Naranjo, J.E. GeoNetworking based V2V Mesh Communications over WSN. In Proceedings of the 16th International IEEE Conference on Intelligent Transportation Systems (ITSC), The Hague The Netherlands, 6–9 October 2013; pp. 2421–2426.
96. Darbha, S.; Konduri, S.; Pagilla, P. Benefits of V2V Communication for Autonomous and Connected Vehicles. *arXiv* **2018** arXiv:1803.02900.
97. Malik, R.Q.; Alsattar, H.A.; Ramli, K.N.; Zaidan, B.B.; Zaidan, A.A.; Kareem, Z.H.; Ameen, H.A.; Garfan, S.; Mohammed, A. Zaidan, R.A. Mapping and Deep Analysis of Vehicle-to-Infrastructure Communication Systems: Coherent Taxonomy, Datasets Evaluation and Performance Measurements, Motivations, Open Challenges, Recommendations, and Methodological Aspects *IEEE Access* **2019**, *7*, 126753–126772. [CrossRef]
98. Xie, G.; Li, Y.; Han, Y.; Xie, Y.; Zeng, G.; Li, R. Recent Advances and Future Trends for Automotive Functional Safety Design Methodologies. *IEEE Trans. Ind. Inform.* **2020**, *16*, 5629–5642. [CrossRef]
99. Bučko, B.; Michálek, M.; Papierniková, K.; Zábovská, K. Smart Mobility and Aspects of Vehicle-to-Infrastructure: A Data Viewpoint. *Appl. Sci.* **2021**, *11*, 10514. [CrossRef]
100. Kaiwartya, O.; Abdullah, A.H.; Cao, Y.; Altameem, A.; Prasad, M.; Lin, C.T.; Liu, X. Internet of Vehicles: Motivation, Layered Architecture, Network Model, Challenges, and Future Aspects. *IEEE Access* **2016**, *4*, 5356–5373. [CrossRef]
101. Anaya, J.; Merdrignac, P.; Shagdar, O.; Nashashibi, F. Vehicle to Pedestrian Communications for Protection of Vulnerable road Users. In Proceedings of the IEEE Intelligent Vehicles Symposium Proceedings, Dearborn, MI, USA, 8–11 June 2014 pp. 1037–1042.
102. Sewalkar, P.; Seitz, J. Vehicle-to-Pedestrian Communication for Vulnerable Road Users: Survey, Design Considerations, and Challenges. *Sensors* **2019**, *19*, 358. [CrossRef]
103. Wang, X.; Mao, S.; Gong, M.X. An Overview of 3GPP Cellular Vehicle-to-Everything Standards. *GetMobile Mobile Comp. Commun* **2017**, *21*, 19–25. [CrossRef]
104. Abdelkader, G.; Elgazzar, K.; Khamis, A. Connected Vehicles: Technology Review, State of the Art, Challenges and Opportunities *Sensors* **2021**, *21*, 7712. [CrossRef] [PubMed]
105. *1609.2-2016*; IEEE Standard for Wireless Access in Vehicular Environments–Security Services for Applications and Management Messages; Revision of IEEE Std 1609.2-2013. IEEE: Piscataway, NJ, USA, 2016; pp. 1–240.
106. *1609.3-2020*; IEEE Standard for Wireless Access in Vehicular Environments (WAVE)–Networking Services; Revision of IEEE Std 1609.3-2016; IEEE: Piscataway, NJ, USA, 2016; pp. 1–210.
107. Gyawali, S.; Xu, S.; Qian, Y.; Hu, R.Q. Challenges and Solutions for Cellular Based V2X Communications. *IEEE Commun. Surv. Tutor.* **2021**, *23*, 222–255. [CrossRef]
108. *802.11p-2010*; IEEE Standard for Information Technology—Local and metropolitan area networks—Specific requirements—Part 11: Wireless LAN Medium Access Control (MAC) and Physical Layer (PHY) Specifications Amendment 6: Wireless Access in Vehicular Environments; IEEE: Piscataway, NJ, USA, 2010.
109. MacHardy, Z.; Khan, A.; Obana, K.; Iwashina, S. V2X Access Technologies: Regulation, Research, and Remaining Challenges *IEEE Commun. Surv. Tutor.* **2018**, *20*, 1858–1877. [CrossRef]
110. Nkenyereye, L.; Nkenyereye, L.; Islam, S.M.R.; Choi, Y.H.; Bilal, M.; Jang, J.W. Software-Defined Network-Based Vehicular Networks: A Position Paper on Their Modeling and Implementation. *Sensors* **2019**, *19*, 3788. [CrossRef] [PubMed]
111. Chen, S.; Hu, J.; Shi, Y.; Zhao, L.; Li, W. A Vision of C-V2X: Technologies, Field Testing, and Challenges With Chinese Development *IEEE Internet Things J.* **2020**, *7*, 3872–3881. [CrossRef]
112. Garcia, M.H.C.; Molina-Galan, A.; Boban, M.; Gozalvez, J.; Coll-Perales, B.; Şahin, T.; Kousaridas, A. A Tutorial on 5G NR V2X Communications. *IEEE Commun. Surv. Tutor.* **2021**, *23*, 1972–2026. [CrossRef]
113. Lazar, R.G.; Militaru, A.V.; Caruntu, C.F.; Pascal, C.; Patachia-Sultanoiu, C. Real-time data measurement methodology to evaluate the 5G network performance indicators. *IEEE Access* **2023**, *11*, 43909–43924. [CrossRef]
114. Muhammad, M.; Safdar, G.A. 5G-based V2V broadcast communications: A security perspective. *Array* **2021**, *11*, 100084.
115. Khan, M.J.; Khan, M.A.; Malik, S.; Kulkarni, P.; Alkaabi, N.; Ullah, O.; El-Sayed, H.; Ahmed, A.; Turaev, S. Advancing C-V2X for Level 5 Autonomous Driving from the Perspective of 3GPP Standards. *Sensors* **2023**, *23*, 2261. [CrossRef] [PubMed]
116. Ploeg, J.; Scheepers, B.T.M.; van Nunen, E.; van de Wouw, N.; Nijmeijer, H. Design and experimental evaluation of cooperative adaptive cruise control. In Proceedings of the 14th International IEEE Conference on Intelligent Transportation Systems (ITSC), Washington, DC, USA, 5–7 October 2011; pp. 260–265.

117. Jiang, D.; Chen, Q.; Delgrossi, L. Optimal Data Rate Selection for Vehicle Safety Communications. In Proceedings of the 5th ACM International Workshop on VehiculAr Inter-NETworking, VANET'08, San Francisco, CA, USA, 15 September 2008; Association for Computing Machinery: New York, NY, USA, 2008; pp. 30–38.
118. Chanfreut, P.; Keijzer, T.; Ferrari, R.M.; Maestre, J.M. A Topology-Switching Coalitional Control and Observation Scheme with Stability Guarantees. *IFAC-PapersOnLine* **2020**, *53*, 6477–6482. [CrossRef]
119. Long, H.; Khalatbarisoltani, A.; Hu, X. MPC-based Eco-Platooning for Homogeneous Connected Trucks Under Different Communication Topologies. In Proceedings of the IEEE Intelligent Vehicles Symposium, Aachen, Germany, 4–9 June 2022; pp. 241–246.
120. Pauca, O.; Maxim, A.; Caruntu, C.F. Communication topologies evaluation for a vehicle merging into a platoon on highway. In Proceedings of the 30th Mediterranean Conference on Control and Automation (MED), Vouliagmeni, Greece, 28 June–1 July 2022; pp. 957–962.
121. Dunbar, W.B.; Caveney, D.S. Distributed Receding Horizon Control of Vehicle Platoons: Stability and String Stability. *IEEE Trans. Autom. Control.* **2012**, *57*, 620–633. [CrossRef]
122. Kianfar, R.; Falcone, P.; Fredriksson, J. A Distributed Model Predictive Control Approach to Active Steering Control of String Stable Cooperative Vehicle Platoon. *IFAC Proc. Vol.* **2013**, *46*, 750–755. [CrossRef]
123. Pauca, O.; Lazar, M.; Caruntu, C.F. Hybrid Cost Function Distributed MPC for Vehicle Platoons with Stability and String Stability Properties. *IEEE Control. Syst. Lett.* **2023**, *7*, 2257–2262. [CrossRef]
124. Maxim, A.; Pauca, O.; Caruntu, C.F. Distributed Model Predictive Control and Coalitional Control Strategies—Comparative Performance Analysis Using an Eight-Tank Process Case Study. *Actuators* **2023**, *12*, 281. [CrossRef]

sensors

MDPI

Article

Urban Advanced Mobility Dependability: A Model-Based Quantification on Vehicular Ad Hoc Networks with Virtual Machine Migration

Luis Guilherme Silva [1], Israel Cardoso [1], Carlos Brito [1], Vandirleya Barbosa [1], Bruno Nogueira [2], Eunmi Choi [3], Tuan Anh Nguyen [4,*], Dugki Min [5,*], Jae Woo Lee [6,*] and Francisco Airton Silva [1]

1 Coordination of the Information Systems Course, Technical College of Teresina (CSHNB), Federal University of Piauí (UFPI), Picos 64049-550, Piauí, Brazil; luis.e@ufpi.edu.br (L.G.S.); israel.araujo@ufpi.edu.br (I.C.); carlosvictor@ufpi.edu.br (C.B.); vandirleya.barbosa@ufpi.edu.br (V.B.); faps@ufpi.edu.br (F.A.S.)
2 Instituto de Computação, Federal University of Alagoas (UFAL), Maceió 57072-900, Alagoas, Brazil; bruno@ic.ufal.br
3 School of Software, College of Computer Science, Kookmin University, Seoul 02707, Republic of Korea; emchoi@kookmin.ac.kr
4 Konkuk Aerospace Design-Airworthiness Research Institute (KADA), Konkuk University, Seoul 05029, Republic of Korea
5 Department of Computer Science and Engineering, College of Engineering, Konkuk University, Seoul 05029, Republic of Korea
6 Department of Aerospace Information Engineering, Konkuk University, Seoul 05029, Republic of Korea
* Correspondence: anhnt2407@konkuk.ac.kr (T.A.N.); dkmin@konkuk.ac.kr (D.M.); jwlee@konkuk.ac.kr (J.W.L.)

Citation: Silva, L.G.; Cardoso, I.; Brito, C.; Barbosa, V.; Nogueira, B.; Choi, E.; Nguyen, T.A.; Min, D.; Lee, J.W.; Silva, F.A. Urban Advanced Mobility Dependability: A Model-Based Quantification on Vehicular Ad Hoc Networks with Virtual Machine Migration. *Sensors* **2023**, *23*, 9485. https://doi.org/10.3390/s23239485

Academic Editors: Constantin Caruntu and Ciprian Romeo Comşa

Received: 13 November 2023
Revised: 26 November 2023
Accepted: 27 November 2023
Published: 28 November 2023

Abstract: In the rapidly evolving urban advanced mobility (UAM) sphere, Vehicular Ad Hoc Networks (VANETs) are crucial for robust communication and operational efficiency in future urban environments. This paper quantifies VANETs to improve their reliability and availability, essential for integrating UAM into urban infrastructures. It proposes a novel Stochastic Petri Nets (SPN) method for evaluating VANET-based Vehicle Communication and Control (VCC) architectures, crucial given the dynamic demands of UAM. The SPN model, incorporating virtual machine (VM) migration and Edge Computing, addresses VANET integration challenges with Edge Computing. It uses stochastic elements to mirror VANET scenarios, enhancing network robustness and dependability, vital for the operational integrity of UAM. Case studies using this model offer insights into system availability and reliability, guiding VANET optimizations for UAM. The paper also applies a Design of Experiments (DoE) approach for a sensitivity analysis of SPN components, identifying key parameters affecting system availability. This is critical for refining the model for UAM efficiency. This research is significant for monitoring UAM systems in future cities, presenting a cost-effective framework over traditional methods and advancing VANET reliability and availability in urban mobility contexts.

Keywords: Vehicular Ad Hoc Networks (VANETs); dependability modeling; Stochastic Petri Nets (SPN); virtual machine migration; network reliability and availability

1. Introduction

Vehicular Ad Hoc Networks (VANETs) represent a significant advancement in wireless communication technology. These networks are pivotal in enhancing vehicular connectivity, thereby fostering a safer and more efficient environment for all traffic participants. GSMA estimates indicate that approximately 20% of the global vehicular fleet, estimated at 1.5 billion, are internet-connected, contributing substantially to data generation [1]. Projections suggest that by 2027, there will be an annual growth rate of around 17%, resulting in 367 million connected vehicles.

Typically, a VANET incorporates a static infrastructure component, known as a Roadside Unit (RSU), positioned alongside thoroughfares. Vehicles interface with this infrastructure through an onboard unit (OBU) [2]. It is presumed that each vehicle is outfitted with sensors to collect environmental data. The OBU processes these data and engages in communication with other vehicles or RSUs, either directly or indirectly. Additionally, RSUs have the capability to connect to the internet, thereby facilitating vehicular access to various services [3].

VANETs find application in diverse traffic-related domains, encompassing areas such as network security [4], traffic management [5], and parking space optimization [6]. Nonetheless, managing the Quality of Service (QoS) within VANETs poses a multifaceted and critical challenge. Key challenges include ensuring network availability and reliability, addressing latency, and managing traffic, as well as grappling with issues of poor connectivity, limited flexibility, and scalability constraints. The physical distance from Edge processing and storage centers can also introduce considerable communication delays [7–9].

The literature review reveals a paucity of studies employing Stochastic Petri Nets (SPN) in the context under consideration in this research. The proposed model incorporates crucial factors such as availability and reliability, utilizing VM migration across the network to link various RSUs and implementing an Edge Computing-based data processing system. This model facilitates a comprehensive analysis of the key parameters, aiming to refine architectures to address the integration challenges of VANETs with Edge Computing. The principal contributions of this research are as follows:

- Development of an SPN model to assess the reliability and availability of VANET-based VCC architectures, factoring in stochastic elements to emulate realistic scenarios. This aims to enhance the robustness of vehicular network environments, ensuring dependable and consistent performance.
- Execution of case studies using the proposed models, offering a blueprint for other researchers in applying these models. These case studies focus on identifying and analyzing primary parameters influencing system availability, providing preliminary insights into the critical variables affecting system reliability, and facilitating enhancements and optimizations.
- Conducting a sensitivity analysis of the SPN model components, identifying parameters with significant influence on system availability. This analysis enhances the understanding of the model and aids in its optimization.

The structure of this work is organized as follows: Section 2 introduces essential concepts foundational to this study. Section 3 describes the architecture forming the basis of the model. Section 4 details the developed SPN model. Section 5 presents two sensitivity analyses performed on the SPN model. Section 6 elaborates on the outcomes of the case study. Finally, Section 7 concludes the paper and outlines directions for future work.

2. Background

This section succinctly delineates the core concepts fundamental to this research, primarily focusing on SPNs. An elucidation of experimental design and sensitivity analysis will follow. These concepts are critical for comprehending the methodologies and techniques employed in the formulation of this article. Understanding these foundational principles is essential for appreciating the intricacies of the proposed models and analyses within the context of this study. Related works of the above discussions are provided in Table 1.

2.1. Stochastic Petri Nets

A Petri Net is a combined graphical and mathematical representation that effectively models systems and processes undergoing continuous changes and concurrent operations. This modeling is particularly valuable for systems characterized by simultaneous multi-action scenarios [10]. The present study employs a sophisticated form of Petri Nets, termed SPNs, which are distinguished by their ability to incorporate randomness and probabilistic behaviors [11]. SPNs are comprised of three primary elements that delineate the system's states or conditions: transitions (denoting potential system events or actions), tokens (symbolizing entities within the system), and each token's potential association with a specific resource [12].

The functionality of SPNs hinges on two essential types of connections: (I) Input Arcs, which are preconditions for triggering transitions, necessitating specific tokens' presence at designated locations; and (II) Output Arcs, which dictate the subsequent transferal of tokens upon a transition's activation [13]. Transitions in SPNs are categorized into timed transitions, obeying stochastic distributions [14], and immediate transitions that occur instantaneously upon activation. Additionally, inhibitor arcs play a pivotal role in controlling token flow between locations, with tokens being allocated to particular places within the system.

Crucially, SPNs utilize guard conditions to define the specific prerequisites for transition activations. These conditions often incorporate random variables, thus introducing a probabilistic dimension. The fulfillment of these guard conditions triggers transitions according to predefined rate functions, effectuating a change in the system's state [15]. Figure 1 visually explicates the core components of an SPN model, providing a comprehensive understanding of its structure and functionality.

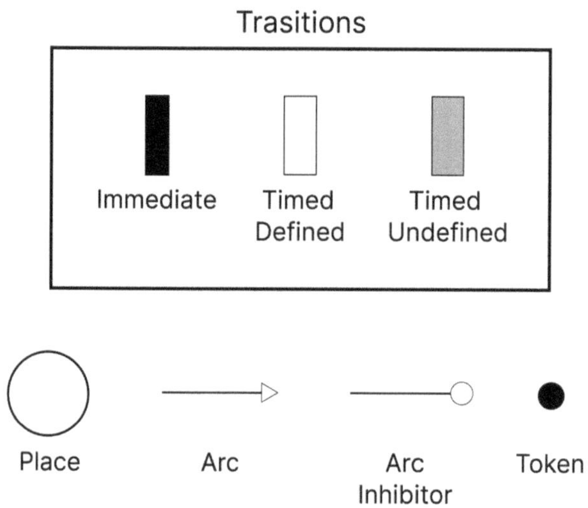

Figure 1. SPN components.

2.2. *Sensitivity Analysis with DoE*

Design of Experiments (DoE) is an extensively utilized methodology in research and development for enhancing processes, products, and systems. It entails the meticulous planning and execution of controlled experiments to gather pertinent and substantial data [16]. The initial step in DoE involves defining the experiment's objective, followed by identifying the variables or factors that could influence the outcome. Subsequently, an experimental plan is formulated, which includes determining the levels for each factor and designing the experiments to extract significant insights. The execution of the experiment is aligned with this plan, leading to the collection and statistical analysis of the results.

DoE equips system designers with the ability to discern the most influential variables, understand their interactions, and fine-tune the conditions to achieve optimal outcomes with minimal experimental iterations [17]. Complementing DoE, sensitivity analysis serves as a pivotal technique for examining how variations in the input parameters or model attributes influence the outputs or results [18]. This analysis is essential for assessing the robustness of a system or model in the face of uncertainties or changes in parameters [19].

Through sensitivity analysis, it becomes feasible to identify which variables significantly affect outputs and which have lesser impacts, thereby guiding the allocation of resources and efforts towards efficient system optimization. Both DoE and sensitivity analysis play crucial roles in research, engineering, and strategic decision making, providing pathways to more effective solutions, resource and time conservation, and the acquisition of valuable insights for the enhancement of processes and systems. This efficiency is achieved as they necessitate fewer experiments to glean significant information [20,21].

Table 1. Related works.

Reference	Contribution	Assessment Method	Metrics	Multiple RSUs	Sensitivity Analysis
[22]	A performance modeling of media access control (MAC).	Simulation	Performance	No	No
[23]	Threat-Oriented Authentication Approach for Secure Communication.	Simulation	Performance	No	No
[24]	Modeling that integrates the transmission of the 802.11p system and the queuing process.	Simulation	Performance	No	No
[25]	A mobile agent-based information dissemination scheme in the VANET environment.	Simulation	Performance	No	No
[26]	A mobile agent migration mechanism based on location simulation experiments in the VANET environment.	Simulation	Performance	No	No
[27]	A TCP Context Migration Scheme (TOMS) method for enhancing data services in vehicular networks.	Simulation	Performance	Yes	No
[28]	Detection of anomalies, loss of messages with conventional and VEC techniques.	Simulation	Availability	Yes	No
[29]	A container-based virtualization and live migration framework for the in-vehicle ad hoc network.	Measurement	Performance	Yes	No
[30]	Provide a classification of security requirements, characteristics and security challenges.	Measurement	Does not have	Yes	No
[31]	A seamless handover system in a software-defined network (SDN) framework.	Measurement	Performance	Yes	No
[32]	BaaS (Broadcast as a Service) transmission is proposed for VANET to disseminate data efficiently to network vehicles using cloud computing.	Measurement	Performance	Yes	No
[33]	It presents a model for the connectivity patterns of chains of vehicles traveling on a highway.	Markov Model	Availability	No	No
[34]	Analytical model based on Stochastic Petri Net (SPN) theory for assessment of Vehicular Ad Hoc Network infrastructures.	SPN Model	Performance	No	No
[35]	Use SDN to improve the allocation and migration of microservices in Vehicular Fog Networks (VFN).	Measurement	Performance	Yes	No
This work	Modeling an architecture with multiple RSUs and migration to assess system availability.	SPN Model	Availability	Yes	Yes

2.2.1. Simulation-Based Methods

In the initial category of research, simulation served as the primary method for evaluation. The study by [22] focused on assessing the performance of Medium Access Control (MAC) protocols within VANETs. Ref. [23] introduced a threat-oriented authentication strategy designed to bolster secure communication in vehicle-to-vehicle (V2V) and vehicle-to-infrastructure (V2I) interactions, utilizing a combination of encryption keys. The research conducted by [24] highlighted deficiencies in the 802.11 system, particularly the absence of backoff and binary exponential retransmission mechanisms, which were found to adversely affect the QoS during periods of intense traffic.

Further, Ref. [25] proposed an innovative mobile agent migration mechanism. This mechanism, rooted in network location analytics, was employed to simulate a VANET environment, exploring its practical applications. In parallel, the studies by [26,27] utilized the TCP Context Migration Scheme (TOMS), a novel approach aimed at enhancing data services within vehicular networks. This scheme entailed the proactive establishment of TCP connections, managed by a mobile TCP proxy that assumed the role of a cluster leader.

Lastly, the investigation by [28] delved into the application of Vehicle Edge Computing (VEC) alongside conventional anomaly detection techniques. This approach was targeted at identifying and quantifying message loss, thereby evaluating the extent of fault coverage within these networked systems.

2.2.2. Measurement-Based Methods

In the second category, the selected studies employed measurement as their primary evaluation methodology. Ref. [29] developed a container-based virtualization architecture, facilitating dynamic migration within ad hoc vehicular networks, particularly in VANETs. This innovation aimed to enhance flexibility and responsiveness in these networks. Meanwhile, Ref. [30] undertook the task of classifying security requirements, identifying key characteristics, and delineating the challenges related to security within similar VANET scenarios.

In another significant contribution, Ref. [31] introduced a seamless transition system that leverages the capabilities of SDN and Media Independent Handover (MIH). This system was designed to dynamically modify the topology of VANETs, enhancing their adaptability and efficiency. Additionally, Ref. [32] proposed a novel concept termed Broadcast as a Service (BaaS), specifically tailored for VANETs. This solution aimed to efficiently disseminate data across networked vehicles utilizing cloud computing technologies. Lastly, the work of Ref. [35] applied SDN to improve the management and migration of microservices in Vehicular Fog Networks (VFNs), taking into account the dynamic nature of vehicular nodes.

2.2.3. Modeling-Based Methods

The third classification encompasses studies that utilized modeling as their core evaluation technique. The research of [33] presented a model to describe the connectivity patterns among vehicles on highways. This model is crucial for the development of protocols and applications in VANETs, tailored to their specific connectivity traits. Similarly, [34] adopted an analytical approach rooted in SPN theory. This approach was used to assess the infrastructure of VANETs, taking into consideration the mobility of the network and its inherent limitations. Notably, this study modeled the service fees of RSUs using exponential distributions.

2.2.4. Contributions of This Work in Relation to Others

This study introduces an SPN model that evaluates the impact of VM migration across multiple RSUs within vehicular networks. A review of the literature reveals the scarcity of studies addressing this specific scenario within the VANET context. The adoption of such an approach is economically advantageous, as it enables the analysis of availability and reliability without necessitating a physical infrastructure for testing.

The review further indicates that systems modeling, as employed in this study, provides a more predictive and comprehensive understanding compared to measurement and simulation methods. This is achieved by simplifying the representation of critical system elements. In contrast, measurement and simulation tend to rely on observational data, which may not fully encapsulate the complexity of the system. Most of the reviewed studies focused primarily on performance evaluation, with limited attention to metrics such as availability, reliability, or downtime. Moreover, there is a noticeable gap in studies exploring the cooperation between multiple RSUs and the use of sensitivity analysis to determine the impact of Mean Time To Failure (MTTF) and Mean Time To Repair (MTTR) parameters on system performance.

3. Evaluated Architecture

In this section, the envisaged architecture for integrating VANETs is elucidated. The foundational scenario, as illustrated in Figure 2, encompasses an array of RSUs, with their respective coverage zones depicted as green and red circles. The operational dynamics of this architecture are as follows:

- (1) *Active RSU Coverage and Vehicle Interaction*: Vehicles in transit enter the coverage area of active RSUs (represented by green circles), wherein these RSUs facilitate communication and gather data from the vehicles.
- (2) *Response to RSU Failure*: In the event of an RSU malfunction, leading to a disruption in data collection, a contingency protocol is activated.
- (3) *VM Migration for Uninterrupted System Availability*: To ensure continued system functionality, an allocation of data from VMs is performed that will be transferred to the subsequent RSU within the network after an RSU fails.
- (4) *Data Management*: Subsequent to collection, all data are transmitted to an Edge Server for storage and further processing.

The RSUs in this architecture are equipped with advanced communication technologies, potentially including 5G [36] or Lora [37], enabling interaction with vehicular systems. These units are strategically placed along roadways to form a dependable communication infrastructure, essential for the efficient operation of the VANET system. This arrangement guarantees a seamless data flow and operational continuity of the system, even amidst individual RSU failures, thereby augmenting the reliability and resilience of the VANET infrastructure.

In this segment, the technical composition of the VANETs is detailed, emphasizing the integration of On-Board Units (OBUs) in each vehicle. These OBUs are imbued with communication capabilities, enabling the transmission and reception of messages. Additionally, they are outfitted with GPS tracking devices, facilitating the sharing of precise, real-time locational data [38]. The infrastructure is based on the assumption that all RSUs maintain a connection to a private Edge server via a high-speed wireless link, such as 5G; 5G offers faster data transmission speeds and reduced latency, key features for system efficiency [39].

The selection of these technologies plays a significant role in influencing the overall system availability [40]. The choice of communication technology emerges as a vital consideration in the model's design process. The architecture's design is inherently scalable, allowing for the integration of a variable number of RSUs to meet the specific requirements of the deployment area. The depicted scenario in the figure showcases four groups of RSUs, but this configuration can be dynamically adjusted to suit varying demands.

A key attribute of this architecture is its fault detection capability in RSUs. When an RSU is compromised, indicated by red in Figure 2, it triggers the migration of VMs to the next cluster of RSUs. This migration is essential for maintaining operational continuity and ensuring system availability, thereby minimizing disruptions in communication and data processing.

Data processing in this system is executed at the Edge (Edge Computing), which enhances communication efficiency and reduces latency. The proximity of RSUs to the vehicles permits a portion of the data processing to occur locally, thus enhancing the system's real-time responsiveness. The proposed architecture aims to establish an effective communicative link between vehicles and cloud infrastructure, with a strong focus on reliability, scalability, and the migration of VMs to guarantee uninterrupted operations. The ensuing section will delve into the model used to assess the reliability and availability of this system.

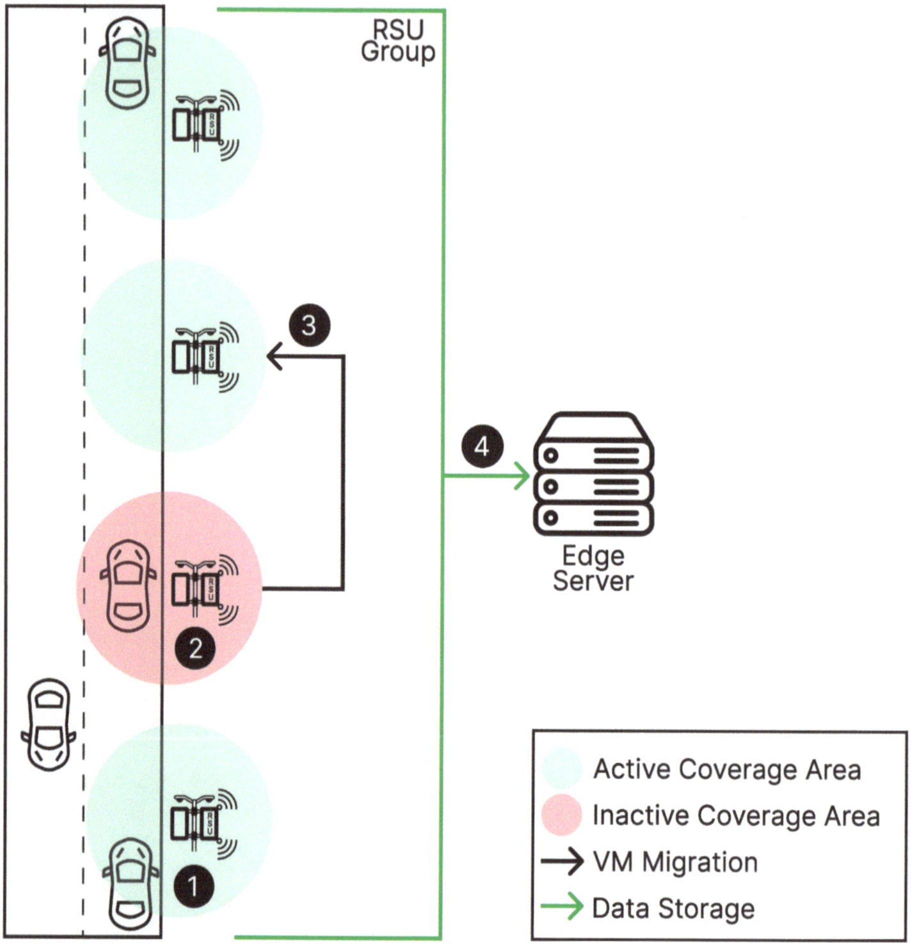

Figure 2. Base architecture.

4. Proposed Model

This section delineates the models applied in this study, which are constructed based on the architecture outlined in the preceding section. It further details the reliability and representative availability models in scenarios both with and without the implementation of migration strategies. All the models and simulations in this study were conducted using the Mercury Tool [41].

4.1. System Reliability Model

The model for analyzing the reliability of VANETs is depicted in Figure 3. Within this context, reliability is defined as the conditional probability that a system will continue functioning over a time interval [0, t], provided that it was operational at the inception of this interval (t = 0) . The presented model (Figure 3) bears resemblance to the model in Figure 4, with a notable distinction: it excludes the MTTR transitions that would facilitate the recovery of components in the event of a failure. This exclusion is a critical aspect, as it directly impacts the system's ability to self-recover post-failure, thereby influencing the overall reliability assessment of the VANET system under study.

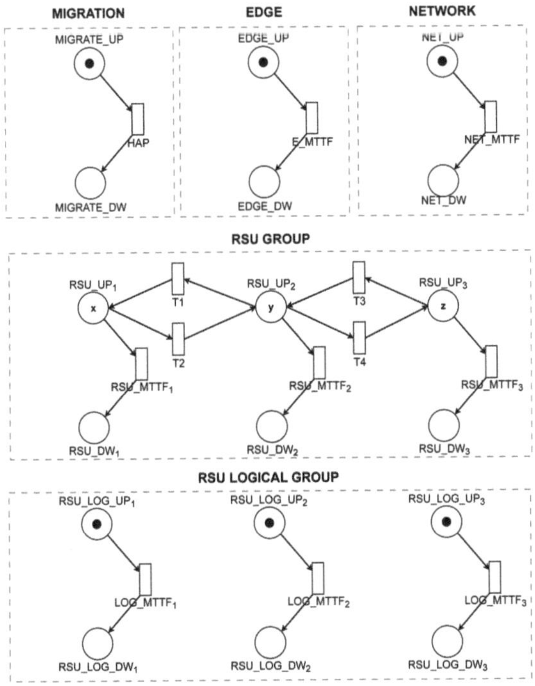

Figure 3. Reliability model.

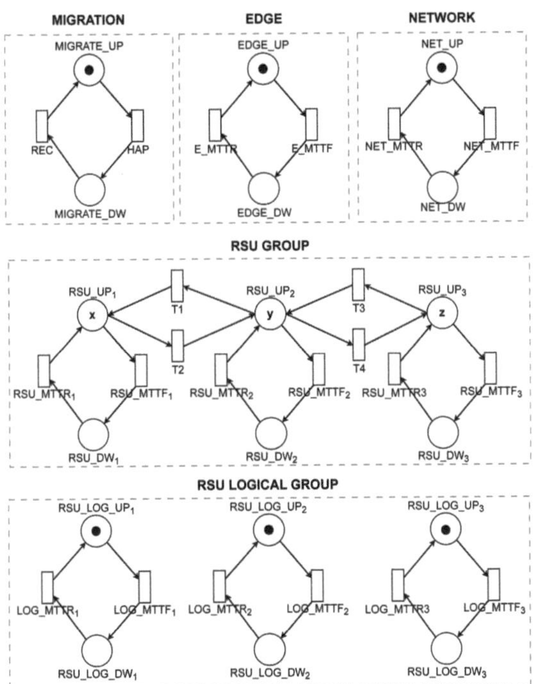

Figure 4. Availability model with migration.

In the RSU GROUP within the VANET system, the RSUs can host varying numbers of VMs, symbolized by the variables x, y, and z. The reliability of the model under consideration can be quantitatively assessed using Equation (1). This metric is defined as the complement of the probability of failure of any component in the system. Specifically, it is represented as one minus the probability that the RSU Group, RSU Logical Group, Network, and Edge components play all operations simultaneously:

$$R = 1 - P((\#EDGE\_U > 0) \text{ AND } (\#NET\_UP > 0) \text{ AND } ((\#RSU\_UP1 + \#RSU\_UP2) + \#RSU\_UP3) = (n \cdot \text{TOKENS}) \text{ AND } (\#RSU\_LOG\_UP1 = 1 \text{ OR } \#RSU\_LOG\_UP2 = 1 \text{ OR } \#RSU\_LOG\_UP3 = 1)) \quad (1)$$

In this model, the variable " P" is utilized to determine the probability that the system will become unavailable or fail. By applying this equation, it is possible to generate a curve that effectively illustrates how the system's reliability diminishes over time. This curve is a crucial analytical tool, as it provides a visual representation of the system's reliability, enabling the identification of trends and potential vulnerabilities over a specified time frame. Such an analysis is vital for understanding the robustness of the system and for making informed decisions regarding maintenance, upgrades, or other interventions to enhance the system's reliability.

4.2. Availability Model with Migration

Table 2 systematically details the elements of the availability model, employing tokens to represent the count of operational VMs within each RSU. In this model, the transitions labeled *RSU_MTTF* and *RSU_MTTR* are integral for facilitating the exchange of information among different RSU groups via VMs. For the effective operation of an RSU Group, it is imperative that the quantity of tokens in the *RSU_UP* state aligns with the initially set value.

The availability of an RSU is contingent upon the presence of tokens in the *RSU_LOG_UP* state and their absence in the *RSU_LOG_DW* state. Concurrently, the operational state of the NETWORK is indicated by the distribution of tokens: tokens in the *NET_UP* state signify an active network, while those in *NET_DW* denote an inactive state. This model underscores the reliance on token distribution for depicting the dynamic status of each RSU and the overall network, thus providing a comprehensive view of the system's availability and the efficacy of the migration strategy.

Table 2. Description of the main components of the model.

Type	Components	Description
Places	*MIGRATE_UP*	Migration of VMs between RSUs is available
	MIGRATE_DW	Migration of VMs between RSUs is unavailable
	EDGE_UP	Data processing at the Edge is available
	EDGE_DW	Data processing at the Edge is unavailable
	NET_UP	The network connecting the RSUs is available
	NET_DW	The network connecting the RSUs is unavailable
	RSU_UP1, RSU_UP2, RSU_UP3	The physical RSUs are available
	RSU_DW1, RSU_DW2, RSU_DW3	The physical RSUs are unavailable
	RSU_LOG_UP1, RSU_LOG_UP2, RSU_LOG_UP3	The logical RSUs are available
	RSU_LOG_DW1, RSU_LOG_DW2, RSU_LOG_DW3	The logical RSUs are unavailable
Transitions	*E_MTTR*	Represents the MTTR of the system's Edge computing
	E_MTTF	Represents the MTTF of the system's Edge computing
	NET_MTTR	Represents the MTTR of the system's network
	NET_MTTF	Represents the MTTF of the system's network
	RSU_MTTR1, RSU_MTTR2, RSU_MTTR3	Represents the MTTR of the system's RSUs
	RSU_MTTF1, RSU_MTTF2, RSU_MTTF3	Represents the MTTF of the system's RSUs
	LOG_MTTR1, LOG_MTTR2, LOG_MTTR3	Represents the MTTR of the logical part of the RSUs
	LOG_MTTF1, LOG_MTTF2, LOG_MTTF3	Represents the MTTF of the logical part of the RSUs
	T1, T2, T3, T4	Transitions between RSUs in the system

The transitions labeled *NET_MTTF* and *NET_MTTR* are pivotal in the management of the network's operational dynamics. Concurrently, the functioning of the EDGE component relies on the *E_MTTF* and *E_MTTR* transitions, which govern the activation (*EDGE_UP*) and deactivation (*EDGE_DW*) states. The migration process, a key element for ensuring uninterrupted system operation, is regulated through *MIGRATE_UP* (activation) and *MIGRATE_DW* (deactivation) markers. These markers facilitate the transfer of VMs between RSUs, an action critical for maintaining system availability.

Figure 4 presents the proposed SPN model, which is constructed based on the previously outlined scenario. This model encompasses various components such as RSU GROUP, RSU LOGICAL GROUP, NETWORK, EDGE, and MIGRATION. Each of these components is associated with metrics like MTTF and MTTR, which are instrumental in evaluating the availability and reliability of systems and their individual components.

The operational status of the VMs in each RSU is indicated by the presence of tokens in the *RSU_UP* and *RSU_DW* states. Here, *RSU_UP* denotes active RSUs, while *RSU_DW* represents inactive ones. The transitions between the active and inactive states of each Road Service Unit (RSU) are controlled by the *RSU_MTTF* and *RSU_MTTR* transitions. Within the RSU Group, each unit contributes to the collective functionality by sharing information via VMs. For the RSU Group to function effectively, it is essential that the number of tokens in *RSU_UP* aligns with its pre-established initial value.

Table 3 illustrates the implementation of guard conditions in the transitions T1, T2, T3, and T4, which are situated between the RSUs designated as UP1, UP2, and UP3. These guard conditions are essential for regulating the migration of VMs in scenarios where a logical component of the RSU becomes unavailable. When this logical component is subsequently reactivated and resumes normal operation, the previously migrated VMs are reintegrated into their original RSU. This mechanism of migration and reintegration is pivotal in ensuring the seamless and continuous functioning of the system, as it provides a dynamic response to temporary outages or disruptions within individual RSUs, thereby maintaining overall system integrity and operational continuity.

Table 3. Description of model storage conditions.

Places	Transition	Condition
RSU_UP1, *RSU_UP2*	T2	IF(#*RSU_LOG_DW*2 =0): (N -(#*RSU_UP1*+#*RSU_DW1*)) ELSE(#*RSU_UP1*+(#*RSU_UP1*+#*RSU_UP2*+#*RSU_UP3*))
RSU_UP2, *RSU_UP3*	T4	IF(#*RSU_LOG_DW*3=0): (N-(#*RSU_UP2*+#*RSU_DW2*)) ELSE(#*RSU_UP2*+(#*RSU_UP1*+#*RSU_UP2*+#*RSU_UP3*))

The transition T2 is activated by the guard condition #*RSU_LOG_DW*1=1, which denotes a critical event indicative of the instability or inactivation of RSU 1's logical component. This event initiates the process for VM migration from the compromised RSU. The procedure begins by verifying the operational status of the subsequent RSU, indicated by #*RSU_LOG_DW*2=0. Should this RSU be operational, the migration of VMs is executed accordingly. Subsequently, in the event of a recovery and reactivation of the initially failed RSU, the previously migrated VMs are reintegrated into it. In a parallel scenario, transition T4 is invoked when #*RSU_LOG_Dw*2=1, a condition signaling the deactivation of the logical component of RSU 2. This state necessitates the migration of VMs from RSU 2 to another operational unit, thereby ensuring the continuity of system functionality.

The availability of the model with migration is quantified using Equation (2). This equation computes the probability that the RSU Group, RSU Logical Group, Network, and Edge components are all operational concurrently. In this context, 'P' denotes the probability, while 'TOKENS' refers to the number of tokens present in a specific state or place within the model. This approach to calculating availability is crucial for assessing the effectiveness of the migration strategy in maintaining continuous operation of the system, even in the face of individual component failures or disruptions. The inclusion of these probabilistic measures provides a comprehensive understanding of the system's resilience and its ability to sustain uninterrupted service through dynamic VM migration processes.

$$\begin{aligned} A = P((\#EDGE_U > 0) \text{ AND } (\#NET\_UP > 0) \text{ AND } \\ ((\#RSU\_UP1 + \#RSU\_UP2) + \#RSU\_UP3) = (n \cdot \text{TOKENS}) \text{ AND } \\ (\#RSU\_LOG\_UP1 = 1 \text{ OR } \#RSU\_LOG\_UP2 = 1 \text{ OR } \#RSU\_LOG\_UP3 = 1)) \end{aligned} \quad (2)$$

4.3. SPN Availability Model: Non-Migration Framework

The depicted SPN availability model in Figure 5 delineates the system's functionality in the absence of migration capabilities. This model parallels its counterpart incorporating migration, encompassing components such as MIGRATION, EDGE, NETWORK, RSU GROUP, and RSU LOGICAL GROUP. Integral to each component is an MTTF and a singular MTTR, with the exception of MIGRATION, which is characterized by the Happened attribute (HAP). This attribute simulates potential disasters or system instabilities, leading to the activation of *MIGRATE_DW*, thereby deactivating migration processes.

A notable deviation in this non-migratory model is the omission of transition mechanisms within the RSU GROUP, precluding inter-RSU migration. The activation of the recovery process (REC) is contingent upon the RSU GROUP configuration, which facilitates the identification of operational RSUs via the tokens in *RSU_UP*. The symbol 'N' signifies the possibility of multiple tokens within each RSU, with the quantity of tokens representing the number of operational RSUs.

Conversely, the *RSU_DW* token count reflects the number of non-operational RSUs. The transitions *RSU_MTTF* and *RSU_MTTR* govern the oscillation between the active and inactive states of individual RSUs. Network functionality is ensured when a token resides in *NET_UP* (active network), and it becomes non-functional with a token in *NET_DW* (inactive network). The transitions *NET_MTTF* and *NET_MTTR* regulate these state changes.

Similarly, the cloud's operational status is indicated by the presence of a token in *EDGE_UP*, and its non-operational status is signified by a token in *EDGE_DW*. The transitions *EDGE_MTTF* and *EDGE_MTTR* are instrumental in toggling between the cloud's active and inactive states.

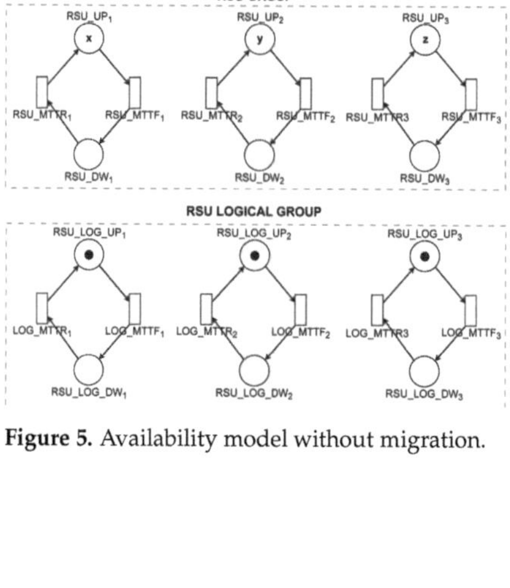

Figure 5. Availability model without migration.

5. Sensitivity Analysis

This research utilizes a Design of Experiments (DoE) approach in conjunction with SPN modeling to derive comprehensive insights into the performance of Edge Computing systems. The variables and their respective levels are consistently applied across models, both incorporating and excluding migration. This methodological consistency is critical to ascertain which variable combinations exert the most significant impact on the system. By exploring these variable combinations across different scenarios, the study robustly underpins the optimization strategies proposed for enhancing the system's availability and reliability.

5.1. Sensitivity Analysis of the System Incorporating Migration

The sensitivity analysis was conducted using the experimental setup illustrated in Figure 2, with a primary focus on the time interval preceding system failure. Table 4 presents a comprehensive enumeration of the variables considered in the Design of Experiments (DoE). This enumeration includes detailed descriptions of each factor and specifies their respective levels. The factors assessed are (a) *EDGE_F*, (b) *NET_F*, (c) *RSU_F*, (d) *RSU_R*, and (e) *LOG_F*. For each factor, evaluations were conducted at both high and low settings to ascertain their impact on the overall system performance. The specific configurations for these factors, as applied in the various experimental scenarios, are thoroughly delineated in the aforementioned table.

Table 4. Design table.

Factor Name	Factor Description	Low Setting	High Setting
EDGE_F	Edge MTTF	125.892	157.365
NET_F	Network MTTF	83,220.0	104,025.0
RSU_F	MTTF Physical Part of RSU	500.0	750.0
RSU_R	MTTR Physical Part of RSU	2.0	3.0
LOG_F	MTTF Logical Part of RSU	168.0	210.0

Table 5 comprehensively catalogs the permutations of factor levels employed in the simulations designed to assess their impact on system availability within an Edge Computing framework. Each row in the table represents a distinct amalgamation of factor values, specifically *EDGE_F*, *NET_F*, *RSU_F*, *RSU_R*, and *LOG_F*. The terminal column of this table quantifies the system availability corresponding to each unique factor combination. This tabulation effectively encapsulates the outcomes of the DOE simulations, facilitating the identification of factor combinations that either significantly influence or minimally affect the system's availability. The structured presentation of these results serves as a pivotal resource for comprehending the dynamics influencing system performance and the relative importance of various system components.

The graph depicted in Figure 6, illustrating the effects of a DOE with migration, provides pivotal insights into the elements most influential on system availability within an Edge Computing environment. Foremost, the MTTF of the physical RSU emerges as the paramount factor, with its impact value approximating 0.70. This underscores the criticality of physical infrastructure reliability in the overall system performance.

Following this, the MTTR of the physical components of RSU, with values oscillating between 0.40 and 0.45, is identified as another crucial determinant in minimizing system failures. This finding accentuates the significance of efficient repair processes in maintaining system integrity.

The interaction between the MTTR of the physical RSU and the MTTF of the logical RSU is also noted to exert a substantial influence on availability. In addition, the chart delineates factors with comparatively lower impacts, such as the interplay between the MTTF of the Edge component and the MTTF of the Network, the MTTF of the logical components, and the interaction between the MTTF of the Network and the MTTR of the physical RSU, each registering impact values below 0.05. While these elements are deemed less consequential in the present analysis, they could acquire greater significance in certain specific scenarios, suggesting the need for a nuanced understanding of different operational contexts in Edge Computing systems.

Table 5. Combination table.

EDGE_F	*NET_F*	*RSU_F*	*RSU_R*	*LOG_F*	**Availability (%)**
125.89	83,220.00	500.00	2.00	168.00	98.35
125.89	83,220.00	500.00	2.00	210.00	97.29
125.89	83,220.00	500.00	3.00	168.00	97.76
125.89	83,220.00	500.00	3.00	210.00	97.41
125.89	83,220.00	750.00	2.00	168.00	98.80
125.89	83,220.00	750.00	2.00	210.00	98.44
125.89	83,220.00	750.00	3.00	168.00	97.93
125.89	83,220.00	750.00	3.00	210.00	98.20
125.89	104,025.00	500.00	2.00	168.00	97.82
125.89	104,025.00	500.00	2.00	210.00	98.02
125.89	104,025.00	500.00	3.00	168.00	98.09
125.89	104,025.00	500.00	3.00	210.00	97.77
125.89	104,025.00	750.00	2.00	168.00	98.62
125.89	104,025.00	750.00	2.00	210.00	98.69
125.89	104,025.00	750.00	3.00	168.00	98.04
125.89	104,025.00	750.00	3.00	210.00	98.61
157.36	83,220.00	500.00	2.00	168.00	97.80
157.36	83,220.00	500.00	2.00	210.00	98.09
157.36	83,220.00	500.00	3.00	168.00	97.16
157.36	83,220.00	500.00	3.00	210.00	97.30
157.36	83,220.00	750.00	2.00	168.00	98.50
157.36	83,220.00	750.00	2.00	210.00	98.71
157.36	83,220.00	750.00	3.00	168.00	98.07
157.36	83,220.00	750.00	3.00	210.00	98.50
157.36	104,025.00	500.00	2.00	168.00	98.16
157.36	104,025.00	500.00	2.00	210.00	98.02
157.36	104,025.00	500.00	3.00	168.00	97.01
157.36	104,025.00	500.00	3.00	210.00	97.92
157.36	104,025.00	750.00	2.00	168.00	99.02
157.36	104,025.00	750.00	2.00	210.00	98.62
157.36	104,025.00	750.00	3.00	168.00	98.06
157.36	104,025.00	750.00	3.00	210.00	98.16

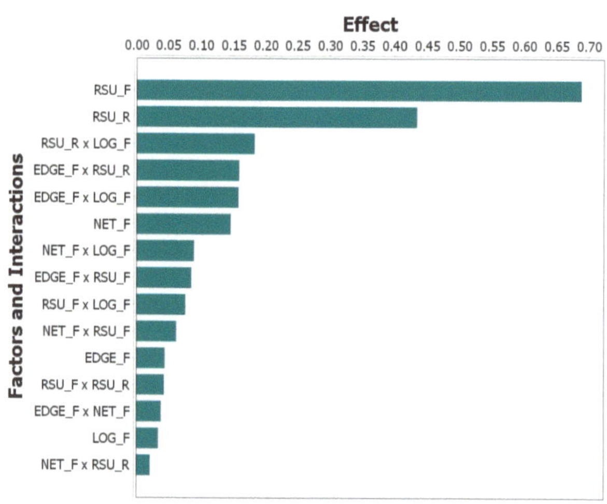

Figure 6. Impact of Different Factors on the System with Migration.

Figure 7 elucidates the intricate interactions between various factors in a migration-inclusive scenario and their collective impact on system availability as assessed through a DOE approach.

In Figure 7a, the interaction between the Mean MTTF of the Edge component and the MTTF of the Network is analyzed. A notable decrease in system availability is observed when the MTTF of the Edge (157.0) is juxtaposed with the Network's MTTF (98.3). This observation is indicative of the system's heightened sensitivity to fluctuations in these parameters, underscoring the critical nature of their balance for optimal system performance.

In Figure 7b, the dynamics between the MTTF of the Network and the MTTF of the logical RSU component are examined. An inverse relationship is discerned here: an escalation in the MTTF of the Edge (from 168.0 to 210.0) correspondingly diminishes the MTTF of the logical RSU. This pattern exemplifies the complex interdependencies within the system, where enhancing the reliability of one component may inversely affect another, thereby impacting overall system availability.

In Figure 7c, the analysis is centered on the interplay between the MTTF of the Network and the MTTR of the physical RSU. An observed increase in the Network's MTTF, along with an enhancement to 210 in the MTTF of the physical RSU component, indicates a notable enhancement in the overall system performance. This result underscores the critical importance of an integrated assessment of both failure and repair durations within the Network and physical components of the RSU. Such a holistic approach is essential for optimizing system availability. These insights collectively contribute to a deeper understanding of system reliability in Edge Computing environments, highlighting the need for a thorough evaluation of the interactions between various system components to achieve optimal system performance.

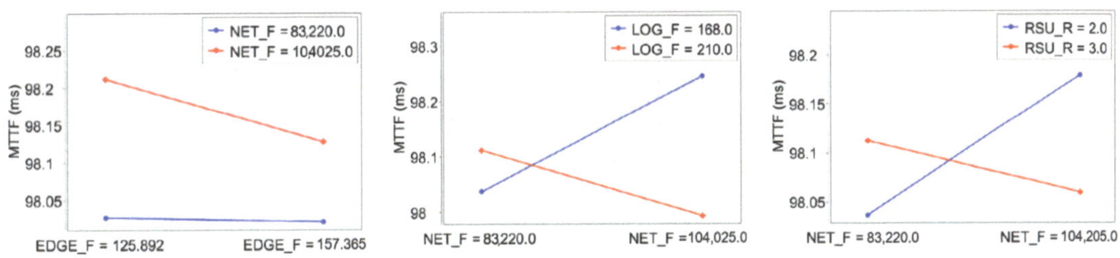

(**a**) MTTF Edge × MTTF Network (**b**) MTTF Network × MTTF RSU Logic (**c**) MTTF Network × MTTR RSU

Figure 7. Interaction between factors in the system with migration.

5.2. Analysis of System Sensitivity in the Absence of Migration

The effect graph in Figure 8, derived from the DOE conducted without the migration feature, elucidates the principal factors impacting system availability in an Edge Computing context. The graph reveals that the MTTF of the logical RSU holds paramount significance, as indicated by its value exceeding 0.90. This underscores the critical role of the reliability of the logical RSU in the overall system.

Secondarily, the MTTR of the RSU physical components also emerges as a consequential factor, exhibiting values around 0.60. This finding emphasizes the importance of efficient repair mechanisms in mitigating system downtime during failures.

Other factors, though less influential, still contribute to the system's performance. These include the MTTF of the Network and the interactions between the MTTF of the Edge component and the MTTF of the logical RSU. Additionally, the interplay between the MTTR of the physical RSU and the MTTF of the logical RSU, each with impact values below 0.1, also bears significance. These findings are instrumental in enhancing system reliability, particularly in scenarios where migrating VMs between RSUs is not feasible. They inform strategic resource allocation decisions and guide interventions aimed at bolstering the overall performance of Edge Computing systems in non-migratory environments. This nuanced understanding of the relative impact of various system components and their interactions is essential for targeted improvements in system robustness and reliability.

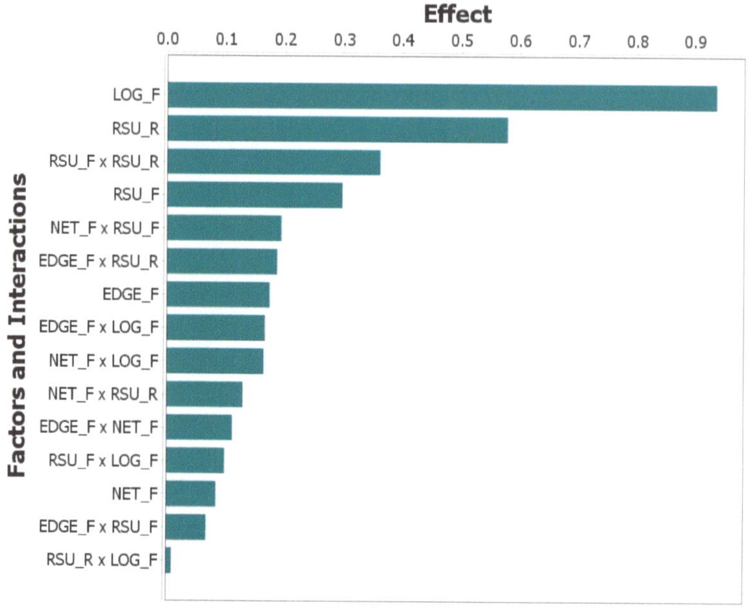

Figure 8. Impact of different factors on the system without migration.

Figure 9 presents an insightful graph of DOE interactions in a context devoid of migration, offering a clear view of how various factor combinations impact system availability in Edge Computing environments.

In Figure 9a, the graph underscores the system's sensitivity to changes in the MTTF of the Edge component relative to the MTTF of the Network. A significant observation here is that an elevation in the Edge's MTTF to 157.0 prompts a slight increase in the Network's MTTF, from 95.0 to 95.3. This shift underscores the substantial influence that the Edge's MTTF exerts on overall system availability.

Figure 9b examines the interplay between the MTTF of the physical RSU and that of its logical counterpart. It is observed that augmenting the reliability of the logical part of the RSU positively influences the MTTF of the physical RSU. This relationship highlights the criticality of integrating these factors to enhance the overall system performance.

Lastly, Figure 9c delves into the interaction between the MTTR and the MTTF of the physical RSU. An increase in the MTTR of the physical RSU is shown to improve the MTTF. However, an increase in MTTF does not significantly impact the MTTR. This finding accentuates the importance of a balanced approach to managing failure and repair times, as this balance is key to optimizing system availability.

Together, these insights from Figure 9 emphasize the complex nature of factor interdependencies in Edge Computing systems, particularly in scenarios where migration is not an option. Understanding these relationships is crucial for the strategic planning and implementation of measures aimed at enhancing the reliability and efficiency of such systems.

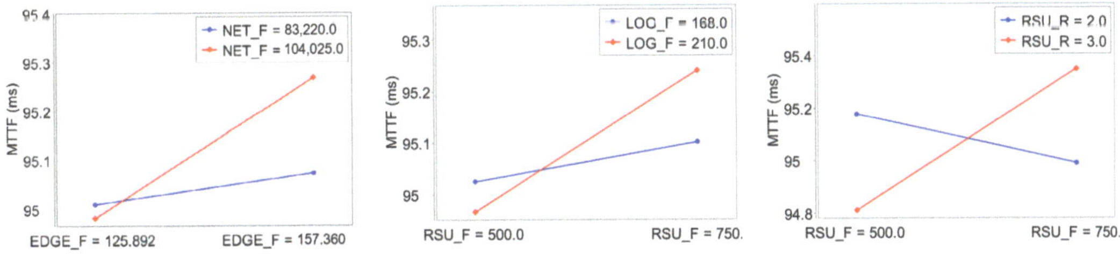

(a) MTTF Edge × MTTF Network (b) MTTF RSU × MTTF RSU Logic (c) MTTF RSU × MTTR RSU

Figure 9. Interaction between factors in the system without migration.

6. Case Study

In this segment, the paper delineates the findings from the analytical evaluation of the models introduced herein. This evaluation encompasses a comprehensive assessment of both the availability and reliability metrics for all the models under consideration. Table 6 provides a detailed account of the parameter values assigned to various system components, with these values meticulously sourced from extant scholarly publications [40,42–44]. The parameters detailed include those pertaining to the RSU group, the MTTF and MTTR for the Edge component, as well as the failure and repair durations associated with the network. Additionally, the table specifies the initial values for the tokens employed in the system, a critical element in the modeling process.

Table 6. Model parameters.

Type	Component	Definition	Value
MTTF	*EDGE_MTTF*	EGDE component failure time	125.89284
	NET_MTTF	Network component failure time	83,220.0
	RSU_MTTF	RSU component failure time	500.0
	RSU_LOG_MTTF	RSU Logical component failure time	168.0
MTTR	*EDGE_MTTR*	EGDE component recovery time	0.913794
	NET_MTTR	Network component recovery time	12.0
	RSU_MTTR	RSU component recovery time	2.0
	RSU_LOG_MTTR	RSU Logical component recovery time	2..0
Variável	TOKENS	Entity representing a state or resource	2.0
	T_M	Migration Time	0.083333

The graph depicting system availability with migration, as shown in Figure 10, delineates a non-linear association between the quantity of VMs utilized and the consequent system availability. Notably, this graph exhibits a convergence of the availability metrics at specific VM counts, namely 8, 16, and 32. This overlapping of data lines suggests that the deployment of eight VMs is sufficient to assure the desired level of availability. Such an observation is pivotal in informing strategies for resource allocation and cost optimization. It implies that beyond a threshold of eight VMs, additional VMs do not significantly enhance system availability, thereby offering a pathway to maximize resource efficiency. This efficient allocation of VMs, without compromising the operational efficacy of the system, is essential in balancing cost effectiveness with system performance.

The research study includes Figure 11, which portrays the graph of system availability in scenarios where migration between RSUs is not implemented. This graph maintains consistency in the range of VMs as used in Figure 10, varying from 2 to 32 VMs in operation. The MTTF for these VMs is set between 100 and 1000 h. A critical observation from the results depicted in this graph is the comparative reduction in system availability when migration is not employed, as opposed to scenarios where migration is feasible.

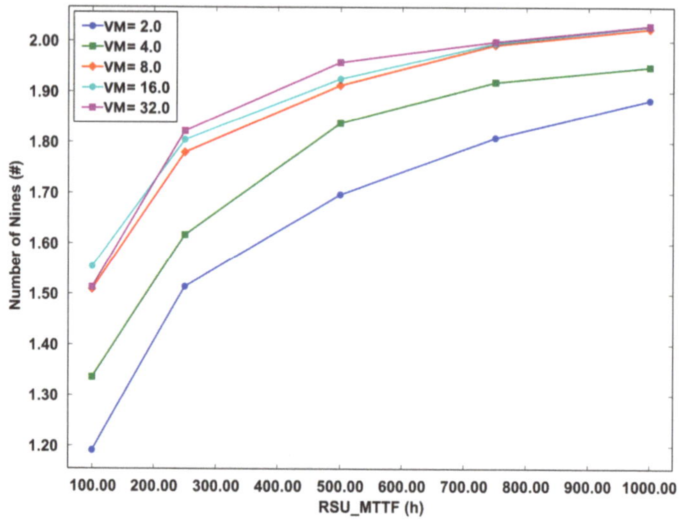

Figure 10. Availability model with migration.

Specifically, the graph shows that system availability commences at '1.00 nines' with the operation of just 2 VMs, and it marginally escalates to '1.36 nines' with the use of four VMs. This pattern of availability underscores the significance of migrating VMs between RSUs in enhancing system reliability. The ability to migrate VMs appears to be a crucial factor in achieving higher availability, as evidenced by the increase in the number of 'nines' in the availability metric. This insight highlights the importance of VM migration as a strategy to bolster the robustness and dependability of the system, especially in contexts where maintaining high availability is paramount. The study thus provides a compelling argument for incorporating VM migration between RSUs as a means to optimize system performance and reliability.

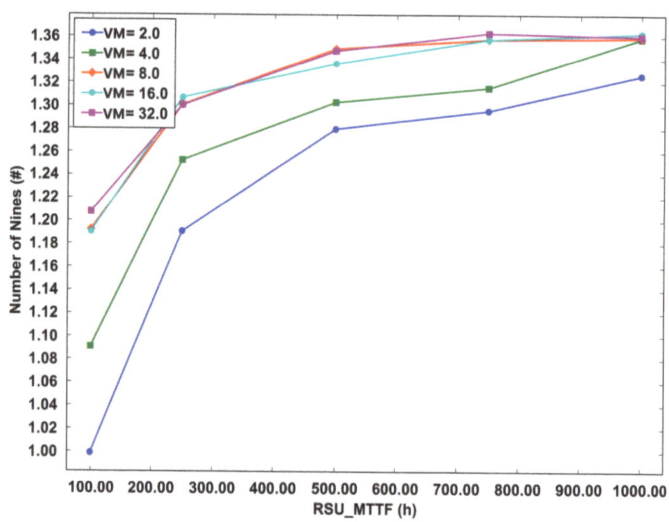

Figure 11. Availability chart—no migration model.

In the context of a system deploying four VMs with a MTTF set at 1000 h, it is feasible to achieve a level of system availability analogous to that obtained with a deployment of 32 VMs. This finding suggests that an augmentation in the number of VMs does not correspondingly result in a substantial increase in system availability. Particularly in the model excluding migration, the availability attributed to the logical component of the RSUs emerges as a predominant factor as discerned from the sensitivity analysis.

This distinction between the models with and without migration underscores the efficacy of the VM migration technique, especially in scenarios characterized by high demand. It accentuates the necessity of integrating VM migration into strategies for computing resource allocation. Figure 12 provides a comparative analysis, juxtaposing the average cases in scenarios of system operation with and without migration. This comparison elucidates the differential impacts of VM migration on system availability, thereby underscoring its critical role in the optimization of computing resources in demanding operational environments.

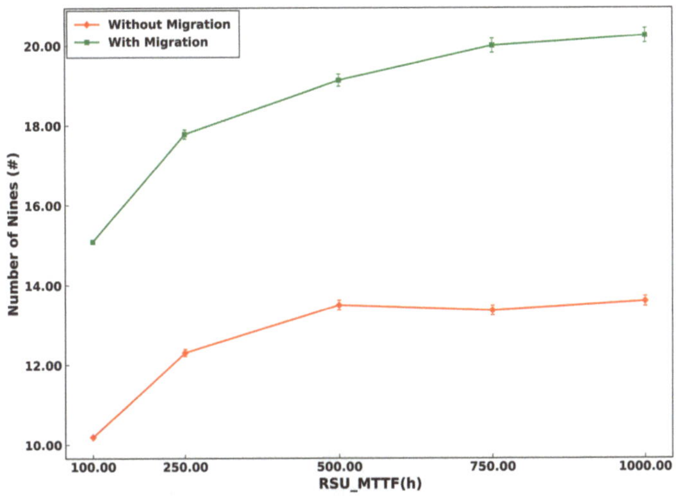

Figure 12. Comparison between the with-migration and without-migration models.

Figure 12 presents an analysis of system availability based on the failure time of RSUs to assess performance in configurations with and without VM migration. In scenarios without migration, the system demonstrates notable availability, achieving approximately `10.00 nines` and increasing slightly to just over `14.00 nines` when the MTTF is set at 100 h. Conversely, in configurations with migration, the availability shows a marked increase as the MTTF is extended, reaching an impressive `20.00 nines` with an MTTF of 100 h. These results clearly indicate that migration exerts a positive influence on system availability, particularly in contexts characterized by extended MTTF periods.

Turning to Figure 13, the reliability graph for the physical component of the system illustrates the variance in the MTTF of the physical aspect of the RSU, with values spanning 168.0, 250.0, and 500.0 over a duration of 800 h. Reliability is a critical metric for evaluating the system's capability to function consistently without failures and interruptions. The data portrayed in this graph establish a direct correlation between the MTTF of the physical component and the overall reliability of the system. This relationship indicates that an increase in the MTTF of the physical part is directly proportional to an enhancement in the system's reliability. This insight is pivotal for understanding the impact of the physical component's robustness on the overall operational stability of the system.

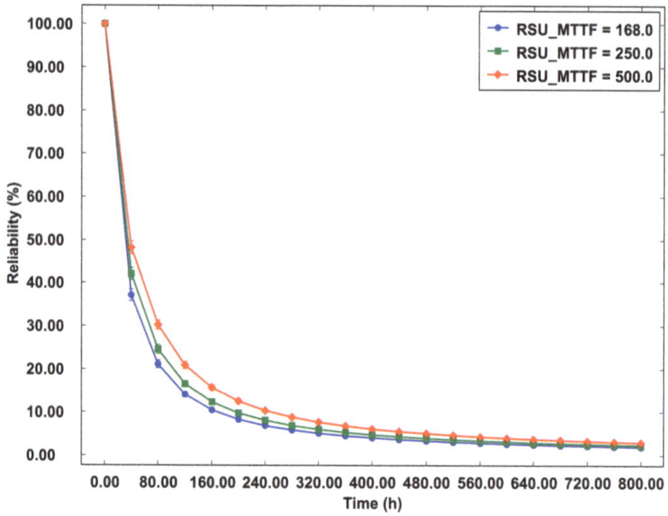

Figure 13. Reliability—physical part of the RSU.

As the MTTF is extended, the system exhibits enhanced robustness and resilience, thereby diminishing the frequency of failures and bolstering reliable operations. Conversely, a system characterized by an MTTF of less than 168 h tends to be more vulnerable to failures and faces challenges in sustaining stable operations. This insight is invaluable for strategizing enhancements to the physical infrastructure of systems in Edge Computing environments. Effective planning in this context encompasses the implementation of both preventive and corrective strategies aimed at augmenting the reliability, quality, and availability of services delivered to end users.

Figure 14 displays the reliability graph for the logical component of the system, with the MTTF varying between 168.0, 250.0, and 500.0 over a span of 800 h. Assessing the MTTF of the logical part is vital to gauge its capacity to sustain adequate operational performance, particularly in Edge Computing contexts where uninterrupted availability is paramount. The data gleaned from this graph offer insights into the reliability trends of the system's logical component relative to the MTTF of VMs.

It is observed that the system with the lowest MTTF of 168 h exhibits a decline in reliability before reaching 80 h of operation. This trend signifies a reduced capability of the system to maintain stability and remain free from failures over a shorter duration. In contrast, the system with the highest MTTF of 500 h demonstrates consistent reliability throughout the initial 80 h of operation, underscoring its superior ability to remain operational and reliable for an extended period before encountering declines in reliability. These findings are crucial in understanding the resilience of the logical component of the system and in guiding decisions related to the management and optimization of Edge Computing systems.

The analysis of the data underscores the heightened significance of the logical component in determining the system's reliability, surpassing the influence of the physical part. Although both aspects are integral, the role of VMs, particularly their efficiency in terms of MTTF, emerges as a critical factor in ensuring uninterrupted system availability. The observation that a system with an elevated MTTF exhibits prolonged stable reliability prior to any decline in performance indicates that the efficacy and dependability of VMs are key determinants of system stability.

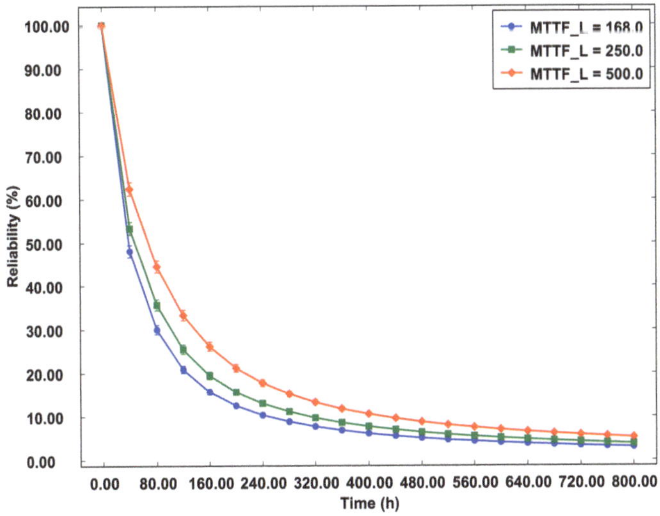

Figure 14. Reliability—logical part of the RSU.

Consequently, it is imperative to place emphasis on strategies aimed at bolstering the reliability of the system's logical aspect. This strategic focus encompasses several key measures:

- `VM management policies`: The development and implementation of comprehensive management policies for VMs are crucial. These policies should be designed to effectively handle resource allocation, scaling, and migration, thereby enhancing system performance.
- `Performance monitoring`: Rigorous and continuous monitoring of system performance is essential. This enables the early identification and resolution of potential issues, thereby maintaining the system's operational integrity.
- `Maintenance practices`: The adoption of appropriate and systematic maintenance practices is vital in ensuring the optimal functioning of VMs. Regular maintenance activities, including updates and troubleshooting, are necessary for sustaining system health and efficiency.

Implementing these strategies will significantly enhance the availability and quality of the services rendered by the system. Such enhancements not only contribute to the system's resilience and operational efficiency but also to the overall user experience in Edge Computing environments. Therefore, prioritizing improvements in the logical part of the system is not merely a technical imperative but a strategic approach to achieving superior service delivery.

7. Conclusions

This research meticulously analyzed system availability within networks of RSUs using SPN and the DOE methodology, focusing on environments with and without VM migration. The findings successfully met the study's objectives, providing insightful revelations about the impact of critical factors on system performance in both migration scenarios. A significant aspect highlighted by the study is the critical role of VM migration in VANETs, which proves to be fundamental in optimizing system availability. This process not only ensures a balanced distribution of workload but also substantially minimizes system downtime. The study identifies the MTTF of physical RSUs and the MTTR of the logical components of RSUs as pivotal determinants of system availability. These findings emphasize the necessity of enhancing these components to secure reliable and efficient operation in VANET contexts.

For future research directions, a more granular investigation of the identified factor interactions is suggested. This should include considering a broader range of variables and conducting empirical experiments in real-world environments to further validate and refine the proposed model. Additionally, exploring alternative approaches to VM migration and incorporating advanced load-balancing strategies are proposed as avenues for further study. These investigations are expected to provide deeper insights and enable additional improvements in system availability and performance, especially in the complex and evolving landscape of Edge Computing. This comprehensive approach to future research will not only enhance the understanding of system dynamics in VANETs but also contribute to the development of more resilient and efficient vehicular network systems.

Author Contributions: Conceptualization, F.A.S. and T.A.N.; methodology, F.A.S. and T.A.N.; software, L.G.S. and I.C.; validation, C.B., V.B. and B.N.; formal analysis, L.G.S.; investigation, F.A.S. and B.N.; resources, E.C. and D.M.; data curation, L.G.S., I.C., C.B. and V.B.; writing—original draft preparation, L.G.S., I.C., C.B. and V.B.; writing—review and editing, F.A.S., B.N. and T.A.N.; visualization, I.C., C.B. and V.B.; supervision, D.M. and J.W.L.; project administration, T.A.N.; funding acquisition, D.M. and J.W.L. All authors have read and agreed to the published version of the manuscript.

Funding: This research was partially supported by Basic Science Research Program through the National Research Foundation of Korea (NRF) funded by the Ministry of Education (No. 2020R1A6A1A03046811). This research was supported by Basic Science Research Program through the National Research Foundation of Korea (NRF) funded by the Ministry of Education (2021R1A2C2094943).

Institutional Review Board Statement: Not applicable.

Informed Consent Statement: Not applicable.

Data Availability Statement: Data are contained within the article.

Conflicts of Interest: The authors declare no conflict of interest.

References

1. Ferreira, W. IBM Alerta Sobre Pressão de Dados dos Carros Conectados. Available online: https://www.telesintese.com.br/ibm-alerta-sobre-pressao-dos-carros-conectados-nas-redes-de-telecom/ (accessed on 9 August 2023).
2. Verma, R. An Efficient Secure VANET Communication Using Multi Authenticate Homomorphic Signature Algorithm. In Proceedings of the 2023 International Conference on Distributed Computing and Electrical Ci rcuits and Electronics (ICDCECE), Ballar, India, 29–30 April 2023; pp. 1–5. [CrossRef]
3. Manivannan, D.; Moni, S.S.; Zeadally, S. Secure authentication and privacy-preserving techniques in Vehicular Ad-hoc NETworks (VANETs). *Veh. Commun.* **2020**, *25*, 100247. [CrossRef]
4. Hussain, R.; Hussain, F.; Zeadally, S. Integration of VANET and 5G Security: A review of design and implementation issues. *Future Gener. Comput. Syst.* **2019**, *101*, 843–864. [CrossRef]
5. Shen, J.; Liu, D.; Chen, X.; Li, J.; Kumar, N.; Vijayakumar, P. Secure real-time traffic data aggregation with batch verification for vehicular cloud in VANETs. *IEEE Trans. Veh. Technol.* **2019**, *69*, 807–817. [CrossRef]
6. Zhang, X.; Li, D.; Wang, J.; Zhang, G.; Jiang, X. Faster parking and less cruise for public parking spot discovery: Modeling and analysis based on timed petri nets. In Proceedings of the 2016 IEEE 13th International Conference on Networking, Sensing, and Control (ICNSC), Mexico City, Mexico, 28–30 April 2016; pp. 1–6.
7. Karabulut, M.A.; Shah, A.F.M.S.; Ilhan, H.; Pathan, A.S.K.; Atiquzzaman, M. Inspecting VANET with Various Critical Aspects—A Systematic Review. *Ad Hoc Netw.* **2023**, *150*, 103281. [CrossRef]
8. Maria Christina Blessy, A.; Brindha, S. Maximizing VANET performance in cluster head selection using Intelligent Fuzzy Bald Eagle optimization. *Veh. Commun.* **2023**, 100660. [CrossRef]
9. Qiong, W.; Shuai, S.; Ziyang, W.; Qiang, F.; Pingyi, F.; Cui, Z. Towards V2I age-aware fairness access: A dqn based intelligent vehicular node training and test method. *Chin. J. Electron.* **2023**, *32*, 1230–1244. [CrossRef]
10. German, R.; Telek, M. Formal relation of Markov renewal theory and supplementary variables in the analysis of stochastic Petri nets. In Proceedings of the 8th International Workshop on Petri Nets and Performance Models (Cat. No. PR00331), Zaragoza, Spain, 8–10 September 1999; pp. 64–73. [CrossRef]
11. Pinheiro, T.; Silva, F.A.; Fé, I.; Oliveira, D.; Maciel, P. Performance and Resource Consumption Analysis of Elastic Systems on Public Clouds. In Proceedings of the 2019 IEEE International Conference on Systems, Man and Cybernetics (SMC), Bari, Italy, 6–9 October 2019; pp. 2115–2120.

12. Zhan, H.; Gu, J. Study of the Normal Generalized Stochastic Petri nets and its Application in Testing System. In Proceedings of the 2006 IEEE Instrumentation and Measurement Technology Conference Proceedings, Sorrento, Italy, 24–27 April 2006; pp. 1123–1128. [CrossRef]
13. Chen, P.; Bruell, S.; Balbo, G. Alternative methods for incorporating non-exponential distributions into stochastic timed Petri nets. In Proceedings of the Third International Workshop on Petri Nets and Performance Models, PNPM89, Kyoto, Japan, 11–13 December 1989; pp. 187–197. [CrossRef]
14. Rodrigues, L.; Endo, P.T.; Silva, F.A. Stochastic Model for Evaluating Smart Hospitals Performance. In Proceedings of the 2019 IEEE Latin-American Conference on Communications (LATINCOM), Salvador, Brazil, 11–13 November 2019; pp. 1–6.
15. Nicol, D.; Miner, A. The fluid stochastic Petri net simulator. In Proceedings of the 6th International Workshop on Petri Nets and Performance Models, Durham, NC, USA, 3–6 October 1995; pp. 214–215. [CrossRef]
16. Van Voorhees, F.; Bahill, A. Sensitivity analysis by design of experiments. In Proceedings of the 1995 International Symposium and Workshop on Systems Engineering of Computer-Based Systems, Tucson, AZ, USA, 6–9 March 1995; pp. 58–65. [CrossRef]
17. Bergès, C.; Song, J.; He, C.; Qian, X.; Ren, J.; Tian, L.; Wen, G.; Zhang, H. Provocation tests, design of experiments and advanced statistical modeling to estimate product sensitivity to a defect: Delamination failure case study for automotive semiconductors. In Proceedings of the 2019 IEEE 26th International Symposium on Physical and Failure Analysis of Integrated Circuits (IPFA), Hangzhou, China, 2–5 July 2019; pp. 1–4. [CrossRef]
18. Dewey, A.; Ren, H.; Zhang, T. Behavioral modeling of microelectromechanical systems (MEMS) with statistical performance-variability reduction and sensitivity analysis. *IEEE Trans. Circuits Syst. II Analog. Digit. Signal Process.* **2000**, *47*, 105–113. [CrossRef]
19. Mach, F. Reduction of Optimization Problem by Combination of Optimization Algorithm and Sensitivity Analysis. *IEEE Trans. Magn.* **2016**, *52*, 7003104. [CrossRef]
20. Kleijnen, J. Sensitivity analysis and optimization in simulation: Design of experiments and case studies. In Proceedings of the WSC '97: Proceedings of the 29th Conference on Winter Simulation, Arlington, VA, USA, 3–6 December 1995; pp. 133–140. [CrossRef]
21. Silva Nunes Junior, M.A.; de Oliveira, T.C.; de Carvalho Filho, J.M.; de Abreu, J.P.G. Design of experiments for sensitivity analysis of voltage sags variables. In Proceedings of the 2012 IEEE 15th International Conference on Harmonics and Quality of Power, Hong Kong, China, 17–20 June 2012; pp. 398–402. [CrossRef]
22. Shah, A.S.; Ilhan, H.; Tureli, U. Modeling and performance analysis of the IEEE 802.11 P MAC for VANETs. In Proceedings of the 2019 42nd International Conference on Telecommunications and Signal Processing (TSP), Budapest, Hungary, 1–3 July 2019; pp. 393–396.
23. Malik, A.; Pandey, B. Security Analysis of Discrete Event Based Threat Driven Authentication Approach in VANET Using Petri Nets. *Int. J. Netw. Secur.* **2018**, *20*, 601–608.
24. Li, B.; Sutton, G.J.; Hu, B.; Liu, R.P.; Chen, S. Modeling and QoS analysis of the IEEE 802.11 p broadcast scheme in vehicular ad hoc networks. *J. Commun. Netw.* **2017**, *19*, 169–179. [CrossRef]
25. Hashimoto, T.; Tsuji, T.; Aoki, J.; Ohta, T.; Kakuda, Y. A Mobile Agent Migration Mechanism for Information Dissemination Scheme in VANETs Considering Entrance and Exit of Mobile Nodes. In Proceedings of the 2015 IEEE International Symposium on Object/Component/Service-Oriented Real-Time Distributed Computing Workshops, Auckland, New Zealand, 13–17 April 2015; pp. 89–94. [CrossRef]
26. Aoki, J.; Hashimoto, T.; Ohta, T.; Kakuda, Y. Experimental Evaluation of Location-Based Mobile Agent Migration Mechanism for Information Dissemination Scheme in VANETs. In Proceedings of the 2015 Third International Symposium on Computing and Networking (CANDAR), Sapporo, Japan, 8–11 December 2015; pp. 271–274. [CrossRef]
27. Jeong, J.; Sheny, Y.C.; Jeong, J.P.; Ohx, T.T.; Jun, J.; Sang, H.S. TOMS: TCP Context Migration Scheme for Efficient Data Services in Vehicular Networks. In Proceedings of the 2017 31st International Conference on Advanced Information Networking and Applications Workshops (WAINA), Taipei, Taiwan, 27–29 March 2017; pp. 360–364. [CrossRef]
28. Weber, J.S.; Ferreto, T.; Zincir-Heywood, N. Exploring Anomaly Detection Techniques for Enhancing VANET Availability. In Proceedings of the 2023 IEEE 97th Vehicular Technology Conference (VTC2023-Spring), Florence, Italy, 20–23 June 2023; pp. 1–7. [CrossRef]
29. Khan, Z.; Abbas, F.; Hamidullah. A Conceptual Framework of Virtualization and Live-Migration for Vehicle to Infrastructure (V2I) Communications. In Proceedings of the 2019 IEEE 11th International Conference on Communication Software and Networks (ICCSN), Chongqing, China, 12–15 June 2019; pp. 590–594. [CrossRef]
30. Goyal, A.K.; Kumar Tripathi, A.; Agarwal, G. Security Attacks, Requirements and Authentication Schemes in VANET. In Proceedings of the 2019 International Conference on Issues and Challenges in Intelligent Computing Techniques (ICICT), Ghaziabad, India, 27–28 September 2019; Volume 1, pp. 1–5. [CrossRef]
31. K, S.; chinnasamy, C. Efficient VANET handover scheme using SSDN by incorporating media independent handover framework. *Meas. Sens.* **2023**, *26*, 100684. [CrossRef]
32. Mchergui, A.; Moulahi, T.; Nasri, S. BaaS: Broadcast as a service cross-layer learning-based approach in cloud assisted VANETs. *Comput. Netw.* **2020**, *182*, 107468. [CrossRef]
33. Cardote, A.; Sargento, S.; Steenkiste, P. On the connection availability between relay nodes in a VANET. In Proceedings of the 2010 IEEE Globecom Workshops, Miami, FL, USA, 6–10 December 2010; pp. 181–185. [CrossRef]

34. Lobo, A.; Matos, R.; Silva, B.; Maciel, P. Expolynomial Modelling for Supporting VANET Infrastructure Planning. In Proceedings of the 2017 IEEE 22nd Pacific Rim International Symposium on Dependable Computing (PRDC), Christchurch, New Zealand 22–25 January 2017; pp. 86–91. [CrossRef]
35. Alvarenga, L.D.C.; Sousa, P.; Costa, A. Allocation and migration of microservices in SDN-based vehicular fog networks. In Proceedings of the 2022 17th Iberian Conference on Information Systems and Technologies (CISTI), Madrid, Spain, 22–25 June 2022; pp. 1–4. [CrossRef]
36. Selvakumar, A.; Ramesh, S.; Manikandan, T.; Michael, G.; Arul, U.; Gnanajeyaraman, R. Microgrid based VANET monitoring and energy management in 5G networks by reinforcement deep learning techniques. *Comput. Electr. Eng.* **2023**, *111*, 108933. [CrossRef]
37. da Rocha Santos, L.C.; Bruschi, S.M.; de Souza, P.S.L.; Ueyama, J.; de Jesus dos Santos, A.; Barbosa, J.S. Performance analysis of a Vehicular Ad Hoc network Using LoRa technology and IoT devices in Amazon Rivers. *Ad Hoc Netw.* **2024**, *152*, 103301 [CrossRef]
38. kumar Pulligilla, M.; Vanmathi, C. An authentication approach in SDN-VANET architecture with Rider-Sea Lion optimized neural network for intrusion detection. *Internet Things* **2023**, *22*, 100723. [CrossRef]
39. Shafi, M.; Molisch, A.F.; Smith, P.J.; Haustein, T.; Zhu, P.; De Silva, P.; Tufvesson, F.; Benjebbour, A.; Wunder, G. 5G: A tutorial overview of standards, trials, challenges, deployment, and practice. *IEEE J. Sel. Areas Commun.* **2017**, *35*, 1201–1221. [CrossRef]
40. Araújo, G.; Rodrigues, L.; Oliveira, K.; Fé, I.; Khan, R.; Silva, F.A. Vehicular cloud computing networks: Availability modelling and sensitivity analysis. *Int. J. Sens. Netw.* **2021**, *36*, 125–138. [CrossRef]
41. Maciel, P.; Matos, R.; Silva, B.; Figueiredo, J.; Oliveira, D.; Fé, I.; Maciel, R.; Dantas, J. Mercury: Performance and dependability evaluation of systems with exponential, expolynomial, and general distributions. In Proceedings of the 2017 IEEE 22nd Pacific Rim International Symposium on Dependable Computing (PRDC), Christchurch, New Zealand, 22–25 January 2017; pp. 50–57.
42. Audu, G.A.; Bhattacharya, S.; Muhtar, A.; Qazi, B.; Elmirghani, J.M. Reliability and quality of service of an off-grid wind powered roadside unit in a motorway vehicular environment. *Veh. Commun.* **2017**, *9*, 176–187. [CrossRef]
43. Andrade, E.; Nogueira, B. Dependability evaluation of a disaster recovery solution for IoT infrastructures. *J. Supercomput.* **2020**, *76*, 1828–1849. [CrossRef]
44. Melo, C.; Matos, R.; Dantas, J.; Maciel, P. Capacity-oriented availability model for resources estimation on private cloud infrastructure. In Proceedings of the 2017 IEEE 22nd Pacific Rim International Symposium on Dependable Computing (PRDC), Christchurch, New Zealand, 22–25 January 2017; pp. 255–260.

sensors

Article

Improving Turn Movement Count Using Cooperative Feedback

Patrick Heyer-Wollenberg *, Chengjin Lyu, Ljubomir Jovanov, Bart Goossens and Wilfried Philips

TELIN-IPI, Ghent University–imec, St-Pietersnieuwstraat 41, B-9000 Ghent, Belgium; chengjin.lyu@ugent.be (C.L.); ljubomir.jovanov@ugent.be (L.J.); bart.goossens@ugent.be (B.G.); wilfried.philips@ugent.be (W.P.)

* Correspondence: patrick.heyerwollenberg@ugent.be

Abstract: In this paper, we propose a new cooperative method that improves the accuracy of Turn Movement Count (TMC) under challenging conditions by introducing contextual observations from the surrounding areas. The proposed method focuses on the correct identification of the movements in conditions where current methods have difficulties. Existing vision-based TMC systems are limited under heavy traffic conditions. The main problems for most existing methods are occlusions between vehicles that prevent the correct detection and tracking of the vehicles through the entire intersection and the assessment of the vehicle's entry and exit points, incorrectly assigning the movement. The proposed method intends to overcome this incapability by sharing information with other observation systems located at neighboring intersections. Shared information is used in a cooperative scheme to infer the missing data, thereby improving the assessment that would otherwise not be counted or miscounted. Experimental evaluation of the system shows a clear improvement over related reference methods.

Keywords: Turn Movement Count (TMC); cooperative vision; vehicle count; smart intersection; traffic analysis

Citation: Heyer-Wollenberg, P.; Lyu, C.; Jovanov, L.; Goossens, B.; Philips, W. Improving Turn Movement Count Using Cooperative Feedback. *Sensors* **2023**, *23*, 9772. https://doi.org/10.3390/s23249772

Academic Editors: Constantin Caruntu and Ciprian Romeo Comşa

Received: 6 October 2023
Revised: 7 December 2023
Accepted: 8 December 2023
Published: 12 December 2023

1. Introduction

Turn Movement Count (TMC) is the task of counting how many vehicles perform each of the possible movements at an intersection in a specific time period. It has been widely used in the applications of infrastructure planning, smart cities, and traffic optimization. Existing automated traffic analysis systems often underperform compared to human annotators, but they are able to annotate much larger datasets for extended periods of time. Despite these systems' successes, they tend to miscount turn movements in situations with many simultaneously visible vehicles. These vehicles occlude each other, making it impossible to correctly identify the origin or destination of the vehicle. If a vehicle is occluded for most of its trajectory, current methods that rely on a single source of information cannot overcome this problem.

More recently, systems that implement multiple cameras at a single intersection have been introduced, with the goal of overcoming some of these problems in cases when a vehicle can be seen by at least one camera at all times. However, to ensure that all vehicles are in view, greater numbers of cameras are required and the computational cost is higher due to the larger amount of data to be processed. Although increasing the number of cameras observing a single intersection is not a cost-effective solution, traffic surveillance systems that observe the entry and exit streets at most intersections are already deployed in most cities. Other methods from the literature rely on information provided directly from vehicles. The available information varies depending on the vehicle types transiting through the area and the implemented technology. Information sources range from RFID tags that simply inform the system that a certain vehicle is present, to highly sophisticated systems common on autonomous vehicles. These latter systems obtain information on traffic environments from onboard sensors such as cameras, radars, and LIDAR.

In this work, we propose a system that utilizes existing camera infrastructure to perform TMC cooperatively. This system consists of multiple independent observational systems that collect and process local data at different locations *in parallel* and share relevant information between nearby intersections. The shared data provides a broader context to the observations at any given intersection, either by confirming previous motion estimations or by providing information about the vehicle's movements before entering and after leaving the intersection. We postulate that with this additional information, a cooperative system should be capable of correctly evaluating a vehicle's movement under conditions in which current systems would fail, given the same sensor configuration.

2. Related Work

Over the years, various innovative approaches have been proposed to solve the turn movement count problem using video analysis. Most of the methods described below employ a variation in the Detect-Track-Count paradigm, which consists of a set of *sequential* processes.

2.1. Detection

The Detection step consists of determining the regions of interest (ROI) in the image containing vehicles. Early methods used a variation in background subtraction to determine the ROIs, such as those presented in [1,2], by removing parts of the image that have not changed significantly over a certain number of frames. As these methods require multiple steps to be performed in sequence, they are often slower and more computationally expensive than more recent approaches.

More recent approaches rely on deep learning techniques to detect vehicles in the scene, such as in the works of [3–5]. These techniques have demonstrated better performance in detecting vehicles. In addition, these approaches are able to process algorithms such as YOLO [6] faster, a technique that provides accurate regions of interest from a single pass over the image. While improvements in detection speed and accuracy have contributed to a better TMC performance, they do not address high occlusion scenarios.

2.2. Tracking

The Tracking step of the process obtains the trajectories of previously detected vehicles as they move across the intersection. The most naive tracking methods, such as those presented by [7,8], extract the distances between the current and previously detected object positions. The trajectory is therefore described as a series of points in a sequence of images in which the vehicle has been detected. As these systems are frame-to-frame based, they require a high frame rate to track vehicles, given that the association of trajectory to a specific vehicle becomes inaccurate at low frame rates.

More sophisticated methods rely on a combination of visual features' re-identification (described further below in Section 2.4) and the current position, to associate a trajectory to a vehicle. Such methods can correctly determine the location of an identified vehicle in a new frame, instead of relying on the proximity of detected positions to associate the tracks. Therefore, they allow for more robust trajectory assignment. This can be seen in the works of Liang et al. and Wojke et al. [9,10]. This approach reduces the uncertainty of the tracker while allowing for lower frame rates to be used, at the expense of higher computational complexity.

Other tracking systems, such as the one presented by Li et al. [11], essentially merge detection and tracking into a single process. This tracker passes a prediction of probable ROIs based on the current trajectory, providing the detector algorithm with a distribution of probable ROIs on which to perform detection. While this solution is the most complex of those discussed here, it is also the most reliable for continuous tracking, once a vehicle has been detected and tracking has started.

2.3. Movement Assignment and Counting

Movement assignment and counting is the last step taken via existing TMC methods. Based on the trajectory obtained from the tracking system, a turn movement is selected and counted based on a preexisting list of possible moves. Two main approaches exist for selecting the turn movement.

The first method consists of determining the entry and exit points of the vehicle by performing an intersection test between the vehicle's trajectory and a predefined region of the intersection; the entry point is defined as the first region where the vehicle was detected, and the exit point corresponds to the last tracked position. This type of TMC can be seen in works such as [1,12–14]. When the detection algorithm is slow, this type of TMC becomes unreliable—the entry point will be incorrectly identified, and in cases where the vehicle is occluded while in the exit region, missed assignments will occur.

The second method consists of comparing the entire trajectory with previously known trajectories from the annotated data. These methods, as presented in [15–17], tend to be computationally expensive, since comparing whole trajectories is more complex than performing point-region intersection tests. The biggest issue in trajectory comparison is the precision of the tracker and the accuracy of the reference trajectory, since minor tracking deviations can generate confusion during evaluation.

2.4. Re-Identification

Re-identification consists of determining if an object of interest is detected—in this case a vehicle—and if it is the same as a previously detected vehicle. When an object has been detected, a series of identifying features are collected and compared to all previously collected feature sets. If the similarities between the features surpass a certain threshold, the object is considered the same as the one generating the initial feature set. While re-identification is not an essential part of TMC, some methods use it to increase the precision of assigning a trajectory to a vehicle. The main limitation of current vehicle re-identification methods is that the similarities of different vehicles tend to surpass the threshold for positive ID. That is, the feature set used to describe a previously observed vehicle may be similar enough to that of a different vehicle and cause incorrect identification, confusing the second vehicle for the first one.

This problem can be partially solved by increasing the similarity threshold required to match a newly detected vehicle to an existing ID. However, this comes with the main drawback of decreasing the number of vehicles correctly identified.

2.5. Other Data Sources

Although in this work we focus on conducting TMC using camera-based systems, other methods perform traffic analysis using data collected from different sources. These methods include systems based on radar, such as in [18,19], while others use 3D data captured via LIDAR, such as [20,21]. In recent years, a new concept has emerged to combine existing roadside sensors, such as those already described, with vehicle-mounted sensors such as LIDAR. Since autonomous driving vehicles already include such technologies, as described in [22], this integration would deliver more relevant information to the traffic system without increasing the cost of the infrastructure. The inclusion of these additional traffic data sources presents a great opportunity for research. However, a main limitation barring a wider application of these concepts is the limited types of data used in existing public infrastructure, where these kinds of systems are to be deployed.

3. Methodology

We propose a cooperative feedback approach to address the vehicle occlusion problem affecting traffic analysis systems. Our approach gathers information from multiple points in surrounding areas. This information is used to assign a turn movement to a vehicle in cases where the assignment could not be made using information obtained only locally. We propose this approach based on the fact that the movement of a vehicle through an intersection is not an isolated event limited to the intersection, but rather part of a complex series of movements through an environment of interconnected roads, observed via traffic monitoring systems at multiple intersections. We represent the interconnections of roads as a directed multigraph, where each intersection is considered a node and each of the lanes connecting these intersections is an edge, as shown in Figure 1 where the main intersection C is connected to A, B, D using directional edges representing the traffic flow. We also consider all possible combinations of incoming and outgoing edges as the list of possible turn movements at that node, without considering the legality of such moves.

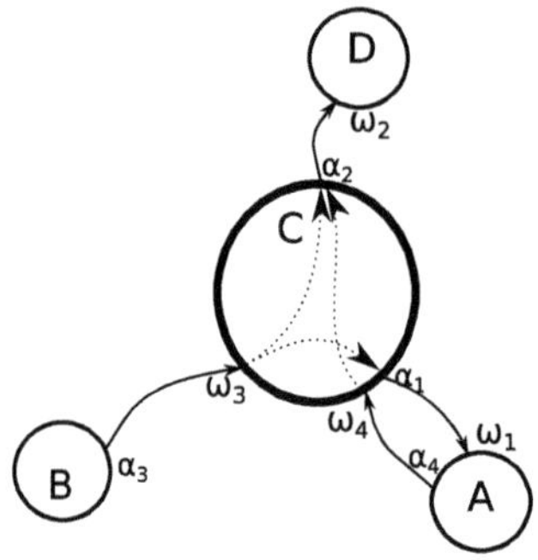

Figure 1. Visual representation of an intersection as a node C in a connected graph, with neighboring intersections A, B, D. Intersection C is connected with entry points w_3, w_4, and exit points a_1, a_2 Dotted lines represent possible movements through the node, representing the legal turn movements at that intersection.

To determine how the information provided by the other intersections influences a TMC system, we propose a method formally described in Section 3.1. In Section 3.2, we describe the software implemented to test this system.

3.1. Formalization

Given the previously described assumptions, the problem can be formalized as follows:

- $U = \{u_v | v = 1, \ldots, V\}$, a set of objects of interest u_v (vehicles in this case) where V denotes the cardinality of the set.
- $\Omega = \{\omega_j | j = 1, \ldots, J\}$, a set of incoming edges' ends ω_j (entry points to intersections in this case), where J denotes the cardinality of the set.
- $A = \{\alpha_k | k = 1, \ldots, K\}$, a set of outgoing edges' ends α_k (exit points to intersections in this case) where K denotes the cardinality of the set.
- $\Gamma = \{\gamma_i | \gamma_i \in \Omega \bigcup A\}$, the set of edges' ends γ_i (accesses, whether entries or exits, to intersections in this case).
- A crossing $P(c_l) : \Omega \rightarrow A$ is a function linking entries to exits according to the rule of association $P(c_l) = \{(\omega_j, \alpha_k) | \omega_j \in \Omega, \alpha_k \in A\}$, where a set of ordered pairs between entries ω_j and α_k exits on a crossing. Note that this allows us to model most types of intersections. We enforce the constraint that any given access γ_i can only be part of, at most, one crossing $p(c_l)$, noted γ_i^l, but within that intersection, it may participate in several ordered pairs, i.e., we assume free ends. We refer to an ordered pair (ω_j^l, α_k^l) as a *pathlink* p_{jk}^l; hence, a crossing is a set of intranodal pathlinks.
- $C = \{P(c_l) | l = 1, \ldots, L\}$, the set of intersections as described by its pathlinks $P(c_l)$, with L as the cardinality of the set.
- A road network can be represented by a directed multigraph $\mathcal{G} = (C, E)$ of intersections and $E = C \times C$ and streets. Now, the incoming edges to a node form Ω, and the outgoing edges are in A. In other words, an edge is an ordered pair $e = (\alpha_k, \omega_j)$, with Γ as all the edges' endpoints. Not all intersections will be monitored, but this

is circumstantial. Furthermore, it is not critical to know the whole network $\mathcal{G}$, and it suffices to know C.

- $S = \{s_r^i | r = 1, \ldots, R, i : \gamma_i \in \Gamma\}$, a set of observations obtained from the video analysis (acquired with some corresponding camera) s_r^i looking at some intersection access γ_i, with R denoting the cardinality of the set. If a camera can monitor more than one access in one or more intersections, this can be represented by as many s_r^i instances as required. If any camera does not monitor some access, there will not be a data source for that access. An access γ_i may be observed via none, one, or more cameras. These observations will be noted as $s_r^i[t_n]$ with $n = 1, \ldots, T$ set of timestamps.

At any given time t_n of $\{T\}$, a vehicle u_v may be detected using an observation algorithm s_r^i. A function f_{detect} in Equation (1):

$$\begin{aligned} f_{detect} &: \{T\} \times S \times U \rightarrow Bool = \{True, False, NA\} \\ b &= f_{detect}(t_n, s_r^i, u_v) \end{aligned} \tag{1}$$

indicates when the vehicle u_v is being detected via some observer s_r^i at the time t_n, $s_r^i[t_n]$. If the status of the detection b is *True*, this indicates that the vehicle is being detected. The outcome *False* indicates that the object is not being detected, while NA indicates that the detection cannot be confirmed (e.g., due to occlusions). Naturally, because s_r^i monitors access γ_i and given the constraint that one access can only be part of, at most, one intersection $p(c_l)$, if $f_{detect}(t_n, s_r^i, u_v) = True$, we also know that object u_v was at node $p(c_l)$ at time t_n.

The problem can be stated as follows: Given a time t, an object u_v, an entry access ω_j at intersection $p(c_l)$, noted ω_j^l, and knowledge of the situation at the intersection C and observation S, *determine* the (most likely) exit access α_k in the same crossing $p(c_l)$, noted α_k^l t with g as the node traversing function, which is unknown.

$$\underset{\alpha_k^l}{\text{argmax}}\ P_r(g(t_n, u_v, \omega_j; C, S) = \alpha_k) \tag{2}$$

In other words, determine the most likely outgoing end of the pathlink (ω_j^l, α_k^l), followed by the vehicle u_v traveling road network $\mathcal{G}$ at time t_n exploiting the info in S. This is solved using the Nelder–Mead algorithm [23] by iteratively adjusting the shape of a graph to find the lowest or highest point, depending on the goal. Shape optimization continues until passing a convergence threshold or reaching a stopping condition.

3.2. The Algorithm

The proposed solution is organized as a modular platform that executes different algorithms in parallel. Each algorithm is implemented as an independent plugin, sharing information using a shared memory whiteboard model. An illustration of the proposed algorithm is shown in Figure 2.

3.2.1. Data Acquisition

This module obtains the most recent unprocessed video frame from the device and associates a timestamp to it, then saves the frame to memory and shares it with the other plugins using the shared whiteboard. The protocols supported by this plugin are *RTSP* [24], *TrafiSense2 Dual* thermal camera, *iDS uEye* camera, and local video files or image sequences.

3.2.2. Detection

This plugin relies on the YOLO version 4 implementation provided by OpenCV, [25] using the parameters shown in Table 1 trained on the COCO dataset [26] and considering only the traffic-related objects from its multi-class output. The algorithm determines the position and bounding boxes of the vehicles in the scene, as illustrated in Figure 3. Once the bounding box is determined, it is cropped and shared along with its view space coordinates.

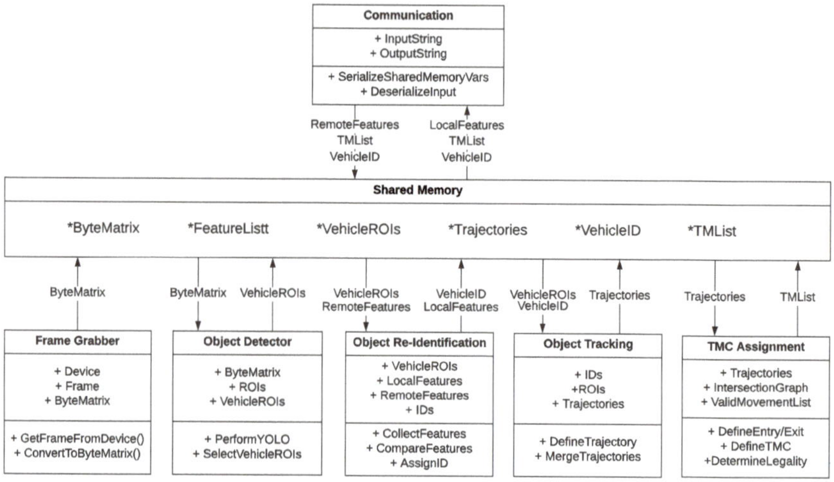

Figure 2. Illustrative data exchange among different modules of the architecture proposed. All modules are executed in parallel and data exchange is bidirectional and asynchronous.

Table 1. Parameters used in the training of YOLO using the COCO dataset.

Parameter	Value	Parameter	Value	Parameter	Value
batch	64	momentum	0.9	learning_rate	0.001
subdivisions	16	decay	0.0005	burn_in	1000
width	608	saturation	1.5	max_batches	500,200
height	608	exposure	1.5	policy	steps
channels	3	hue	0.1	steps	400,000, 450,000

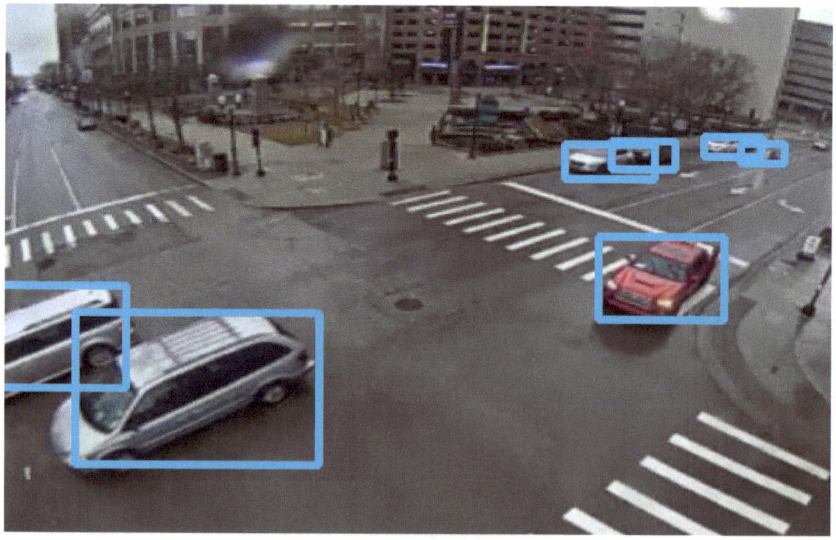

Figure 3. Visual representation of resulting bounding boxes defined by the detection algorithm.

3.2.3. Feature Collection and Re-Identification

This plugin extracts recognizable features from the bounding box using the ORB algorithm [27]. A sample of the feature match progress is shown in Figure 4. A set of newly extracted features are compared to either the previously known features collected via this processing node, or to those received from other locations via the network, using the *Hamming distance* defined in Equation (3) where the features a from the current camera image are compared to features b stored from previously collected features, as well as where a_i and b_i are the individual features.

$$d_{hamming}(a,b) = \sum_{i=0}^{n-1}(a_i \oplus b_i) \quad (3)$$

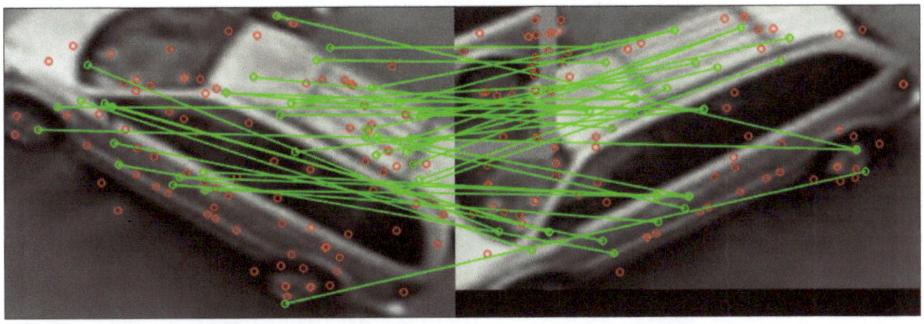

Figure 4. Feature matching during the re-identification process. *Red circles* features detected via the ORB algorithm. *Green lines* matching features.

If there is a match, the vehicle is considered re-identified and the known ID is assigned. If the comparison does not pass the threshold, the vehicle is regarded as *unknown* and a new ID is assigned. Once the ID is assigned, the system shares it, appending its features to the existing feature set.

3.2.4. Tracking

This module follows the vehicle through the scene. The position obtained via the detector is projected to a top-view representation. In addition, the position of each detected object is tracked using a Kalman tracker in combination with the ID provided by the re-identification algorithm. Once the top-view trajectory has been determined for all cameras, the trajectories are merged based on their associated ID, using the Frechet distance as shown in Algorithm 1 and in Figure 5. This algorithm measures the similarity between two curves that maintain a certain proximity by recursively calculating the Euclidean distance of the points that belong to the curves. If the resulting distance is below a certain threshold, the trajectories are merged using point averaging. Thus, the resulting trajectory is stored as an observation S for the intersection C. The stored trajectories are then used by the turn movement assignment algorithm to select both the most probable route taken and the appropriate TMC.

3.2.5. Turn Movement Assignment

The system determines the most probable trajectory (ω_j^l, α_k^l) for a vehicle u_v based on its entry point ω_j and the observations collected via the tracking module using Equation (2). This is performed by comparing the observed trajectory, aligning the sequences in a non-linear manner to a series of predefined turn movements, as shown in Figure 6, and finding the optimal match by stretching or compressing one of the trajectories to match the other, as described in Algorithm 2. The trajectory that requires the smallest change is considered the best match to the known legal turn movement.

Algorithm 1 MergeTrajectories

```
function MERGETRAJECTORIES (splines, threshold)
    trajectory ← ∅
    if splines is not empty then
        for i ← 0 to length(splines) − 1 do
            for j ← i + 1 to length(splines) do
                distance ← CalculateEuclideanDistance(splines[i], splines[j])
                if distance < threshold then
                    mergedSpline ← AveragePoints(splines[i], splines[j])
                    trajectory ← trajectory ∪ {mergedSpline}
                end if
            end for
        end for
    end if
    return trajectory
end function
```

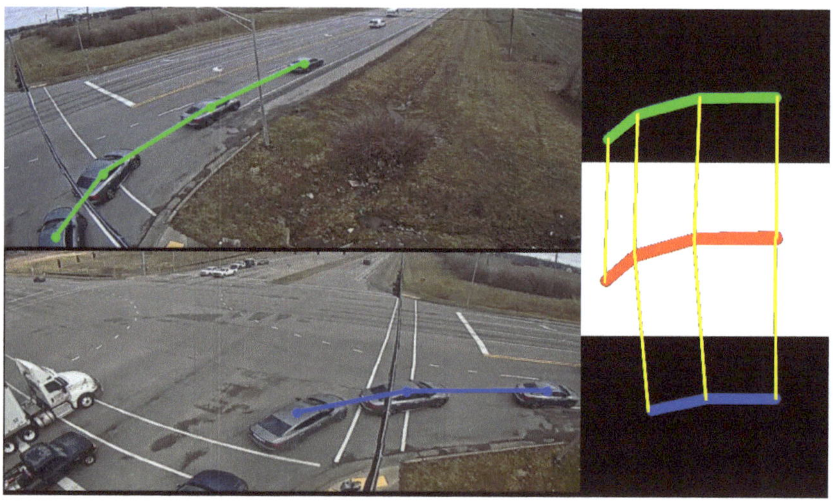

Figure 5. Tracking and smoothing process. Image-space trajectories projected to top-view *blue and green lines* and combined to obtain the final trajectory in top-view space *red line*.

Algorithm 2 TrajectorySimilarityEvaluation

```
function TSS(A, B): float
    n ← length of A
    m ← length of B
    DP ← a 2D array of size (n + 1) × (m + 1)
    for i ← 1 to n do
        for j ← 1 to m do
            cost ← distance between A[i] and B[j]
            DP[i][j] ← cost + min(DP[i − 1][j], DP[i][j − 1], DP[i − 1][j − 1])
        end for
    end for
    return DP[n][m]
end function
```

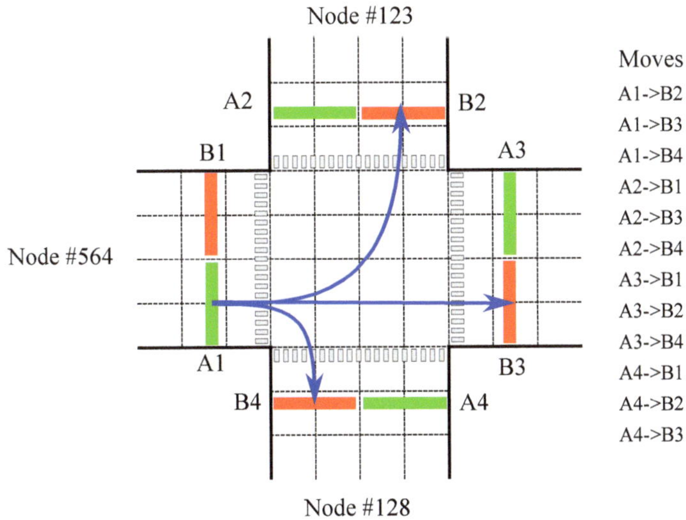

Figure 6. Visual representation of the proposed scene model at one intersection. Blue arrows indicate three legal turn movements at this specific intersection starting from A1: left (A1→B2), straight (A1→B3), and right (A1→B4).

Once the most probable trajectory is assigned, it is stored along with the vehicle ID and used for future comparisons with the trajectories obtained by human observers. In addition, the computed TMC is communicated to other nodes on the road network $\mathcal{G}$ to be used in further computations.

3.2.6. Communication

This module collects all available information shared by the other modules of its node and sends it to other nodes on the network. It also receives incoming data from other nodes and integrates it into the node's shared whiteboard. As the main component of the cooperative system, this module allows the node access to information about the environment that the locally connected sensors cannot collect. The information exchange uses a low-latency broadcasting messaging protocol that sends messages to all the other nodes simultaneously. In addition, the communication system also sends messages at each *vehicle entry event*, when a vehicle has been re-identified via a new node, and *vehicle exit event*, when the vehicle has left an intersection. The data communicated is user-defined; to reduce the amount of data exchange in our experiments, data sharing was limited to: *originating node, timestamp, re-identification ORB features, vehicle IDs, and entry/exit point*.

In addition, this module continuously reviews the events provided by the other systems and compares the entry/exit events to determine if the vehicle has been re-identified via another node. If re-identification occurs, the turn movement assignment is confirmed, thereby giving a higher certainty of a correct assessment.

4. Experiments

To evaluate the influence the proposed cooperation scheme has on the overall performance of a TMC system, we conducted a comparison between four scenarios, using the same dataset. The scenarios were a no-cooperation setup, a partial TMC confirmation, a complete TMC confirmation, and a partial blackout setup. We assessed the accuracy of TMC assignments compared to the ground truth manual annotations. We hypothesized that the TMC system would have the highest performance when maximum information was shared between the different nodes.

4.1. Experimental Setup

To correctly assess how the knowledge of its surroundings affects the result of a TMC algorithm, the dataset used should provide the context of surrounding intersections. To the best of our knowledge, no such dataset was available. A new dataset was captured using publicly accessible traffic cameras provided by the Lexington-Fayette Urban County Government in Lexington, KY, USA. The dataset was recorded from a total of 90 intersections, each equipped with four cameras pointing in the general direction of the legs of the intersection. Recordings were 1280 × 720 pixels, collected for 30 min at 25 frames per second. One issue we had to address is that the cameras in this system were destined for human traffic monitoring. Therefore, these cameras have not been calibrated or configured for software analysis. Sample images captured at one of the intersections used in this dataset can be seen in Figure 7. To overcome the problem of calibrations, the views were manually projected to align with the observable ground plane.

Figure 7. Illustrative frames of video captured at one of the intersections and approximated locations of all the cameras recorded.

Since the full context was not available for all intersections, a subset was selected. We chose intersections whose adjacent intersections were also recorded. The videos from these intersections were manually labeled by assigning an identifier to each leg of the intersection, enumerating each possible movement to be performed, and finally, assigning one of the possible movements to each vehicle in the scene. The data from the adjacent intersections were labeled and assigned the same unique ID to each vehicle in all scenes. As the experiment did not allow for on-site analysis, the experiments were performed offline on a single computer. To best replicate real-world conditions where each intersection would be processed via a single *edge* device such as an Intel NUC or a similar low-power unit, each node was executed on a single thread of the host unit, limiting the RAM available to 1 GB and with no GPU acceleration. Meanwhile, the network connection was set to 10 Mbps with an average simulated latency of 100–150 ms, which is similar to the latency found on low-cost public infrastructure solutions. For each experiment, the system simulated five processing nodes simultaneously, i.e., one main central node and four secondary nodes. The results reported in Section 4.3 only consider the results of the central node. The secondary nodes are used to provide context to the main node to increase its accuracy, but are not used directly for the evaluation. In this way, the systems are compared under the most similar conditions possible.

4.2. Experimental Scenarios

These scenarios were designed to test our hypothesis that "A cooperative system will surpass a non-cooperative system under the same conditions". Specifically, we selected the three most circulated intersections, each in a traditional crossroad configuration (i.e., four-legged symmetric intersection). For each intersection, the central intersection area acted as the main point of interest, and the four surrounding nodes served as cooperative inputs. An illustration of a typical four-legged intersection can be seen in Figure 8. The four scenarios are described next.

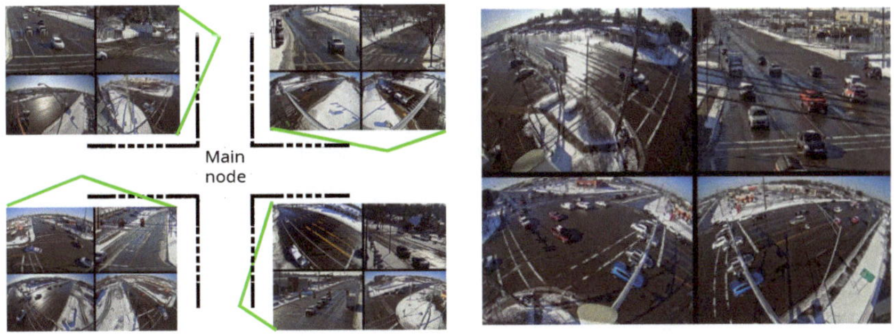

Figure 8. Illustrative distribution of a common four-legged intersection with one central node and four secondary nodes that communicate.

Scenario 1: No cooperation. Cooperation is entirely disabled. Information provided by other nodes is *not* used to determine the vehicle's movement. To maintain a similar workload of the system, we left communication enabled, but configured the main node to ignore any information provided by the other nodes of the network. This reflects the basic scenario of how most TMC systems are tested. That is, a standalone system that is only capable of using locally collected data. This also reflects the fundamental behavior of non-cooperative systems typically found in the literature. Therefore, we consider this scenario as a baseline for the comparison of a cooperative system, as all other conditions remain equal.

Scenario 2: Partial TMC confirmation. In this scenario, the data from other nodes is only used if the entry/exit points cannot be determined using local data. This scenario represents the main cases where the vehicle is occluded during the initial or final parts of its trajectory. This can be reduced to the following steps:

(1) If entry and exit points are defined:
- Store trajectory and exit.

(2) If no entry point is defined:
- Request re-identification.
- If the re-identification is positive:
  - Assign the vehicle entry point based on the location where the vehicle was initially identified.
- If the re-identification is negative:
  - Assign the vehicle entry point based on the closest available point.

(3) If no exit point is defined:
- Await a positive re-identification reported by another node.
- If the re-identification is positive:
  - Assign the exit point to the leg of the intersection connected to that node.
- If the re-identification is negative:
  - Assign the vehicle exit point based on the closest available point.

We limit the cooperation in this scenario to undetermined cases. Therefore, the system will not perform any correction to incorrectly assessed movements. Consequently, incorrect assignments made on the central node, not related to an incomplete trajectory, will remain incorrect even if other nodes perform correct re-identification.

Scenario 3: Complete TMC confirmation. In this scenario, we take full advantage of the cooperative mode of the system. Each node performs the turn movement assignment using the local data, while also confirming with other nodes following the steps listed below:

(1) Each node performs the turn movement assignment using the local data.
(2) Confirm the entry point by:

- Performing a re-identification of the vehicle.
- Determining if the assigned entry point coincides with the previously identified node.

(3) Confirm the exit point by:

- Awaiting a positive re-identification reported by another node.
- Determining if the assigned exit point coincides with the subsequently identified node.

(4) If more than one node presents an entry/exit:

- Maintain the locally obtained trajectory and exit as probable misidentification occurred.

(5) In case of any discrepancy between local and remote ID:

- Re-assign the points if the re-identification confidence level is above the threshold.
- Maintain original points if the re-identification confidence level is below the threshold (assumed misidentification).

In addition, the node also adds vehicle IDs with a turn movement when two neighboring nodes indicate the route to the intersection as the exit point and entry point, respectively. Therefore, the node manages to assign probable turn movements even to occluded vehicles based on the information provided by its peers.

Scenario 4: Partial blackout. In this scenario, the system is configured with the same cooperative capabilities as in scenario 3—*Complete*. However, the locally obtained data is blocked in order to simulate a failure of the cameras. This is a common situation when traffic accidents damage the sensing infrastructure. Traffic behavior analysis becomes more important in these cases, as the conditions at the intersection produce unexpected traffic flows. In order to make its turn movement assignment with these limiting conditions, the node depends entirely on the cooperative aspect of the system and the data provided by the surrounding nodes. Under this scenario, the system should be able to correctly assess those cases where a vehicle is identified by two of the adjacent nodes and assign the turn movement based on this. While this scenario represents an extreme case where a system that only uses local information is completely incapable of performing correctly, it presents an opportunity to determine if cooperation is beneficial in cases of technical failures, such as the one illustrated in Figure 9.

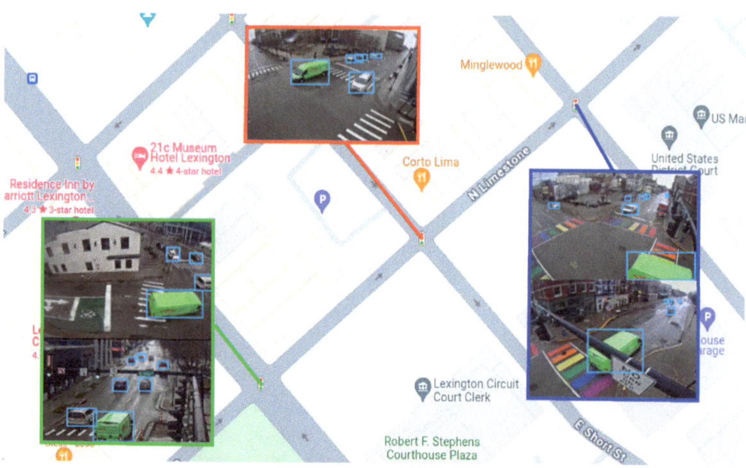

Figure 9. Illustrative example of the cooperation of the system. In this scenario, it is possible to infer the trajectory of the vehicle across the middle intersection (red) from the information provided by the neighboring nodes (green, blue), even if the camera at the central intersection was obstructed.

The architecture is configured in such a way that the plugins of the platform can be activated based on the cooperative level being used in each tested scenario, which can be seen in Algorithm 3. This is used to dynamically change how each node of the network uses the context provided by the other nodes and what information is to be shared using the communication plugin.

Algorithm 3 Simplified plugin call pseudocode

```
trajectory ← GET_REDUCED_TRAJECTORY()
if cooperationlevel = 1 then
    if not EntryPoint then
        Assign closest entry point to trajectory
    end if
    if not ExitPoint then
        Assign closest exit point to trajectory
    end if
end if
if cooperationlevel = 2 then
    if not EntryPoint then
        trajectory.Entry ← REIDENTIFYENTRY(features)
    end if
    if not ExitPoint then
        trajectory.Exit ← REIDENTIFYEXIT(features)
    end if
end if
if cooperationlevel = 3 then
    if REIDENTIFYENTRY(features) > threshold then
        trajectory.Entry ← NewEntry
    end if
    if REIDENTIFYEXIT(features) > threshold then
        trajectory.Exit ← NewExit
    end if
end if
TM ← TRAJECTORYSIMILARITYEVALUATION(trajectory)
ASSIGNTMC(TM)
```

4.3. Experimental Results

The system was applied to three intersections, with a total of 890 manually labeled turn movements.

As shown in Table 2 and in Figure 10, Scenario 3—*Complete*—achieved the highest average at 95.65% $\pm$ 1.55 correctly identified movements and reached a performance of 97% on Intersection 3, outperforming the other scenarios by 4% to 70%. On scenario 4—*Partial Blackout*, the system was capable of correctly assigning on average 30.67% $\pm$ 4.04 of the turn movements. Systems without a cooperative feature would be incapable of this task because the local data has been blocked, where such systems would identify zero vehicles.

Table 2. Percentage of Correctly Assigned Turn Movements for the three intersections under the four scenarios.

	Intersection A	Intersection B	Intersection C	Average	STD
No-Coop	87%	82%	91%	86.67%	4.51
Partial	91%	86%	93%	90.00%	3.61
Complete	**94%**	**96%**	**97%**	**95.65%**	**1.55**
P-Blackout	30%	35%	27%	30.67%	4.04

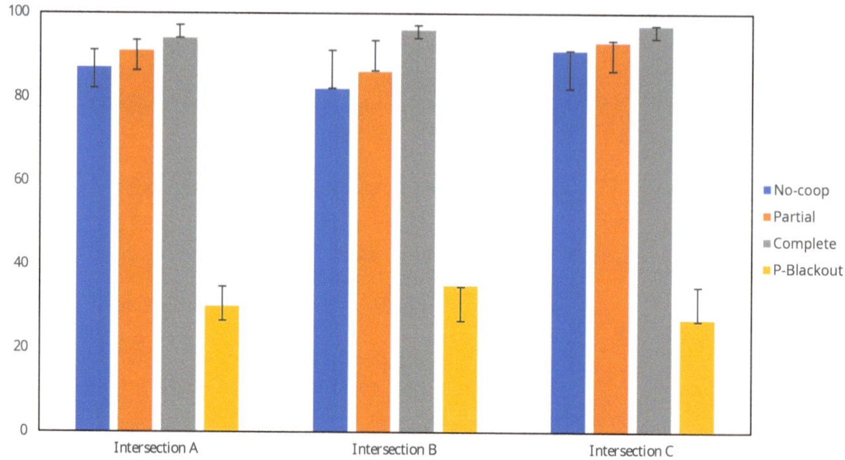

Figure 10. Correctly assigned turn movements in different scenarios.

Table 2 also shows that the standard deviation (STD) is smaller in more cooperative scenarios. This shows a clear tendency for an overall higher performance when the system shares more information.

These results can also be contrasted in Table 3, showing a clear increase in the F1 score when the system uses cooperative mode. The clear advantage of the use of cooperation in scenario 4 *P-Blackout* can also be seen, in contrast to a standard system, where the score without cooperation would be zero. It should also be noted that in scenarios 2 *Partial* and 3 *Complete*, cooperation reduces the false negative rate without inducing a large increase in the false positive rate, demonstrating that cooperation does not increase incorrect turn movement assignments.

$$\text{IDF1} = \frac{2 \times \text{TP}}{2 \times \text{TP} + \text{FP} + \text{FN}} \tag{4}$$

Table 3. IDF1 score of the TMC under different scenarios on a total of 890 different turn movements. **IDF1** *score obtained using Equation* (4); **TP** *true positive;* **FP** *false positive;* **FN** *false negative.*

	IDF1	TP	FP	FN
No-Coop	0.7120	55.28	24.49	20.22
Partial	0.7464	59.55	21.01	19.43
Complete	**0.7799**	**63.93**	21.91	**14.15**
P-Blackout	0.4786	31.46	**9.77**	58.76

Further testing is required due to the relatively small dataset used in these experiments. Nevertheless, we note a tendency for performance improvements when using cooperative methods compared to non-comparative ones. Moreover, this system is capable of correctly inferring the information of an intersection without directly observing it.

5. Conclusions

While the TMC results of the proposed system, when functioning in non-cooperative mode, are similar to the published results in the literature, the integration of cooperation significantly improves the TMC performance. Additionally, it is noteworthy that by integrating a cooperative scheme into the algorithm, it is possible to perform TMC under adverse conditions where one part of the system is incapable of collecting data on its own. This system will be more robust and reliable in conditions where the correct assessment of movements is the most important factor when making decisions about infrastructure improvements.

The main contribution of a cooperative system, compared to non-cooperative systems, is that the proposed system is capable of determining the movements of vehicles even when no direct observation of the intersection is possible. As the cooperation of the system does not depend on any particular detection or tracking algorithm, the proposed system can be integrated with other existing TMC methods, providing additional certainty. The experiments described here demonstrate that the system is highly scalable, as each node is only required to both analyze the data provided by the cameras at one intersection and communicate with the directly neighboring intersection, e.g., 3–5 neighbors in most common urban scenarios. This scalability at a relatively low cost, compared with a centralized solution, allows for ad hoc expansion of the system. We expect to expand this research to evaluate the system in real-world scenarios using real *edge* hardware solutions. We also plan to integrate additional sensing capabilities such as thermal cameras, radar, and LIDAR, as these sensors provide additional information, expanding the possibilities of better infrastructure planning. We are also exploring the possibility of expanding the system to observe larger areas, such as highways and interstates, to perform long-term vehicle tracking. Once these sensing technologies have been integrated into the system, we plan to deploy the complete system in real-world scenarios to improve the local capabilities of traffic analysis.

Author Contributions: Methodology, P.H.-W.; Writing—original draft preparation, P.H.-W. and C.L.; Writing—review and editing, L.J., B.G. and W.P.; Supervision, B.G. and W.P. All authors have read and agreed to the published version of the manuscript.

Funding: This work was funded by EU Horizon 2020 ECSEL JU research and innovation programme under grant agreement 876487 (NextPerception) and by the Flemish Government (AI Research Program).

Institutional Review Board Statement: Not applicable.

Informed Consent Statement: Not applicable.

Data Availability Statement: The data and code can be requested by contacting the corresponding author.

Acknowledgments: We wish to extend our sincere gratitude to Asli Kumcu for her invaluable assistance, particularly her support and expertise in editing.

Conflicts of Interest: The authors declare no conflict of interest.

Abbreviations

The following abbreviations are used in this manuscript:

ID	Identification
LIDAR	Light Detection and Ranging
RFID	Radio Frequency Identification
ROI	Region Of Interest
RTSP	Real Time Streaming Protocol
STD	Standard Deviation
TMC	Turn Movement Count
YOLO	You Only Look Once

References

1. Barcellos, P.; Bouvié, C.; Escouto, F.L.; Scharcanski, J. A novel video based system for detecting and counting vehicles at user-defined virtual loops. *Expert Syst. Appl.* **2015**, *42*, 1845–1856. [CrossRef]
2. Bas, E.; Tekalp, A.M.; Salman, F.S. Automatic vehicle counting from video for traffic flow analysis. In Proceedings of the IEEE Intelligent Vehicles Symposium, Istanbul, Turkey, 13–15 June 2007; IEEE: Piscataway, NJ, USA, 2007; pp. 392–397.
3. Lan, S.; Ren, Z.; Wu, Y.; Davis, L.S.; Hua, G. SaccadeNet: A fast and accurate object detector. In Proceedings of the IEEE/CVF Conference on Computer Vision and Pattern Recognition, Virtual, 14–19 June 2020; pp. 10397–10406.
4. Lin, T.Y.; Goyal, P.; Girshick, R.; He, K.; Dollár, P. Focal loss for dense object detection. In Proceedings of the IEEE International Conference on Computer Vision, Venice, Italy, 22–29 October 2017; pp. 2980–2988.

5. Ren, S.; He, K.; Girshick, R.; Sun, J. Faster r-cnn: Towards real-time object detection with region proposal networks. In Proceedings of the Advances in Neural Information Processing Systems, Montreal, QC, Canada, 7–12 December 2015; Volume 28.
6. Redmon, J.; Divvala, S.; Girshick, R.; Farhadi, A. You only look once: Unified, real-time object detection. In Proceedings of the IEEE Conference on Computer Vision and Pattern Recognition, Las Vegas, NA, USA, 26 June–1 July 2016; pp. 779–788.
7. Bewley, A.; Ge, Z.; Ott, L.; Ramos, F.; Upcroft, B. Simple online and realtime tracking. In Proceedings of the IEEE International Conference on Image Processing, Phoenix, AZ, USA, 25–28 September 2016; IEEE: Piscataway, NJ, USA, 2016; pp. 3464–3468.
8. Fang, J.; Meng, H.; Zhang, H.; Wang, X. A low-cost vehicle detection and classification system based on unmodulated continuous wave radar. In Proceedings of the IEEE Intelligent Transportation Systems Conference, Bellevue, WA, USA, 30 September–3 October 2007; IEEE: Piscataway, NJ, USA, 2007; pp. 715–720.
9. Liang, C.; Zhang, Z.; Lu, Y.; Zhou, X.; Li, B.; Ye, X.; Zou, J. Rethinking the competition between detection and reid in multi-object tracking. *arXiv* **2020**, arXiv:2010.12138.
10. Wojke, N.; Bewley, A.; Paulus, D. Simple online and realtime tracking with a deep association metric. In Proceedings of the IEEE International Conference on Image Processing, Beijing, China, 17–20 September 2017; IEEE: Piscataway, NJ, USA, 2017; pp. 3645–3649.
11. Li, W.; Xiong, Y.; Yang, S.; Deng, S.; Xia, W. Smot: Single-shot multi object tracking. *arXiv* **2020**, arXiv:2010.16031.
12. Bui, N.; Yi, H.; Cho, J. A vehicle counts by class framework using distinguished regions tracking at multiple intersections. In Proceedings of the IEEE/CVF Conference on Computer Vision and Pattern Recognition Workshops, Seattle, WA, USA, 14–19 June 2020; pp. 578–579.
13. Wang, W.; Gee, T.; Price, J.; Qi, H. Real time multi-vehicle tracking and counting at intersections from a fisheye camera. In Proceedings of the IEEE Winter Conference on Applications of Computer Vision, Waikoloa, HI, USA, 5–9 January 2015; IEEE: Piscataway, NJ, USA, 2015; pp. 17–24.
14. Dai, Z.; Song, H.; Wang, X.; Fang, Y.; Yun, X.; Zhang, Z.; Li, H. Video-based vehicle counting framework. *IEEE Access* **2019**, *7*, 64460–64470. [CrossRef]
15. Shokrolah Shirazi, M.; Morris, B.T. Trajectory prediction of vehicles turning at intersections using deep neural networks. *Mach Vis. Appl.* **2019**, *30*, 1097–1109. [CrossRef]
16. Yu, L.; Feng, Q.; Qian, Y.; Liu, W.; Hauptmann, A.G. Zero-virus: Zero-shot vehicle route understanding system for intelligent transportation. In Proceedings of the IEEE/CVF Conference on Computer Vision and Pattern Recognition Workshops, Seattle, WA, USA, 13–19 June 2020; pp. 594–595.
17. Liu, Z.; Zhang, W.; Gao, X.; Meng, H.; Tan, X.; Zhu, X.; Xue, Z.; Ye, X.; Zhang, H.; Wen, S.; et al. Robust movement-specific vehicle counting at crowded intersections. In Proceedings of the IEEE/CVF Conference on Computer Vision and Pattern Recognition Workshops, Seattle, WA, USA, 13–19 June 2020; pp. 614–615.
18. Nassereddine, H.; Santiago-Chaparro, K.R.; Noyce, D.A. Framework for quantifying right-turn-on-red conflicts from existing radar-based vehicle detection infrastructure. *Transp. Res. Rec.* **2022**, *2676*, 556–567. [CrossRef]
19. Czyżewski, A.; Kotus, J.; Szwoch, G. Estimating traffic intensity employing passive acoustic radar and enhanced microwave doppler radar sensor. *Remote Sens.* **2019**, *12*, 110. [CrossRef]
20. Zhang, M.; Fu, R.; Morris, D.D.; Wang, C. A framework for turning behavior classification at intersections using 3D LIDAR. *IEEE Trans. Veh. Technol.* **2019**, *68*, 7431–7442. [CrossRef]
21. Zhang, J.; Xiao, W.; Coifman, B.; Mills, J.P. Vehicle tracking and speed estimation from roadside lidar. *IEEE J. Sel. Top. Appl. Earth Obs. Remote. Sens.* **2020**, *13*, 5597–5608. [CrossRef]
22. Abbasi, R.; Bashir, A.K.; Alyamani, H.J.; Amin, F.; Doh, J.; Chen, J. Lidar point cloud compression, processing and learning for autonomous driving. *IEEE Trans. Intell. Transp. Syst.* **2022**, *24*, 962–979. [CrossRef]
23. Singer, S.; Nelder, J. Nelder-mead algorithm. *Scholarpedia* **2009**, *4*, 2928. [CrossRef]
24. Schulzrinne, H.; Rao, A.; Lanphier, R. Real Time Streaming Protocol (RTSP); Technical Report; 1998. Avaliable online: https://www.rfc-editor.org/info/rfc2326 (accessed on 5 October 2023).
25. Bradski, G. The OpenCV Library. *Dr. Dobb J. Softw. Tools* **2000**, *25*, 120–123.
26. Lin, T.Y.; Maire, M.; Belongie, S.; Hays, J.; Perona, P.; Ramanan, D.; Dollár, P.; Zitnick, C.L. Microsoft coco: Common objects in context. In Proceedings of the European Conference on Computer Vision, Zurich, Switzerland, 6–12 September 2014; Springer: Cham, Switzerland, 2014; pp. 740–755.
27. Rublee, E.; Rabaud, V.; Konolige, K.; Bradski, G. ORB: An efficient alternative to SIFT or SURF. In Proceedings of the International Conference on Computer Vision, Barcelona, Spain, 6–13 November 2011; IEEE: Piscataway, NJ, USA, 2011; pp. 2564–2571.

sensors

MDPI

Article

The Impact of Line-of-Sight and Connected Vehicle Technology on Mitigating and Preventing Crash and Near-Crash Events

Eileen Herbers [1,2,*], Zachary Doerzaph [1,2] and Loren Stowe [1]

[1] Virginia Tech Transportation Institute, Virginia Tech, Blacksburg, VA 24060, USA
[2] Department of Biomedical Engineering and Mechanics, Virginia Tech, Blacksburg, VA 24060, USA
* Correspondence: eherbers@vtti.vt.edu

Abstract: Line-of-sight (LOS) sensors developed in newer vehicles have the potential to help avoid crash and near-crash scenarios with advanced driving-assistance systems; furthermore, connected vehicle technologies (CVT) also have a promising role in advancing vehicle safety. This study used crash and near-crash events from the Second Strategic Highway Research Program Naturalistic Driving Study (SHRP2 NDS) to reconstruct crash events so that the applicable benefit of sensors in LOS systems and CVT can be compared. The benefits of CVT over LOS systems include additional reaction time before a predicted crash, as well as a lower deceleration value needed to prevent a crash. This work acts as a baseline effort to determine the potential safety benefits of CVT-enabled systems over LOS sensors alone.

Keywords: advanced driver assistance systems; naturalistic driving data; automated driving systems; connected vehicles

Citation: Herbers, E.; Doerzaph, Z.; Stowe, L. The Impact of Line-of-Sight and Connected Vehicle Technology on Mitigating and Preventing Crash and Near-Crash Events. *Sensors* **2024**, *24*, 484. https://doi.org/10.3390/s24020484

Academic Editors: Constantin Caruntu and Ciprian Romeo Comşa

Received: 14 December 2023
Revised: 6 January 2024
Accepted: 8 January 2024
Published: 12 January 2024

1. Introduction

New vehicles are being equipped with a multitude of sensors to be used in advanced driver assistance systems (ADAS) and/or automated driving systems (ADS) to develop an understanding of their environment. These environmental sensors can generally be characterized as line-of-sight (LOS) sensors because they rely on information directly captured by the sensors' field of view. However, to increase the amount of information available and to expand the sensed field of view, it is beneficial to use additional shared data from other vehicles and the infrastructure. By leveraging evolving communication systems, data shared over connected vehicle technologies (CVT) may provide a variety of performance benefits to transportation. This new level of collaborative communication has the potential to develop a collective perception of a vehicle's environment, which could directly improve safety as events unfold.

The purpose of this research is to act as a baseline attempt to measure the potential safety impact that advanced sensors and communication methods can make in real-world crashes and near-crashes. Therefore, in this research LOS sensors are used to describe any sensors that use vision-based technology for object detection (such as cameras, RADAR, or LiDAR). These sensors are most often used in advanced driver assistance systems (ADAS) and automated driving systems (ADS). CVT may have additional sensors that are used in sharing or receiving information with other vehicles or infrastructure. Currently, the implementation of CVT is mostly found in simulation testing [1].

Previous work has characterized some of the potential advantages of LOS and CVT, such as increasing traffic speed or flow rate without any negative impact on traffic safety, improving individual mobility, providing environmental impact reduction benefits through reduced fuel use and better efficiency, and preventing/mitigating fatal and injury-causing crashes [2–6]. Although some of these potential advantages can be readily predicted through macrosimulation techniques, quantifying the actualized safety benefit of LOS sensors or CVT requires a more nuanced approach. This is because the specific factors leading up to a police-reported crash are generally unknown and can vary greatly between one another [7]. Generally, the prospective safety impact of more advanced vehicle-sensing technology, like LOS or CVT, is projected by estimating the number or percentage of police-reported crashes that could have been avoided if the vehicles involved were equipped with additional sensors. For example, through a meta-analysis model of 73 different studies, it was found that up to 48.07% of crashes in the US could have been prevented if all of the vehicles involved had CVT or were automated vehicles [8]. Additionally, through the analysis of GES crash records from 2005–2008, CVT or LOS sensors could conceivably have prevented 32.99% of crashes, and 47% of rear-end crashes in the US [9]. According to a preliminary study by the National Highway Traffic Administration (NHTSA), CVT could eventually prevent or mitigate about 80% of non-alcohol-related crashes [10]. These studies use crash aggregates and crash types to assume how many crashes could have been prevented if there was 100% market penetration of these technologies.

Studies like these provide a general estimate of the types of crashes that could be mitigated or avoided with the implementation of these sensors and technologies, but each crash is unique, and the actual impact of LOS and CVT may be affected by additional factors. It becomes more difficult to accurately predict the safety impact of these systems on the microscopic scale [11]. One way to do so is to evaluate the performance of current ADAS systems (which contain LOS sensors). For example, a partnership between automakers and NHTSA used real-world vehicle data from 47 million ADAS-equipped vehicles to determine that vehicles with automatic emergency braking (AEB) are 49% less likely to strike another vehicle in a rear-end crash [12]. Another way to do so is by introducing LOS or CVT sensors into simulation studies. A study in which intersection crashes were re-simulated predicted that an intersection-specific ADAS could prevent 25–59% of crashes [13]. Another simulation study using a bottom-up microscopic simulation approach to predict macroscopic statistics found that 24–87% of fatal crashes could have been avoided in scenarios involving vehicles with ADAS/ADS systems compared to fully manual driving scenarios [14].

In developing simulations to assess CVT effectiveness, machine learning models have been used to simulate specific events which can then be tested on the road. One study developed a long short-term memory model to predict vehicle trajectories to simulate a cut-in maneuver in a V2V environment, which was superior to traditional collision-warning models [15]. Another study developed a road safety information system using naturalistic data from connected vehicles on Korean highways to assess how connected vehicles could affect traffic safety and flow. However, this study used a macroscopic model for each section of the highway, and suggests a more microscopic calibration to assess actual crash risk [16]. Although one study used Doppler shift to assess a collision-avoidance system that specifically used only wireless communication without any LOS sensors [17], another study developed a high-level fusion of LOS sensors and wireless vehicle communication data to predict the trajectories of conflict with vehicles and pedestrians and found that this fusion enabled higher driver and pedestrian safety [18]. This fusion method is more similar to how the sensors are viewed within LOS and CVT systems in this research. However, an important piece of information that is missing from these simulations and could be beneficial in predicting the actual impact of more advanced vehicle sensors is the actual vehicle kinematic signatures before and during some of these safety-critical events (SCEs).

In this paper, a physics-based model was developed to simulate real-world crash and near-crash scenarios using naturalistic data from the Second Strategic Highway Research Program Naturalistic Driving Study (SHRP2 NDS). Naturalistic driving data provide a wealth of information before, during, and after SCEs and baseline scenarios. Especially important for this research, real-time vehicle data of near-crashes were captured, which enabled us to analyze SCEs that are not found in national crash datasets. These events were reconstructed so that the benefit of LOS sensors and CVT could be compared to the baseline scenarios that did not have either of those technological benefits. Four different crash configurations were studied, and the system activation time and resulting required deceleration to avoid these crash and near-crash events were calculated. The research in this report intends to add to the body of knowledge around the potential quantitative safety impact of vehicles equipped with LOS sensors (ADAS-equipped vehicles) and the probable added benefit of CVT systems (vehicle-to-everything (V2X)). This work was part of a larger Safety Through Disruption (Safe-D) University Transportation Center (UTC) report [19].

2. Materials and Methods

The following section includes a description of the dataset, and how events were identified to be included in the data extraction. A flowchart of this method can be found in Appendix B. The data extraction included pulling de-identified data from VTTI's data enclave to obtain the Global Positioning System (GPS) positions of the subject vehicle to generate a map image of the location without the linked GPS coordinates. The subject vehicle in this research is used to describe the "host vehicle", or the vehicle containing the DAS. The target vehicle is used to describe the vehicle that conflicted with the subject vehicle (i.e., it was the other vehicle in the crash or near-crash). The subject vehicles and the target vehicles were then manually tracked throughout the event to determine their trajectories and speed. This information was used to recreate the events and calculate crash-specific variables that could be used to calculate the impact that ADS technologies could have had on the outcome of these events.

2.1. Dataset

The Second Strategic Highway Research Program Naturalistic Driving Study (SHRP2 NDS), funded by the U.S. Federal highway Administration (FHWA), is the largest naturalistic driving study that has been undertaken to date. The SHRP2 database consists of over 5.5 million trips driven by 3542 drivers across 6 collections sites in the continental United States. These sites hosted from 150 to 450 participants each and included rural sights such as central Pennsylvania, and more populated urban areas such as Seattle, Washington. VTTI developed a data acquisition system (DAS) to support the research questions and objectives of the SHRP2 NDS program, which included compiling a dataset that could be used to support future data mining activities such as this one [20]. SHRP2 was used because of the availability of a "breadcrumb" trail of the GPS location, speed, acceleration, and other timeseries data [21]. The DAS facilitated the collection of the following data of interest to this study:

- video data of the forward view;
- subject vehicle speed;
- subject vehicle yaw and yaw rate;
- GPS latitude and longitude.

These variables were chosen because they could be used to recreate events of interest within a simple physics-based model. However, the DAS collected a variety of additional variables such as multiple video angles, machine vision, accelerometer data in all 3 axes, driver cell phone use, vehicle network data, and more. The study was conducted in accordance with the Declaration of Helsinki and approved by the Institutional Review Board of Virginia Tech (IRB #18-957 23 October 2018).

2.2. Event Identification

To correctly compare the potential benefit of CVT over LOS sensors, a subset of the crash and near-crash events from SHRP2 were identified. These events included ones where the view of target vehicle was obstructed so that the LOS sensor would not be able to perceive the target vehicle prior to an imminent potential conflict. Therefore, the capability of an LOS system would be limited, while a CVT system could provide a benefit to the operation of the associated safety system. Therefore, the conflict object, or target vehicle, was out of sight for the majority of the time leading up to the event. Figure 1 illustrates an example of such an event. Although this constitutes a strong selection bias, these specific events were chosen because they showed the most promise to fulfill the purpose of determining how the addition of LOS and CVT systems could mitigate or prevent real-world crash and near-crash scenarios.

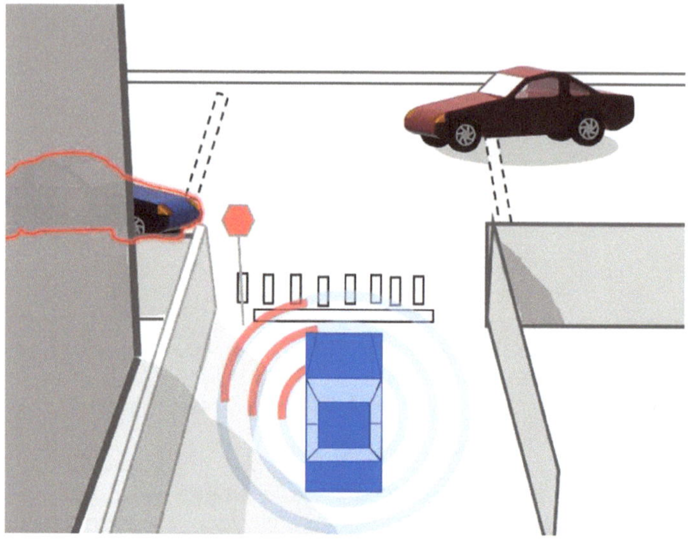

Figure 1. Example of a visual obstruction event.

From initial data mining, 594 events showed initial promise to be relevant to this research because there was a view-obstructing object involved within the incident and the necessary data elements were available. These events were defined by the SHRP2 dataset and included crashes and near-crashes. A crash was defined as "any contact that the subject vehicle has with an object" and a near-crash was defined as "any circumstance that requires a rapid evasive maneuver by the subject vehicle, or any other vehicle, pedestrian, cyclist, or animal, to avoid a crash" [20].

The candidacy of each event was then rated by its relevancy to the project and its ability to be reconstructed; those that were not good candidates included events with insufficient video, unpredictable conflict object maneuvers, host driver error, and more. Insufficient video included video with heavy precipitation, video with insufficient lighting, or video that was unavailable. Unpredictable conflict object maneuvers included any vehicles or animals that made erratic movements that would be difficult to reproduce with a simple physics-based model. Thus, 18 crashes and 162 near-crashes were identified that promised a strong ability to be recreated for the purposes of this project.

After these events were determined, they were reviewed once again to classify the leading cause of conflict. Table 1 provides a summary of the obstruction type for the events with strong candidacy.

Table 1. Events with strong candidacy by obstruction type.

Obstruction Type	Near Crash	Crashes
vehicle	126	11
bend in path (lateral or vertical)	9	1
none (small agent)	0	1
fog	0	1
other (e.g., building, vegetation)	27	4

2.3. Data Extraction

To obtain a better picture of the environment during a crash or near-crash event, a birds-eye view of the location of the event was created. The corresponding Google map image was extracted for each event and overlayed with the relative location of the subject vehicle. To keep the locations anonymous, the actual GPS coordinates of the subject vehicle were kept in the VTTI data enclave and converted to pixel locations, relative to the map image. The other data taken from SHRP2 were the instantaneous kinematic details of the subject vehicle and the front-facing video of the subject vehicle. Together, these four pieces of data were used to reconstruct the events in a physics-based model.

2.4. Event Reconstruction

2.4.1. Identifying Important Timepoints

The first step in recreating the event was to superimpose the subject vehicle trajectory over the map image. The accuracy of automotive-grade GPS is not always good enough to directly overlay on the map. For example, in one case, the subject vehicle appeared to be offset from the road and driving in the grass a few meters to the side. Therefore, the relative kinematic information collected from the DAS was used to generate the trajectory given a set of initial conditions. Points were marked individually on the map for the subject vehicle's initial location and the kinematic information from the subject vehicle was used to recreate its trajectory.

A graphical user interface (GUI) was developed to complete the following two tasks:

1. Determining the impact proximity frame (timestamp) by watching the event video. This is the approximate timestamp in which the subject vehicle and target vehicle come into contact (or near contact for near-crash events). This is later referred to as the conflict time.
2. Then, two frames (timestamps) are identified within the video that correspond to two locations of the subject vehicle on the map. The frames are chosen based on the ability to accurately place the concurrent subject vehicle position and heading on the map (i.e., lane markings, buildings, trees, etc.).

2.4.2. Calculating Subject Vehicle Trajectory

Once two positions, headings, and corresponding timestamps were identified, the vehicle's trajectory throughout the event was calculated given the vehicle kinematic data that were extracted previously from the DAS. This was performed by using the vehicle's starting position (x_n), heading (θ_n), and speed (v_n) in a basic iterative trajectory formula shown in Equation (1).

$$x_{n+1} = x_n + v_n \cos \theta_n \quad (1)$$

This produced a photo of the event trajectory superimposed on the corresponding map as shown in Figure 2. These photos were reviewed to determine if it would be beneficial to modify and repeat earlier steps for any events that had an unexpected trajectory. If the trajectory was unexpected (Figure 2, right), the associated trace factor and theta shifter were internally developed to determine more accurate positions and headings for the subject vehicle in the first step. The trace factor is the ratio between the trajectory distance of the event (as determined from the DAS) and the distance calculated from the vehicle positions chosen during the video review. The theta shifter is the difference in heading (degree).

2.4.3. Determining Locations of Objects of Interest

The GUI generated a .MAT file that contained the positions of the subject vehicle, view-obstructing objects, and the target vehicle (or conflict object). To generate the locations of the objects of interest, the subject vehicle front camera and the position of the subject vehicle on the map were displayed at corresponding timestamps.

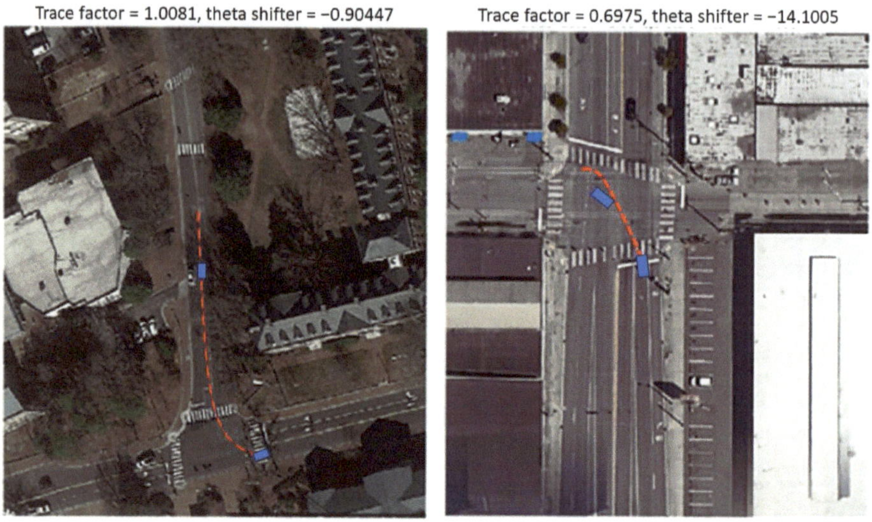

Figure 2. Review step showing an example of correctly calculated trajectory (**left**) and incorrectly calculated trajectory (**right**).

First, the impact proximity frame (conflict time) was brought up and the location of the target vehicle was identified on the map. Then, the event was reversed by 4 frames at a time until the frame in which the target vehicle was no longer visible was reached. At this frame, each object that could be obstructing the view of the subject vehicle was identified and its location was marked. Then, the event continued 4 frames at a time and the corresponding locations of each view obstructing object and the target vehicle was marked. These locations were marked up until the conflict time.

This step went quickly if the position and heading of the subject vehicle was accurate (as determined by the previous two steps), there was only one target vehicle, and only one stationary view-obstructing object. This step took considerable effort if there were many view-obstructing objects that move throughout each frame, there were multiple target vehicles, or the subject vehicle position was not accurate.

2.5. Physics-Based Model

Each event now had the relative locations and headings on the subject vehicle, the target vehicle(s), and the view-obstructing object(s) at certain time frames of the event. Linear interpolation was used to fill in the locations of these objects for the missing timestamps. This allowed for different parameters to be manipulated to simulate different scenarios that stem from one event. For the purposes of this project, three scenarios were simulated. The first scenario acts as the base case, which used the data in the original event reconstruction. The second scenario acts as though the subject vehicle has line-of-sight (LOS) technology. For simplicity in calculation, the subject vehicle detects the target vehicle when an uninterrupted line can be drawn from the centroid of each vehicle as shown in Figure 3. Since LOS sensors differ in range and width, this simplification allows for consistent calculation. We expect that most sensors will need to view much of the vehicle in order to correctly detect it, and this was a simple way to exemplify this expectation. The third scenario describes when the subject vehicle and target vehicle are using CVT (i.e., the subject vehicle knows the location, speed, acceleration, and trajectory of the target vehicle). These assumptions were used to calculate how environmental sensors (LOS) and information-sharing between traffic participants and smart infrastructures (CVT) could impact vehicle safety in these crash and near-crash scenarios. Using these data, two pieces of information were calculated for each scenario: (1) the activation time in which a potential conflict is identified in both

CVT and LOS ADS and (2) the required deceleration of the subject vehicle to prevent a crash in both CVT and LOS ADS.

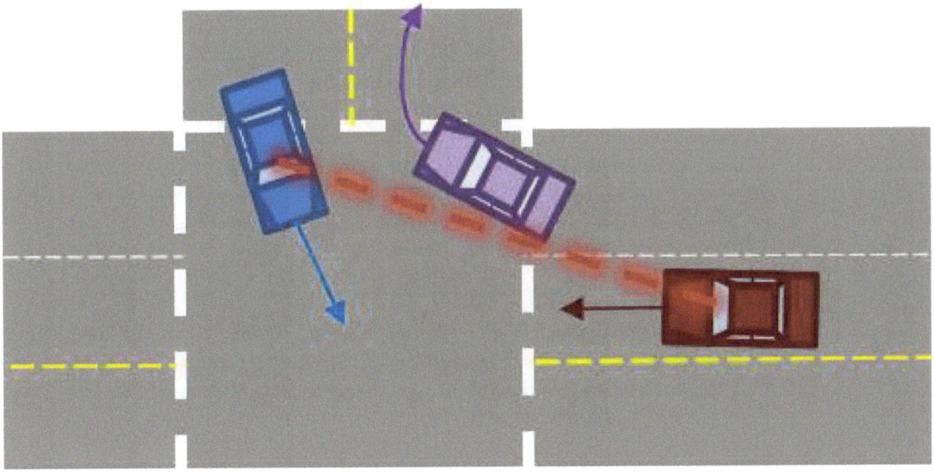

Figure 3. The centroid of the target vehicle (blue) is within the line of sight of the subject vehicle (red) around the visual obstruction object (purple), which determines LOS activation.

2.5.1. Activation Time before Conflict Calculation

To calculate these two pieces of information, the relative speed, heading, and locations of the target vehicle and view-obstructing objects were needed. The instantaneous speed, v_n, was calculated by taking a simple derivative of the pixel location of each object over time. The pixel distance was converted to distance in meters by using the zoom factor used to produce the map image (i.e., the zoom factor, zf, was a fraction that correlated the number of pixels to a distance in meters).

Then, in each frame, the current speed, location, and heading of the subject vehicle and target vehicle were used to determine the expected trajectory of each vehicle (Equation (1)). A conflict was identified if the vehicle centroids came within 4 m of each other at some point within their predicted trajectories.

The first time a potential conflict was identified in the data, defined when the CVT system would activate. The LOS activation timepoint would occur once the subject vehicle could "see" the target vehicle (as shown in Figure 3). Therefore, a LOS system would activate either concurrently, or after a CVT system. By taking the difference between those two vectors, it could be determined how much earlier a CVT system could notify a driver of a potential conflict over an LOS system. The system activation time before the conflict time is essentially the commonly used safety surrogate measure, time-to-collision (TTC). However, because the real-world conflict time was known, we used this predicted system activation time relative to the actual conflict time.

2.5.2. Required Deceleration

With the distance between each vehicle (d), the predicted point of collision, the time until collision (t), and the speed of both vehicles (v_S and v_T), the minimum required deceleration ($-a$) to prevent a crash can be calculated by Equation (2).

$$-a = \frac{d - |v_S - v_T|t}{t^2} \tag{2}$$

The collision avoidance strategy assumes that the subject vehicle does not swerve and the conflict is avoided with braking only; the driver (or vehicle) does not have to perform any other evasive maneuvers. Additionally, a simplifying assumption is made such that the target vehicle does not accelerate or swerve once a conflict is identified.

3. Results and Discussion

As discussed earlier in Section 2.2 Event Identification, 18 crashes and 162 near-crash events were reconstructed. After analyzing these 180 events, 68 events still possessed with usable data. Furthermore, 112 events were excluded due to incorrect satellite images (e.g., major construction since the date of the event, incorrect GPS data points in SHRP2) or missing kinematic data within the time of interest for our project. The resulting 68 events were separated into four crash configuration categories, and the following two values were calculated: (1) the difference between CVT and LOS activation time, and (2) the minimum required deceleration to avoid a collision.

3.1. Crash Configuration Categories

To organize the events and find significance within the values calculated, events were categorized into four crash configurations. These four categories were based off the General Estimates System (GES) accident type diagram and can be found in Appendix A, with the number of events within each configuration shown in Table 2.

Table 2. Crash configuration categories and the number of events analyzed.

Crash Configuration	Number of Events
left turn across path	40
perpendicular	8
rear-end	14
turn into same direction	6

Even though rear-end collisions are the most frequent type of crash in the US [22], the study did not include a proportional number of these events. This is because they are generally not caused by a visual obstruction, but rather due to driver distraction or lack of driver awareness. Although the events in this research make up a relatively low sample size for each crash configuration, it is important to note that the left turn across path configurations made up a significant number of crash and near-crash events that could be mitigated by LOS and CVT systems within this sample. Therefore, it could be beneficial to focus on left turn across path scenarios in future work involved with assessing the safety of technologically advanced vehicles.

3.2. Activation Time before Conflict

The activation time represents the amount of time between when each system detected an imminent conflict and the actual time of conflict. This concept somewhat represents the common safety surrogate measure, time-to-collision (TTC), but is calculated slightly differently here since the time of the actual conflict is known. Figure 4 shows the difference in activation time between CVT and LOS sensors. Each dot represents the actual value for each scenario, the X represents the mean value, the horizontal line represents the median value, the box encompasses the first and third quartile, and the whiskers extend to the maximum value that is within 1.5 times the inner quartile range.

Across all scenarios, the CVT system activated 0.51 ± 0.15 s before a LOS sensor detects the target vehicle on average. This means that CVT could provide an additional $\frac{1}{2}$ second of reaction time over LOS systems. Additionally, since all crash and near-crash events are generally classified together as safety-critical events (SCEs), an additional $\frac{1}{2}$ second could allow for earlier activation of forward collision-warning systems or automated emergency braking in these scenarios [20].

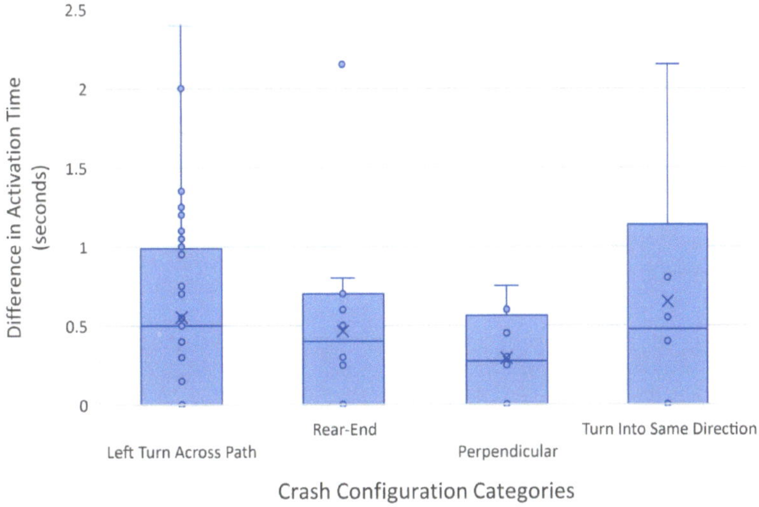

Figure 4. Box and whisker plot of the difference in activation time (seconds) between LOS and CVT systems separated by crash configuration categories. Each circle represents the actual value for each scenario, the X represents the mean value, the horizontal line represents the median value, the box encompasses the first and third quartile, and the whiskers extend to the maximum value that is within 1.5 times the inner quartile range.

3.3. *Required Deceleration*

The required deceleration is a value calculated to determine the minimum deceleration necessary to avoid a crash or near-crash if the vehicle began braking at the system activation time. Figure 5 is a box and whisker plot of the deceleration values required to prevent a potential conflict. Similar to Figure 4, the X represents the mean value, the horizontal line represents the median value, the box encompasses the first and third quartile, the whiskers extend to the maximum value that is within 1.5 times the inner quartile range, and each dot represents the actual value.

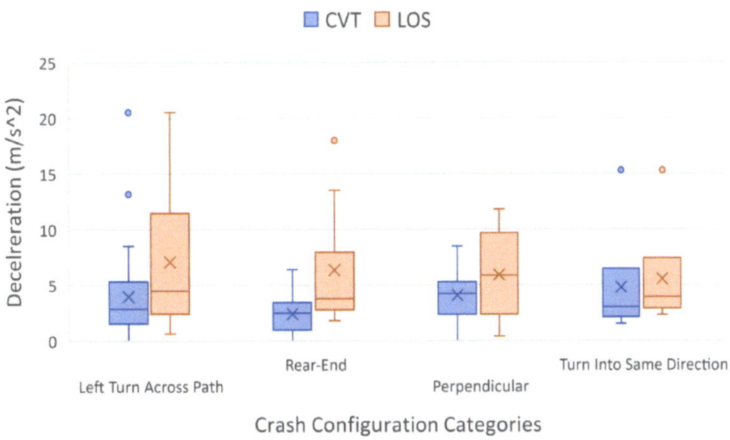

Figure 5. Box and whisker plot of mean required deceleration (negative acceleration) separated by crash configuration category and sensor type. The X represents the mean value, the horizontal line represents the median value, the box encompasses the first and third quartile, the whiskers extend to the maximum value that is within 1.5 times the inner quartile range, and the circles represents any values outside of the whisker range.

Across all four crash configuration categories, the average required deceleration of CVT systems vs. the LOS systems were 3.79 m/s^2 and 6.22 m/s^2, respectively. Generally, a deceleration value over 0.45 g (4.41 m/s^2) is considered hard braking [23]. Therefore, CVT systems could reduce the need for hard braking, and reduce the average deceleration required by 2.43 m/s^2 to avoid the conflict by braking alone.

Some events required excessive deceleration values. Figure 6 shows the number of events binned by their respective required deceleration to avoid a conflict (anything above 14 m/s^2 is shown as 14+). In general, a deceleration value of less than 1 g (9.8 m/s^2) is reasonable for most modern light vehicles equipped with crash-avoidance systems [24]. This figure shows that more LOS events (than CVT events) require an acceleration value of more than 1g as depicted by the black dashed line. Additionally, 91.2% of the CVT events and 75.0% of the LOS events analyzed required a deceleration value less than 1g, implying that a vehicle equipped with LOS features alone could prevent 75.0% of conflicts within this dataset, while a connected vehicle could prevent 91.2% of conflicts.

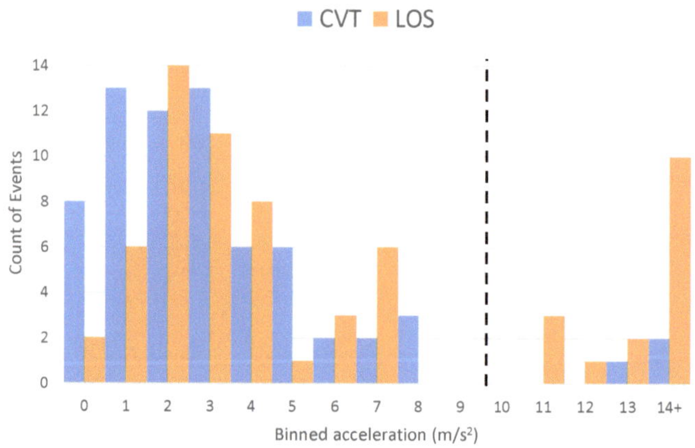

Figure 6. Count of events binned by required acceleration to avoid the crash or near-crash separated by sensor type.

4. Conclusions

In this research, crash and near-crash scenarios from the Second Strategic Highway Research Program Naturalistic Driving Study (SHRP2 NDS) were simulated via a physics-based model to calculate the potential safety benefit of line-of-sight (LOS) sensors and connected vehicle technology (CVT). Previous work has predicated the potential safety impacts of LOS and CVT through an estimation of the types of police-recorded crashes that could have been avoided if these sensors and systems were in place, or by simulating different crash scenarios with these technologies in place. A missing piece has been the use of real-time kinematic data of vehicles during a crash, as well as using other safety-critical events, such as near-crashes, to analyze how additional LOS or CVT could perform in these scenarios. This research acts as a baseline attempt to measure the potential safety impact that advanced sensors and communication methods can provide in real-world safety-critical events (SCEs).

This project looked at four different crash configurations, including left turn across path, rear-end, perpendicular, and turn into same direction. The turn across left path configurations contained the largest number of crash and near-crash scenarios that could be addressed through LOS sensors or CVT systems. Therefore, it could be beneficial to focus more attention on the accuracy of these sensors specifically for left-turn maneuvers.

On average, the CVT system would activate 0.51 $\pm$ 0.15 s before a LOS sensor detects the target vehicle. This means that CVT could provide about an additional half-second of reaction time over LOS systems. In future research, determining how this calculated value might change at different speeds could greatly affect the added safety benefit of some of these sensors. On average, the required deceleration of CVT systems vs. the LOS systems to avoid a conflict were 3.79 m/s^2 and 6.22 m/s^2, respectively. Additionally, 91.2% of the CVT events and 75.0% of the LOS events analyzed required a deceleration value less than 1 g.

From the required deceleration interpretation, it is expected that any event which required a deceleration value greater than 9.8 m/s^2 could have resulted in a crash. However, only 4.4% of the actual events that were analyzed resulted in a collision; the remaining 95.6% were near-crashes. This is because one or more of the vehicles involved performed an evasive maneuver in addition to braking, which is often how near-crashes are categorized [25]. The simulated event then calculated a required deceleration that was higher than the baseline event to avoid the collision because it did not include a swerving maneuver. These near-crashes were used for this research to increase the sample size and they can be used as a potential surrogate measure to crashes in similar scenarios [26]. Although the near-crashes did not result in an actual police-reported crash, these are important to use in determining the potential safety impact of automated driving systems since these were events in which the driver performed a successful evasive maneuver. However, since the results from the simulated LOS and CVT systems only included braking as an evasive maneuver, further research could look into how swerving could be used to avoid some of these conflicts. This would be especially beneficial for ADS development. Most of the drivers in these near-crash events were able to avoid the crash with a combination of swerving and braking, so it would also be beneficial to see if vehicles with additional sensors and more advanced driving assistance systems could do the same.

Although this research includes only a small sample of SCEs, this work demonstrates how certain safety-surrogate measures can be used to measure the potential safety impact of more advanced sensors and communication methods. It would also be beneficial to calculate these same surrogate measures with a larger dataset for use in different baseline scenarios without a visual obstruction to compare the results. The events analyzed were specifically chosen because CVT is most likely to have an impact in scenarios in which LOS sensors are blocked. However, many SCEs occur when there are no visual obstructions, and CVT has the potential for also avoiding or mitigating these events.

Finally, these simulations assumed that the sensors would have 100% accuracy in determining an imminent conflict. More conservative estimates could be made to account for sensor inaccuracies or additional reaction time within technological systems. This research acts as a baseline sample of how to use real SCEs to predict the potential safety benefit of advanced vehicle sensors, and which events should be focused on for future research.

Author Contributions: Conceptualization, Z.D. and E.H.; methodology, E.H.; software, E.H.; validation, E.H. and L.S.; formal analysis, E.H. and L.S.; resources, E.H. and Z.D.; data curation, E.H.; writing—original draft preparation, E.H.; writing—review and editing, E.H., L.S., and Z.D.; supervision, Z.D. and; project administration, E.H., L.S., and Z.D.; funding acquisition, Z.D. and L.S. All authors have read and agreed to the published version of the manuscript.

Funding: This research was funded by the USDOT program, Safety through Disruption (Safe-D) University Transportation Center (UTC), grant number 69A355177115.

Institutional Review Board Statement: The study was conducted in accordance with the Declaration of Helsinki and approved by the Institutional Review Board of Virginia Tech (IRB #18-957 23 October 2018).

Informed Consent Statement: Informed consent was obtained from all subjects involved in the study.

Data Availability Statement: Data used for this research is available on VTTI's Dataverse here: https://safed.vtti.vt.edu/projects/impacts-of-connected-vehicle-technology-on-automated-vehicle-safety/. (accessed on 11 January 2024).

Acknowledgments: The team would like to acknowledge the support of Ericsson in the project. The team would also like to sincerely thank Thomas Gorman for his insight during the development of the project. The team would also like to thank Eric Glenn for his support for the UTC deliverables and milestones throughout the project.

Conflicts of Interest: The authors declare no conflicts of interest.

Appendix A

This appendix contains Figure A1, which shows the crash configurations used for this research as they compare to the crash configuration from the accident glossary of the 1988-2015 General Estimates System (GES) Analytical User Manual [27].

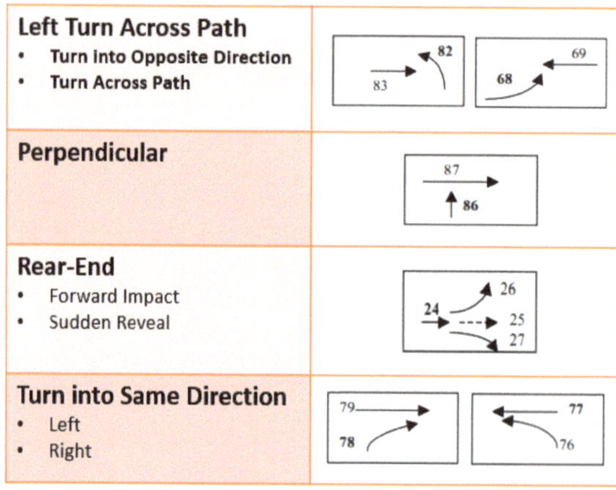

Figure A1. GES crash configurations are organized into four categories relevant for this research.

Appendix B

This appendix contains Figure A2, which depicts the methods flowchart for the data extraction and data analysis. More information about the specific data method can be found in the Safe-D Report that this research is based on [19].

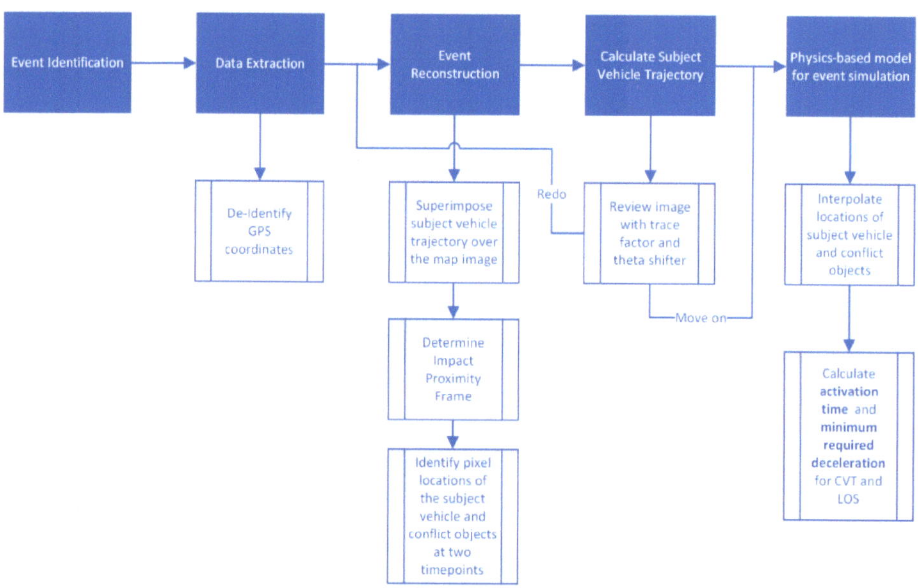

Figure A2. Methods flowchart.

References

1. Wang, B.; Han, Y.; Wang, S.; Tian, D.; Cai, M.; Liu, M.; Wang, L. A Review of Intelligent Connected Vehicle Cooperative Driving Development. *Mathematics* **2022**, *10*, 3635. [CrossRef]
2. Chang, J.; Hatcher, G.; Hicks, D.; Schneeberger, J.; Staples, B.; Sundarajan, S.; Vasudevan, M.; Wang, P.; Wunderlich, K. *Estimated Benefits of Connected Vehicle Applications: Dynamic Mobility Applications, AERIS, V2I Safety, and Road Weather Management Applications*; FHWA-JPO-15-255; FHWA: Washington, DC, USA, 2015.
3. Guériau, M.; Billot, R.; El Faouzi, N.-E.; Monteil, J.; Armetta, F.; Hassas, S. How to assess the benefits of connected vehicles? A simulation framework for the design of cooperative traffic management strategies. *Transp. Res. Part C Emerg. Technol.* **2016**, *67*, 266–279. [CrossRef]
4. NHTSA. Automated Vehicles for Safety. Available online: https://www.nhtsa.gov/technology-innovation/automated-vehicles-safety (accessed on 9 September 2023).
5. Shokravi, H.; Shokravi, H.; Bakhary, N.; Heidarrezaei, M.; Rahimian Koloor, S.S.; Petrů, M. A Review on Vehicle Classification and Potential Use of Smart Vehicle-Assisted Techniques. *Sensors* **2020**, *20*, 3274. [CrossRef] [PubMed]
6. Schiegg, F.A.; Llatser, I.; Bischoff, D.; Volk, G. Collective Perception: A Safety Perspective. *Sensors* **2020**, *21*, 159. [CrossRef]
7. Rolison, J.J.; Regev, S.; Moutari, S.; Feeney, A. What are the factors that contribute to road accidents? An assessment of law enforcement views, ordinary drivers' opinions, and road accident records. *Accid. Anal. Prev.* **2018**, *115*, 11–24. [CrossRef] [PubMed]
8. Wang, L.; Zhong, H.; Ma, W.; Abdel-Aty, M.; Park, J. How many crashes can connected vehicle and automated vehicle technologies prevent: A meta-analysis. *Accid. Anal. Prev.* **2020**, *136*, 105299. [CrossRef] [PubMed]
9. Yue, L.; Abdel-Aty, M.; Wu, Y.; Wang, L. Assessment of the safety benefits of vehicles' advanced driver assistance, connectivity and low level automation systems. *Accid. Anal. Prev.* **2018**, *117*, 55–64. [CrossRef] [PubMed]
10. NHTSA. *Vehicle-to-Vehicle Communication Technology for Light Vehicles*; FMVSS No. 150; NHTSA: Washington, DC, USA, 2016.
11. Abdeen, M.A.R.; Yasar, A.; Benaida, M.; Sheltami, T.; Zavantis, D.; El-Hansali, Y. Evaluating the Impacts of Autonomous Vehicles' Market Penetration on a Complex Urban Freeway during Autonomous Vehicles' Transition Period. *Sustainability* **2022**, *14*, 10094. [CrossRef]
12. Czapp, T.; Chen, C.-L.; Lawrence, S.S.; Wiacek, C. *Real-World Effectiveness of Model Year 2015–2020 Advanced Driver Assistance Systems*; Public Release Case Number 22-3734; Partnership for Analytics Research in Traffic Safety (PARTS): Washington, DC, USA, 2022.
13. Scanlon, J.; Sherony, R.; Gabler, H. Injury Mitigation Estimates for an Intersection Driver Assistance System in Straight Crossing Path Crashes in the US. *Taylor Fr.* **2017**, *18*, S9–S17. [CrossRef]
14. Kitajima, S.; Chouchane, H.; Antona-Makoshi, J.; Uchida, N.; Tajima, J. A Nationwide Impact Assessment of Automated Driving Systems on Traffic Safety Using Multiagent Traffic Simulations. *IEEE Open J. Intell. Transp. Syst.* **2022**, *3*, 302–312. [CrossRef]
15. Lyu, N.; Wen, J.; Duan, Z.; Wu, C. Vehicle Trajectory Prediction and Cut-In Collision Warning Model in a Connected Vehicle Environment. *IEEE Trans. Intell. Transp. Syst.* **2022**, *23*, 966–981. [CrossRef]
16. Jo, Y.; Jang, J.; Park, S.; Oh, C. Connected vehicle-based road safety information system (CROSS): Framework and evaluation. *Accid. Anal. Prev.* **2021**, *151*, 105972. [CrossRef] [PubMed]
17. Guo, Y.; Lee, Y.-H.; Tseng, H.-W.; Yang, C.-F. Simulation for Non-line-of-sight Collision Avoidance Warning System Based on 5G Mobile Car Communication Network. *Sens. Mater.* **2022**, *35*, 723–732. [CrossRef]
18. Baek, M.; Jeong, D.; Choi, D.; Lee, S. Vehicle Trajectory Prediction and Collision Warning via Fusion of Multisensors and Wireless Vehicular Communications. *Sensors* **2020**, *20*, 288. [CrossRef] [PubMed]
19. Herbers, E.; Stowe, L. *Impacts of Connected Vehicle Technology on Automated Vehicle Safety*; 04-120; Virginia Transportation Institute: Blacksburg, VA, USA, 2022.
20. Hankey, J.M.; Perez, M.A.; McClafferty, J.A. *Description of the SHRP2 Naturalistic Database and the Crash, Near-Crash, and Baseline Data Sets*; Virginia Tech Transportation Institute—The Strategic Highway Research Program 2 Transportation Research Board of the National Academies: Blacksburg, VA, USA, 2016.
21. Brown, J.L.; Richard, C.M. *Analysis of SHRP2 Speeding Data: Methods Used to Conduct the Reseearch*; DOT HS 812 793; National Highway Traffic Safety Administration: Washington, DC, USA, 2020.
22. Lee, S.E.; Llanera, E.; Klauer, S.; Sudweeks, J. *Analyses of Rear-End Crashes and Near-Crashes in the 100-Car Naturalistic Driving Study to Support Rear-Signaling Countermeasure Development*; DOT HS 810 846; Virginia Tech Transportation Institute: Blacksburg, VA, USA, 2007.
23. Simons-Morton, B.G.; Ouimet, M.C.; Wang, J.; Klauer, S.G.; Lee, S.E.; Dingus, T.A. Hard Braking Events among Novice Teenage Drivers by Passenger Characteristics. *Driv. Assess. Conf.* **2009**, *2009*, 236–242. [CrossRef]
24. American Automobile Association (AAA). *Automatic Emergency Braking Performance in the Context of Common Crash Scenarios*; American Automobile Association: Heathrow, FL, USA, 2022.
25. Seacrist, T.; Douglas, E.C.; Hannan, C.; Rogers, R.; Belwadi, A.; Loeb, H. Near crash characteristics among risky drivers using the SHRP2 naturalistic driving study. *J. Saf. Res.* **2020**, *73*, 263–269. [CrossRef]

26. Guo, F.; Klauer, S.G.; Hankey, J.M.; Dingus, T.A. Near Crashes as Crash Surrogate for Naturalistic Driving Studies. *Transp. Res. Rec.* **2010**, *2147*, 66–74. [CrossRef]
27. NHTSA. *General Estimates System (GES) Analytical User's Manual*; DOT HS 812 320; U.S. Department of Transportation: Washington, DC, USA, 2016.

sensors

MDPI

Article

Dual-Slope Path Loss Model for Integrating Vehicular Sensing Applications in Urban and Suburban Environments

Herman Fernández [1,*], Lorenzo Rubio [2], Vicent M. Rodrigo Peñarrocha [2] and Juan Reig [2]

[1] Telecommunications Research Group, Pedagogical and Technological University of Colombia, Sogamoso 152211, Colombia

[2] Antennas and Propagation Lab, Universitat Politècnica de València, 46022 Valencia, Spain; lrubio@dcom.upv.es (L.R.); vrodrigo@dcom.upv.es (V.M.R.P.); jreigp@dcom.upv.es (J.R.)

* Correspondence: herman.fernandez@uptc.edu.co; Tel.: +57-313-429-1263

Abstract: The development of intelligent transportation systems (ITS), vehicular ad hoc networks (VANETs), and autonomous driving (AD) has progressed rapidly in recent years, driven by artificial intelligence (AI), the internet of things (IoT), and their integration with dedicated short-range communications (DSRC) systems and fifth-generation (5G) networks. This has led to improved mobility conditions in different road propagation environments: urban, suburban, rural, and highway. The use of these communication technologies has enabled drivers and pedestrians to be more aware of the need to improve their behavior and decision making in adverse traffic conditions by sharing information from cameras, radars, and sensors widely deployed in vehicles and road infrastructure. However, wireless data transmission in VANETs is affected by the specific conditions of the propagation environment, weather, terrain, traffic density, and frequency bands used. In this paper, we characterize the path loss based on the extensive measurement campaign carrier out in vehicular environments at 700 MHz and 5.9 GHz under realistic road traffic conditions. From a linear dual-slope path loss propagation model, the results of the path loss exponents and the standard deviations of the shadowing are reported. This study focused on three different environments, i.e., urban with high traffic density (U-HD), urban with moderate/low traffic density (U-LD), and suburban (SU). The results presented here can be easily incorporated into VANET simulators to develop, evaluate, and validate new protocols and system architecture configurations under more realistic propagation conditions.

Keywords: vehicular ad hoc network (VANET); vehicle-to-everything (V2X); artificial intelligence (AI); internet of things (IoT); path loss models; path loss exponent; 5G; autonomous driving (AD); cooperative autonomous driving (CAD); cooperative sensing; connected and autonomous vehicles (CAVs)

Citation: Fernández, H.; Rubio, L.; Rodrigo Peñarrocha, V.M.; Reig, J. Dual-Slope Path Loss Model for Integrating Vehicular Sensing Applications in Urban and Suburban Environments. *Sensors* **2024**, *24*, 4334. https://doi.org/10.3390/s24134334

Academic Editors: Constantin Caruntu and Ciprian Romeo Comşa

Received: 21 May 2024
Revised: 21 June 2024
Accepted: 2 July 2024
Published: 4 July 2024

1. Introduction

The development of intelligent transportation systems (ITS), vehicular ad hoc networks (VANETs), and autonomous driving (AD) has accelerated in recent years, driven by artificial intelligence (AI), the internet of things (IoT), and their integration with dedicated short-range communications (DSRC) systems and fifth-generation (5G) networks since 5G is emerging as a platform for connecting sensors and vehicles on the road, providing vehicle-to-everything (V2X) services to drivers and pedestrians [1–4]. This is due to the potential of IoT-focused applications of AD and the use of 5G New Radio interfaces to meet these integration requirements given the use scenarios defined in the ITU-M2150-1 recommendation for IMT-2020 systems [5]. These include enhanced mobile broadband (eMBB), ultra-reliable low-latency communication (URLLC), and massive machine-type communication (mMTC). In [6], the authors also described how services such as autonomous vehicles (AV), ITS, V2X, industry 4.0, and smart grid are related to URLLC. This leads to improved mobility conditions in the different road propagation environments, including urban, suburban, rural, and highway, among others. The integration of ITS, VANETs, AD, AV, and V2X with the URLLC scenario has rendered drivers and pedestrians more aware of the need to improve their behavior and decision making in adverse traffic conditions by enabling them to exchange information from different types of sensors widely deployed in vehicles and road infrastructure. However, the wireless data transmission in these systems is affected by the specific conditions of the propagation environment, including the weather, orography, traffic density, and frequency bands used.

The future design of vehicle communication systems has received considerable attention in recent years from the automotive industry, government, and the scientific community as the integration of vehicle-to-vehicle (V2V) and vehicle-to-infrastructure (V2I). These are aimed at proposing better vehicle-reliable safety applications based on DSRC [2], which require high reliability in the connectivity between V2V and V2I communications. In this sense, various types of sensors such as cameras, radars, and speed detectors, among others, have been implemented to support the development of AD and improve mobility conditions in cities and on the road network [4,7,8]. For example, in [4], an integrated sensing and communication system (ISAC) was proposed based on a path loss prediction approach to improve the knowledge of wireless data transmission in vehicular networks The approach predicts the end-to-end path loss distribution using multimodal data collected by millimeter wave (mmWave) radars, laser radars, and cameras. The important role of sensor fusion in intelligent transportation systems was described in [7]. This paper provides a comprehensive overview of the capabilities, the impacts, the planning, and the technological challenges of AVs. In [8], the author introduced several V2X use cases for autonomous driving, where cooperative autonomous driving is categorized into two types: cooperative sensing and cooperative decision. Cooperative sensing focuses on the exchange of sensor information between V2V and V2I. This will further enhance road safety, reduce traffic congestion, and improve travel comfort in urban environments [9,10]. In this sense, it is necessary to develop measurement campaigns in vehicular propagation channels in order to provide more accurate propagation models that can be used to determine the parameters in the characterization of the propagation channel.

Cooperative sensing uses VANET simulators to develop, evaluate, and validate new protocols and system architecture configurations under more realistic propagation conditions. Therefore, it is necessary to propose dual-slope models that allow for more efficient radio planning. Dual-slope models consider a breaking point or critical distance that depends on factors such as the propagation scenario and the frequency band used, among others. This critical distance indicates the end of one wave propagation mechanism and the beginning of another, where more path loss can appear. Dual-slope propagation models allow the generation of results that can be easily incorporated into VANET simulators to optimize the deployment of sensors in the road infrastructure networks.

Government standards bodies have identified specific bands for the development of ITS applications. For example, the Federal Communication Commission (FCC) issued a report and order adopting rules that repurposed the 5850–5895 MHz to expand unlicensed mid-band operations while continuing to dedicate the 5895–5925 MHz for ITS operations [11]. Also, the FCC has proposed that the transition of ITS operations from DSRC-based technology to cellular V2X (C-V2X)-based technology will occur in accordance with a schedule to be determined in a future report and order. In Europe, the European Telecommunication Standard Institute (ETSI) has adopted the DSRC band for ITS applications, allocating 50 MHz (5.875 to 5.925 GHz) [12]. The specific characteristics of both applications (safety and non-safety) require the development and implementation of new communication technologies, where the characterization and modeling of the vehicle channel play a very important role [13,14]. Although progress has been made in the characterization of the V2X radio channel in recent years [15,16], more research is needed that focuses on future vehicle networks that enable interaction with the 5G IoT ecosystem.

Several path loss models have been developed for vehicular network scenarios. These include the one/dual-slope, ray tracing (RT), floating intercept (FI), close-in (CI), free space reference distance, and ABG models. For example, large-scale path loss models for urban environments have been proposed based on the CI, FI, and ABG models under line-of-sight (LOS) and non-line-of-sight (NLOS) conditions in mmWave frequency bands [17]. In [18], the authors presented a shadow fading model for system simulations based on real measurements in urban and highway scenarios. The measurement data were divided into three categories, namely LOS, obstructed line-of-sight (OLOS) by vehicles, and NLOS, and it was observed that vehicles obstructing the LOS induce an additional average attenuation of about 10 dB in the received signal power. In addition, a connection probability in V2V urban scenarios was analyzed based on a dual-slope path loss model on both LOS and OLOS [19]. In [20], path loss models were proposed considering weather conditions, and RT simulations were performed to verify the accuracy of the proposed expression. In [21], the authors adopted a measurement-based dual-slope path loss model to analyze the channel capacity performance for both direct transmission of inter-vehicle and infrastructure-based cooperative communication in VANETs. On the other hand, Wei and Tao analyzed the effects of antenna height on V2I communication in rural areas. They first classified the V2I communication into LOS and NLOS links and then established two-beam ground reflection and integrated models for LOS and NLOS conditions, respectively [22]. In [23], a two-slope model was proposed by performing path loss measurements in a sports utility vehicle (SUV) at 915 MHz and 2.4 GHz. On the other hand, ref. [24] provided an overview of experimentally verified propagation models for wireless sensor networks (WSNs) and quantitative comparisons of propagation models used in WSN research under different scenarios and frequency bands.

Furthermore, the large differences between V2V propagation channels and fixed-to-mobile (F2M) propagation channels, transmitter (Tx) and receiver (Rx) heights, propagation environments, and frequency bands mean that the propagation models developed for the deployment of F2M systems cannot be applied in the performance evaluation and development of future ITS applications over VANETs, thus forcing characterization and modeling of the radio channel at 700 MHz and in the DSRC band at 5.9 GHz for future V2X systems.

This paper presents the vehicular channel characterization by using the dual-slope path loss propagation model in urban with high traffic density (U-HD), urban with low traffic density (U-LD), and suburban (SU) scenarios in the city of Valencia, based on real propagation measurements carried out in 2012 at 700 MHz and 5.9 GHz. The paper is organized as follows: Section 2 describes the measurement system and the main characteristics of the propagation scenarios where the measurements were performed. Section 3 presents the results derived from the analysis of the measurements, where the path loss exponent (PLE) behavior of the received signal is analyzed and classified according to the propagation scenarios. Finally, the conclusions derived from the study are summarized in Section 4.

2. Methodology

2.1. Measurement System

The channel sounder implemented at 700 MHz is shown in Figure 1. This channel sounder consists of an Hewlett HP8648C signal generator in the Tx vehicle transmitting an unmodulated carrier at 700 MHz and a Hewlett Packard HP8590L spectrum analyzer (SA) in the Rx vehicle. A SPAN of zero was selected in the SA to measure received power on 401-point traces. The antennas used are monopoles with a horizontal plane gain of approximately −5.43 dB. For transmission, a power amplifier was used that allowed transmission with an equivalent isotropic radiated power (EIRP) of 26.3 dBm. For reception, an amplifier with a gain of 32.75 dB was used.

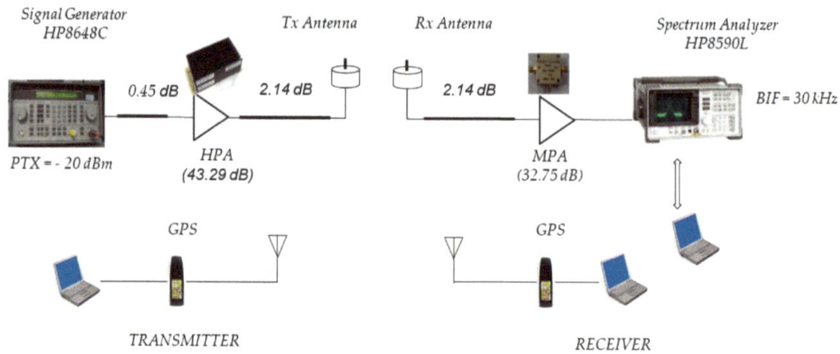

Figure 1. Channel sounder at 700 MHz.

The channel sounder implemented at 5.9 GHz is shown in Figure 2. This channel sounder consists of a Hewlett Packard HP83623A signal generator in the Tx vehicle, which transmits an unmodulated carrier at 5.9 GHz, and the Rohde & Schwartz ZVA24 vector network analyzer (VNA) in the Rx vehicle. The VNA was used in power meter mode, directly measuring the b_2 parameter in traces of 5000 points. The antennas used in Tx/Rx are monopoles at $\lambda/4$ with a gain in the horizontal plane of about -2.56 dB and a scattering parameter S_{11} lower than -22 dB. A power amplifier was used for transmission, allowing transmission with an EIRP of 23.8 dBm. For reception, two amplifiers were used in series with a total gain of 68.12 dB.

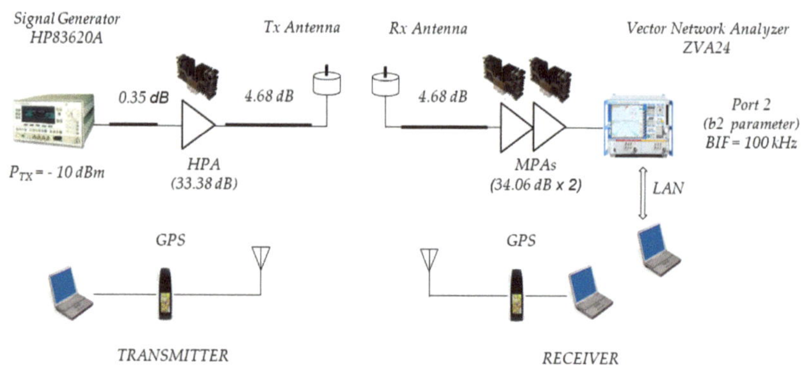

Figure 2. Channel sounder at 5.9 GHz.

The vehicles were equipped with global position systems (GPS) receivers controlled by laptops to obtain information on the time of measurement, relative speed, and Tx-Rx separation distance. All laptops were synchronized in time to correlate the measurements taken by the VNA with the information provided by the GPS receivers.

The vehicles used for the measurements were a Renault (Tx) (Figure 3a) and a Peugeot (Rx) (Figure 3b). The antennas were mounted on the roof of the vehicles at a height of approximately 1.41 m and 1.45 m above the ground for Tx and Rx, respectively. Part of the on-board equipment is also shown in Figure 3. In each of the vehicles, 75 Ah batteries and 12 V DC to 220 V AC inverters were used as the power supply system, allowing an autonomous time of approximately 90 min.

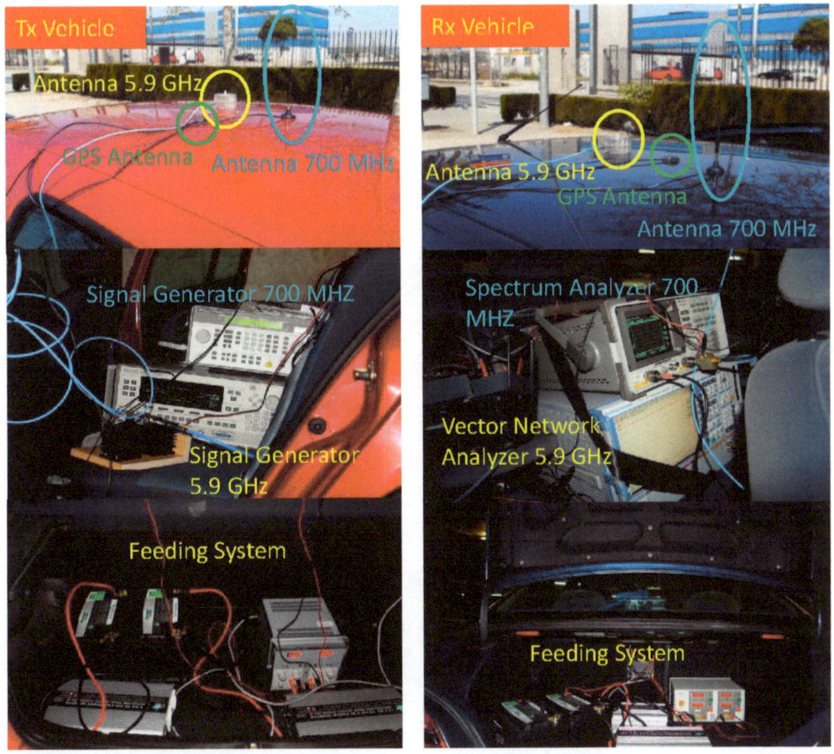

(**a**) Tx vehicle (**b**) Rx Vehicle

Figure 3. Tx/Rx vehicles together with on-board equipment at 700 MHz and 5.9 GHz.

2.2. Scenarios

The type of environment, vehicle speed, and road traffic density determine the propagation characteristics of the vehicle channel. Traffic density and vehicle speed are usually higher in suburban environments. These conditions allowed the selection of the most suitable areas for the measurement campaign to characterize the path loss propagation. A series of avenues and ring roads were found, forming a concentric ring around the city of Valencia, which, according to its characteristics, was defined as a suburban environment and is referred to as scenario 3. Figure 4 illustrates the routes taken in this scenario with a yellow line. Moving inland in the concentric rings, a clearly urban environment with a high road traffic density is presented and referred to as scenario 1. The routes taken are shown with a red line in Figure 4. Within this urban environment, some areas with particular and interesting characteristics were selected. Thus, we analyzed the area of the old town of the city, where narrow streets without a defined shape meet small squares, abrupt intersections, and cobbled pedestrian streets with a large absence of pedestrians. We also utilized a special area within this old town, such as the Plaza del Ayuntamiento. Figure 4 shows in green the routes taken in these urban environments with low road traffic density, which are referred to as scenario 2.

Scenario 1: This is an urban environment with high traffic density, with an average of 44,200 vehicles/24 h. Two avenues in the city of Valencia were chosen for the analysis, namely Avenida del Puerto (one-way traffic) and Avenida Blasco Ibañez (two-way traffic), with four and five lanes in each direction, respectively. Figure 5a shows a view of the measurement scenario.

Figure 4. Scenario 1 is shown with a red line, scenario 2 is shown with a green line, and scenario 3 is shown with a yellow line.

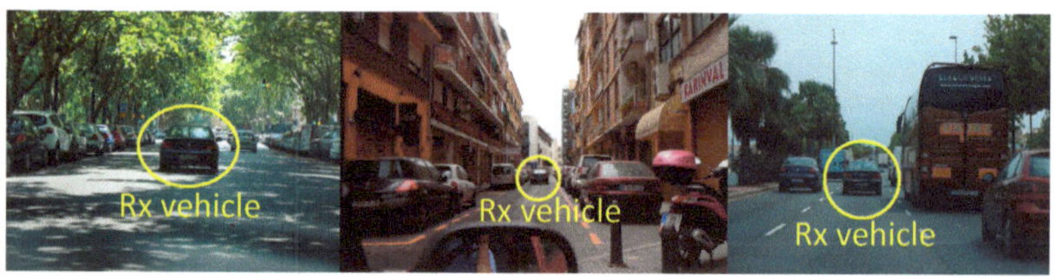

(**a**) (**b**) (**c**)

Figure 5. Measurement scenarios, (**a**) urban high density, (**b**) urban low density, and (**c**) suburban The Rx vehicle can be seen driving in front of the Tx vehicle.

Scenario 2: This is the urban center of the city of Valencia, the old town, with low traffic density at an average of 7500 vehicles/24 h and with narrow streets and one-way traffic. The streets are approximately 10 meters wide, including, in some cases, parking lots and sidewalks on both sides of the street. Figure 5b shows a view of the measurement scenario.

Scenario 3: This scenario is the Ronda Norte in the city of Valencia, Spain, with an average intensity of 71,200 vehicles/24 h, with three and four lanes of traffic in each direction, with wide open spaces on both sides alternating with buildings close to the roadway and medium-height trees along several sections of the avenue. Figure 5c shows a view of the measurement scenario.

It is relevant to note that the measurements were carried out at 700 MHz and 5.9 GHz (DSRC band), with the Tx and Rx vehicles traveling in the same direction several times over the measurement scenario on weekdays between 10:00 and 13:00. Measurements were also carried out under normal driving conditions, with alternations of LOS and NLOS between the Tx vehicle and the Rx vehicle.

3. Link Budget and Dual-Slope Path Loss Measurement Results

The path loss is one of the most significant parameters in radio link design and provides a measure of channel quality. It is expressed as the average level of path loss in dB and varies as a function of the Tx and Rx separation distance. The path loss considers the propagation mechanisms present in the radio channel, such as free space, rejection, diffraction and scattering, the influence of the propagation environments (urban, suburban, rural, and highway), directional characteristics, antenna heights, and Tx and Rx separation distance. In addition, the signal-to-noise ratio (SNR) is inversely related to the path loss; i.e., as the path loss increases, the SNR at the Rx decreases, and the coverage area is reduced.

The average received power in log units (dBm) under free space propagation conditions, denoted by $P_{Rx}(d)$, is given by the following:

$$P_{Rx}(d) = P_{Tx} + G_{Tx} + G_{Rx} - 10log_{10}\left(\frac{4\pi d}{\lambda_c}\right)^2, \quad (1)$$

where P_{Tx} is the transmitted power in dBm, d is the Tx/Rx separation distance; G_{Tx} and G_{Rx} are the transmit and receive antennas' gain expressed in decibels (dB), respectively; and λ_c is the wavelength associated with the carrier frequency f_c. The last term in Equation (1) is the path loss for free space propagation conditions, $PL_{FS}(d)$, expressed in dB.

3.1. Link Budget at 5.9 GHz and at 700 MHz

Figure 6 shows the schematic diagram of the 5.9 GHz V2V measurement system and the components used to perform the link budget.

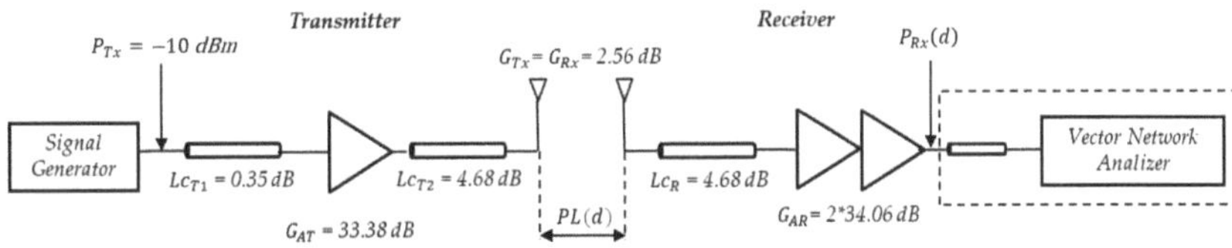

Figure 6. V2V schematic diagram at 5.9 GHz.

According to (1) and taking the received power level $P_{Rx}(d)$ as that related to the VNA input with P_{Tx}= −10 dBm and subtracting the cable losses Lc_{T1} = 0.35 dB, Lc_{T2} = 4.68 dB, and Lc_R = 4.68 dB and then adding the gains of the amplifier G_{AT} = 33.38 dB in Tx and the two cascaded amplifiers G_{AR} = 2 × 34.06 dB in Rx and the Tx and Rx antenna gains G_{Tx} = G_{Rx} = 2.56 dB, the value of $P_{Rx}(d)$ is given by the following:

$$P_{Rx}(d) = P_{Tx} - Lc_{T1} + G_{AT} - Lc_{T2} + G_{Tx} + G_{Rx} - PL(d) - Lc_R + G_{AR}. \quad (2)$$

Therefore, the path loss propagation at 5.9 GHz, expressed in (dB) and denoted by $PL(d)$, is given by the following:

$$PL(d) = 76.67 - P_{Rx}(d). \quad (3)$$

Using the parameter $b_2(f)$ measured by the VNA as the received power value, records were obtained to analyze the behavior of the path loss as a function of d.

The link budget for the 700 MHz test system was performed in the same manner as for the 5.9 GHz test system, as shown in Figure 7.

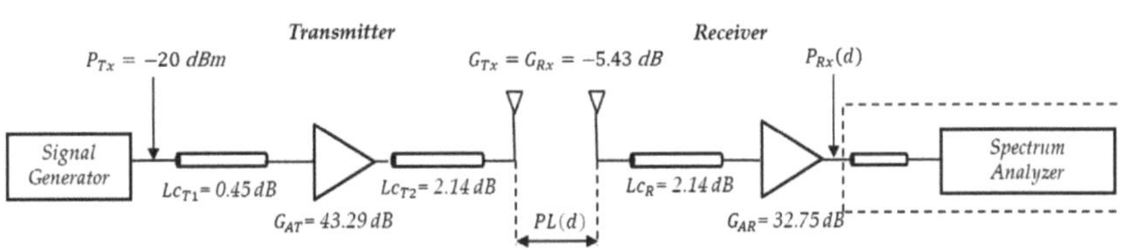

Figure 7. V2V schematic diagram at 700 MHz.

Taking the received power level $P_{Rx}(d)$ with respect to the SA input and also, taking into account a P_{Tx} = −20 dBm and subtracting the cable losses Lc_{T1} = 0.45 dB, Lc_{T2} = 2.14 dB, and Lc_R = 2.14 dB and then adding the gains of the amplifier G_{AT} = 43.29 dB in Tx and the amplifier G_{AR} = 32.75 dB in Rx and the Tx and Rx antenna gains G_{Tx} = G_{Rx} = −5.43 dB gives the same expression as in Equation (2). In this case, the $PL(d)$ at 700 MHz is expressed (in dB) by the following:

$$PL(d) = 40.45 - P_{Rx}(d). \tag{4}$$

Table 1 summarizes the main parameters used to configure the measurement systems.

Table 1. Measurement system configuration parameters at 700 MHz and 5.9 GHz.

Parameter	Channel Sounder at 700 MHz	Channel Sounder at 5.9 GHz
Transmitted power	−20 dBm	−10 dBm
Points number	401	5000
Resolution bandwidth	30 kHz	100 kHz
SPAN	0	0
Dynamic range	80 dB	80 dB

3.2. Dual-Slope Path Loss Model Results

To process the data sets obtained during the measurement campaign, a selection was made according to the propagation scenarios, analyzing the behavior of the path loss propagation as a function of the separation distance between the Tx and Rx. A further process of data filtering was carried out to obtain more accurate values for the parameters of the dual-slope path loss model. According to [25], a linear relationship can be established between the mean path loss expressed in logarithmic units, $PL(d)$, and $log_{10}(d)$ in V2V radio channels, analogous to traditional F2M channels, given by the following:

$$PL(d) = L_0 + 10\gamma log_{10}\left(\frac{d}{d_0}\right) + S, \quad d \geq d_0, \tag{5}$$

where L_0 represents the average path loss propagation at a separation distance d_0, the term $10\gamma log_{10}\left(\frac{d}{d_0}\right)$ refers to the average path loss propagation referring to the Tx/Rx distance d_0, γ is the PLE related to the type of propagation environment, and S is a Gaussian distributed random variable with zero mean and standard deviation σ_S in dB that is used to model long-term fading or shadowing.

However, there are environments in which a dual-slope model can more accurately fit the measured data. A dual-slope model is characterized by a path loss exponent of γ_1 and a standard deviation of σ_{S_1} above a reference distance up to critical distance d_C, a path loss exponent of γ_2, and a standard deviation of σ_{S_2} for a distance higher than the critical distance. Using this model, the average path loss value can be estimated as the following:

$$PL(d)[dB] = \begin{cases} L_0 + 10\gamma_1 log_{10}\left(\frac{d}{d_0}\right) + S, & d_0 \leq d \leq d_C; \\ L_0 + 10\gamma_1 log_{10}\left(\frac{d_C}{d_0}\right) + 10\gamma_2 log_{10}\left(\frac{d}{d_C}\right) + S, & d \geq d_C. \end{cases} \tag{6}$$

On the other hand, the behavior of the path loss propagation based on a dual-slope model was observed for some of the analyzed routes. It should be noted that this behavior did not appear in all propagation scenarios or in the two frequency bands (700 MHz and 5.9 GHz) in which the measurements were made.

Figure 8 shows the path loss as a function of the Tx/Rx separation distance for an urban environment with high traffic density at 700 MHz. A double-slope behavior was observed; thus, applying a linear least squares fit (magenta curve for the first slope and red curve for the second slope in Figure 8) yields values of L_0 = 36.83 dB, γ_1 = 1.65, σ_{S_1} = 4.73 dB, γ_2 = 3.26, and σ_{S_2} = 5.1 dB, with a d_C = 39.67 m.

Similarly, Figure 9 shows the path loss as a function of the Tx/Rx separation distance for an urban environment with low traffic density at 5.9 GHz. We thus obtained L_0 = 59.88 dB, γ_1 = 1.61, σ_{S_1} = 4.00 dB, γ_2 = 4.42, σ_{S_2} = 5.26 dB, and d_C = 134.56 m.

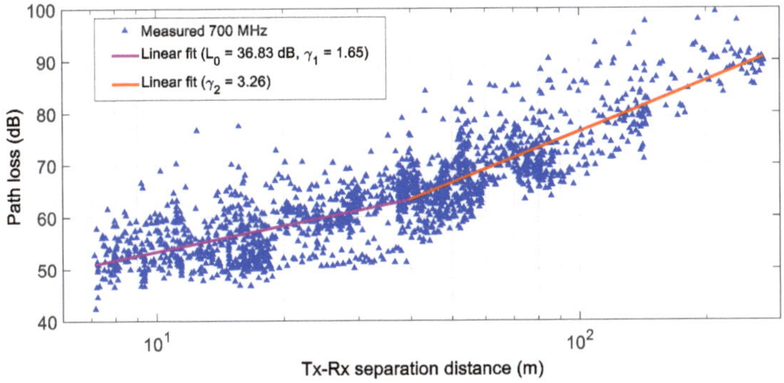

Figure 8. Path loss as a function of Tx/Rx separation distance for an urban environment with high road traffic density at 700 MHz.

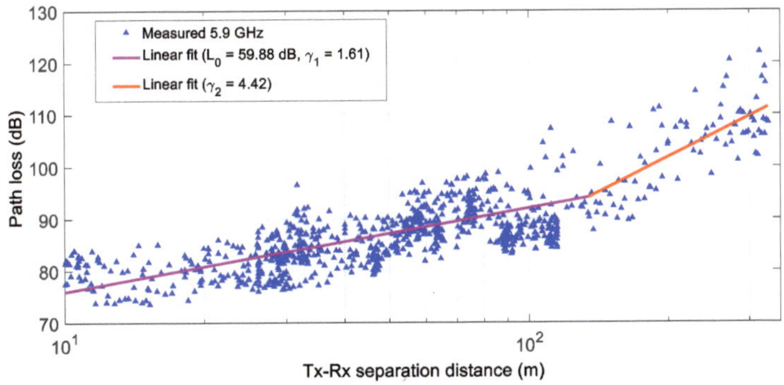

Figure 9. Path loss as a function of Tx/Rx separation distance for an urban environment with low road traffic density at 5.9 GHz.

The mean parameters that characterize the path loss propagation for the dual-slope model are summarized in Table 2. The PLE values for the first slope range from 1.04 to 2.13 and 1.72 to 2.46 at 700 MHz and 5.9 GHz, respectively. The highest PLE values for the first slope occurred in urban environments with low traffic density at 700 MHz and 5.9 GHz. Similarly, the PLE values for the second slope range from 3.43 to 4.92 and from 6.26 to 9.69 at 700 MHz and 5.9 GHz, respectively. In addition, the largest exponents of PLE for the second slope occurred in suburban environments at 700 MHz and 5.9 GHz. Note that for the suburban environment at 700 MHz and 5.9 GHz, constructive interference was observed for the first slope due to the multipath effect. This was also observed for the urban environment with high traffic density at 700 MHz.

Table 2. Parameters obtained at 700 MHz and 5.9 GHz.

Scenario	700 MHz					
	γ_1	L_0 (dB)	σ_{S_1} (dB)	γ_2	σ_{S_2} (dB)	d_C (m)
1 (Urban H-D)	1.45	37.34	4.52	3.43	5.00	35.16
2 (Urban L-D)	2.13	31.42	5.47	3.76	6.66	35.44
3 (Suburban)	1.04	38.78	4.28	4.92	5.67	58.40
Scenario	**5.9 GHz**					
	γ_1	L_0 (dB)	σ_{S_1} (dB)	γ_2	σ_{S_2} (dB)	d_C (m)
2 (Urban L-D)	2.46	45.88	4.49	6.26	5.01	132.42
3 (Suburban)	1.72	52.40	4.80	9.69	5.25	306.04

With respect to the large-scale fading modeled by the parameter σ_S, the received signal suffered the highest variations in urban with low traffic density and suburban environments at 700 MHz and 5.9 GHz. On the other hand, the critical distance ranges from 35.16 m to 58.40 m at 700 MHz, with the largest critical distance obtained in a suburban environment, while the critical distance at 5.9 GHz varies from 132.42 m to 306.04 m for urban with low traffic density and suburban environments, respectively.

It should be noted that in our measurement campaign, there are very few records and scenarios where dual-slope behavior occurred in the path loss propagation analysis. Similarly, only a very small number of researchers have shown results from dual-slope models for vehicular applications [18,26–30]. Table 3 summarizes the values of the dual-slope path loss exponent and the standard deviations of shadowing derived from channel measurement campaigns conducted in different vehicular environments. These results can be used in vehicular network simulators to design, evaluate, and validate new protocols that improve the quality of vehicular communication systems and enable the integration of ITS, AD, and vehicular sensing with the 5G-IoT ecosystem.

Table 3. Large-scale dual-slope path loss model parameters for vehicular environments.

Scenario	γ_1	σ_{S_1} (dB)	γ_2	σ_{S_2} (dB)	d_C (m)	Frequency Band	Related Works
Suburban	2.00–2.10	5.60–2.60	3.80–4.00	4.40–8.40	100	5.9 GHz	Ref. [26]
Highway	1.90	2.50	4.00	0.90	220	5.9 GHz	Ref. [27]
Rural	2.30	3.20	4.00	0.40	226		
Urban intersection 1	1.54	3.64	3.96	4.81	24.5		
Urban intersection 2	1.56	3.64	5.34	4.81	40	5.9 GHz	Ref. [28]
Urban intersection 3	1.53	3.64	4.86	4.81	45		
Urban	0.10	3.03	5.24	7.40	120	500 MHz	Ref. [29]
Suburban	1.47	6.63	6.42	9.22	200		
Urban	1.60	2.20	3.14	4.50	35	725 MHz	Ref. [30]
Scenario	**Conditions**	γ_1	γ_2	σ_S **(dB)**	d_C **(m)**	**Frequency Band**	**Related Works**
Highway	LOS	1.66	2.88	3.95	104	5.6 GHz	Ref. [18]
	OLOS	----	3.18	6.12			
Urban	LOS	1.81	2.85	4.15	104		
	OLOS	1.93	2.74	6.67	104		

The results obtained in this work show the influence of the frequency band and the propagation environment on the estimated values of the path loss propagation and the critical distance. Therefore, when deploying sensor networks for ITS applications in real conditions, the propagation scenario must be considered. According to the results presented in this paper, it is necessary to take into account that urban scenarios with low traffic density have more path loss propagation values for the first slope, up to the critical distance. Similarly, it should be noted that the path loss increased significantly from the critical distance for all propagation environments analyzed in this study at 700 MHz and 5.9 GHz.

4. Conclusions

The results obtained for a dual-slope path loss propagation model in three different environments, i.e., urban with high traffic density, urban with low traffic density, and suburban, are herein presented. These results can be used in vehicular network simulators to design, evaluate, and validate new protocols that improve the quality of vehicular communication systems and enable the integration of ITS, AD, and vehicular sensing with the 5G-IoT ecosystem.

From the results obtained, it should be noted that the parameters of the dual-slope path loss propagation model are influenced by the frequency band. For example, in the results reported in this paper, it was observed that the mean values of the first- and second-slope

PLEs were lower at 700 MHz than at 5.9 GHz for urban with low density and suburban environments studied, where the highest values of the first- and second-slope PLEs were obtained for the urban environments with low traffic density at 5.9 GHz.

Future lines of work will include performing broadband measurement campaigns for applications in the ITS to generate path loss and capacity models that enable the optimal deployment of device and sensor networks in V2X 5G-IoT ecosystem.

Author Contributions: Conceptualization, H.F., L.R., J.R. and V.M.R.P.; methodology, H.F., L.R. and V.M.R.P.; channel measurements, H.F., L.R., J.R. and V.M.R.P.; validation, H.F., L.R. and V.M.R.P.; formal analysis, H.F., L.R., J.R. and V.M.R.P.; investigation H.F., L.R., J.R. and V.M.R.P.; resources, H.F., L.R., J.R. and V.M.R.P.; data curation, H.F., L.R., J.R. and V.M.R.P.; writing—original draft preparation, H.F., L.R., J.R. and V.M.R.P.; writing—review and editing H.F., L.R., J.R. and V.M.R.P.; visualization, H.F., L.R., J.R. and V.M.R.P.; supervision, H.F., L.R., J.R. and V.M.R.P.; project administration, H.F., L.R., J.R. and V.M.R.P.; funding acquisition, H.F., L.R., J.R. and V.M.R.P. All authors have read and agreed to the published version of the manuscript.

Funding: This research was funded in part by the MCIN/AEI/10.13039/501100011033/ through the 362 I+D+i Project under Grant PID2020-119173RB-C21, and by the Pedagogical and Technological University of Colombia (Project Number SGI 3721).

Institutional Review Board Statement: Not applicable.

Informed Consent Statement: Not applicable.

Data Availability Statement: The original contributions presented in the study are included in the article, further inquiries can be directed to the corresponding author.

Acknowledgments: The authors would like to thank B. Bernardo-Clemente and A. Vila-Jiménez for their support in the design and assembly of the channel sounder and D. Balaguer, A. Campuzano, and L. Mondragón for their support in carrying out the measurement campaign.

Conflicts of Interest: The authors declare no conflicts of interest. The funders had no role in the design of the study; in the collection, analyses, or interpretation of data; in the writing of the manuscript; or in the decision to publish the results.

References

1. Guo, R.; Vallati, M.; Wang, Y.; Zhang, H.; Chen, Y.; Wang, F.-Y. Sustainability Opportunities and Ethical Challenges of AI-Enabled Connected Autonomous Vehicles Routing in Urban Areas. *IEEE Trans. Intell. Veh.* **2023**, *9*, 1–4. [CrossRef]
2. Lyu, F.; Zhu, H.; Cheng, N.; Zhou, H.; Xu, W.; Li, M.; Shen, X. Characterizing urban vehicle-to-vehicle communications for reliable safety applications. *IEEE Trans. Intell. Transp. Syst.* **2019**, *21*, 2586–2602. [CrossRef]
3. Lu, R.; Zhang, L.; Ni, J.; Fang, Y. 5G vehicle-to-everything services: Gearing up for security and privacy. *Proc. IEEE* **2019**, *108*, 373–389. [CrossRef]
4. Wei, Z.; Mao, B.; Guo, H.; Xun, Y.; Liu, J.; Kato, N. An Intelligent Path Loss Prediction Approach based on Integrated Sensing and Communications for Future Vehicular Networks. *IEEE Open J. Comput. Soc.* **2024**, *5*, 170–180. [CrossRef]
5. ITU. M.2150: Detailed Specifications of the Terrestrial Radio Interfaces of International Mobile Telecommunications-2020 (IMT-2020). Available online: https://www.itu.int/rec/R-REC-M.2150/en (accessed on 6 June 2023).
6. Muñoz, J.; Mancipe, D.; Fernández, H.; Rubio, L.; Peñarrocha, V.M.R.; Reig, J. Path Loss Characterization in an Outdoor Corridor Environment for IoT-5G in a Smart Campus University at 850 MHz and 3.5 GHz Frequency Bands. *Sensors* **2023**, *23*, 9237. [CrossRef] [PubMed]
7. Nawaz, M.; Tang, J.K.-T.; Bibi, K.; Xiao, S.; Ho, H.-P.; Yuan, W. Robust Cognitive Capability in Autonomous Driving Using Sensor Fusion Techniques: A Survey. *IEEE Trans. Intell. Transp. Syst.* **2023**, *25*, 3228–3243. [CrossRef]
8. Liu, S. Innovations on the Vehicle-to-Everything Infrastructure. In *Engineering Autonomous Vehicles and Robots: The DragonFly Modular-Based Approach*; Wiley-IEEE Press: Piscataway, NJ, USA, 2019.
9. Brambilla, M.; Nicoli, M.; Soatti, G.; Deflorio, F. Augmenting vehicle localization by cooperative sensing of the driving environment: Insight on data association in urban traffic scenarios. *IEEE Trans. Intell. Transp. Syst.* **2019**, *21*, 1646–1663. [CrossRef]
10. Magsinoy, E.R.; Von Galang, G.D.; He, M.L.X.; Inomata, R.C. Cooperative vehicles for monitoring urban roads based on behavioral rules. In Proceedings of the 2019 IEEE 9th Symposium on Computer Applications & Industrial Electronics (ISCAIE), Kota Kinabalu, Malaysia, 27–28 April 2019; IEEE: Piscataway, NJ, USA, 2019; pp. 103–107.
11. Federal Communications Commission. Dedicated Short Range Communications (DSRC) Service. Available online: https://www.fcc.gov/wireless/bureau-divisions/mobility-division/dedicated-short-range-communications-dsrc-service (accessed on 17 May 2024).

12. *ASTM E2213-03*; ASTM E2213-03 Standard Specification for Telecommunications and Information Exchange between Roadside and Vehicle Systems—5 GHz Band Dedicated Short Range Communications (DSRC) Medium Access Control (MAC) and Physical Layer (PHY) Specifications. Available online: https://www.standards.its.dot.gov/Standard/390 (accessed on 9 May 2024).
13. Mecklenbrauker, C.F.; Molisch, A.F.; Karedal, J.; Tufvesson, F.; Paier, A.; Bernado, L.; Zemen, T.; Klemp, O.; Czink, N. Vehicular Channel Characterization and Its Implications for Wireless System Design and Performance. *Proc. IEEE* **2011**, *99*, 1189–1212. (In English) [CrossRef]
14. Karedal, J.; Czink, N.; Paier, A.; Tufvesson, F.; Molisch, A.F. Path Loss Modeling for Vehicle-to-Vehicle Communications. *IEEE Trans. Veh. Technol.* **2011**, *60*, 323–328. (In English) [CrossRef]
15. Zhou, T.; Li, C.; Zhang, W.; Ai, B.; Liu, L.; Liang, Y. Narrow-Beam Channel Measurements and Characterization in Vehicle-to-Infrastructure Scenarios for 5G-V2X Communications. *IEEE Internet Things J.* **2024**, *11*, 16074–16086. [CrossRef]
16. Granda, F.; Azpilicueta, L.; Celaya-Echarri, M.; Lopez-Iturri, P.; Vargas-Rosales, C.; Falcone, F. Spatial V2X traffic density channel characterization for urban environments. *IEEE Trans. Intell. Transp. Syst.* **2020**, *22*, 2761–2774. [CrossRef]
17. Hindia, M.N.; Al-Samman, A.M.; Rahman, T.A.; Yazdani, T. Outdoor large-scale path loss characterization in an urban environment at 26, 28, 36, and 38 GHz. *Phys. Commun.* **2018**, *27*, 150–160. [CrossRef]
18. Abbas, T.; Sjöberg, K.; Karedal, J.; Tufvesson, F. A measurement based shadow fading model for vehicle-to-vehicle network simulations. *Int. J. Antennas Propag.* **2015**, *2015*, 190607. [CrossRef]
19. Chen, R.; Zhong, Z. Analysis on V2V connectivity under dual-slope path loss model in urban scenarios. In Proceedings of the 2014 XXXIth URSI General Assembly and Scientific Symposium (URSI GASS), Beijing, China, 16–23 August 2014; IEEE: Piscataway, NJ, USA, 2014; pp. 1–4.
20. Eldeeb, H.B.; Miramirkhani, F.; Uysal, M. A path loss model for vehicle-to-vehicle visible light communications. In Proceedings of the 2019 15th International Conference on Telecommunications (ConTEL), Graz, Austria, 3–5 July 2019; IEEE: Piscataway, NJ, USA, 2019; pp. 1–5.
21. Chen, R.; Sheng, Z.; Ni, M.; Zhong, Z.; Michelson, D.G. Channel capacity under measurement-based model for cooperative vehicular ad hoc networks. In Proceedings of the 2015 IEEE International Symposium on Antennas and Propagation & USNC/URSI National Radio Science Meeting, Vancouver, BC, Canada, 19–24 July 2015; IEEE: Piscataway, NJ, USA, 2015; pp. 302–303.
22. Li, W.; Hu, X.; Jiang, T. Path loss models for IEEE 802.15. 4 vehicle-to-infrastructure communications in rural areas. *IEEE Internet Things J.* **2018**, *5*, 3865–3875. [CrossRef]
23. Dhanavanthan, B. *Propagation Loss Measurements and Two Slope Modelling in Vehicular Environments for Intelligent Transportation Systems*; SAE Technical Paper, 0148-7191; SAE International: Warrendale, PA, USA, 2017.
24. Kurt, S.; Tavli, B. Path-Loss Modeling for Wireless Sensor Networks: A review of models and comparative evaluations. *IEEE Antennas Propag. Mag.* **2017**, *59*, 18–37. [CrossRef]
25. Rubio, L.; Reig, J.; Fernández, H. Propagation aspects in vehicular networks. In *Vehicular Technologies: Increasing Connectivity*; Intech Open: London, UK, 2011; pp. 376–414.
26. Cheng, L.; Henty, B.E.; Stancil, D.D.; Bai, F.; Mudalige, P. Mobile vehicle-to-vehicle narrow-band channel measurement and characterization of the 5.9 GHz dedicated short range communication (DSRC) frequency band. *IEEE J. Sel. Areas Commun.* **2007**, *25*, 1501–1516. [CrossRef]
27. Cheng, L.; Henty, B.E.; Bai, F.; Stancil, D.D. Highway and rural propagation channel modeling for vehicle-to-vehicle communications at 5.9 GHz. In Proceedings of the 2008 IEEE Antennas and Propagation Society International Symposium, San Diego, CA, USA, 5–11 July 2008; IEEE: Piscataway, NJ, USA, 2008; pp. 1–4.
28. Yang, M.; Ai, B.; He, R.; Ma, Z.; Mi, H.; Fei, D.; Zhong, Z.; Li, Y.; Li, J. Dynamic V2V channel measurement and modeling at street intersection scenarios. *IEEE Trans. Antennas Propag.* **2023**, *71*, 4417–4432. [CrossRef]
29. Huang, J.; Wang, C.-X.; Yang, Y.; Liu, Y.; Sun, J.; Zhang, W. Channel measurements and modeling for 400–600-MHz bands in urban and suburban scenarios. *IEEE Internet Things J.* **2020**, *8*, 5531–5543. [CrossRef]
30. Kryszkiewicz, P.; Sroka, P.; Sybis, M.; Kliks, A. Path loss and shadowing modeling for vehicle-to-vehicle communications in terrestrial TV band. *IEEE Trans. Antennas Propag.* **2022**, *71*, 984–998. [CrossRef]

MDPI AG
Grosspeteranlage 5
4052 Basel
Switzerland
Tel.: +41 61 683 77 34

Sensors Editorial Office
E-mail: sensors@mdpi.com
www.mdpi.com/journal/sensors

www.ingramcontent.com/pod-product-compliance
Lightning Source LLC
Chambersburg PA
CBHW041356210926
51850CB00013B/3095
* 9 7 8 3 7 2 5 8 2 5 0 1 1 *